1/19/99

SELECTED PAPERS ON GEOMETRY

THE
RAYMOND W. BRINK SELECTED MATHEMATICAL PAPERS

Published by
THE MATHEMATICAL ASSOCIATION OF AMERICA

———

Committee on Publications
EDWIN F. BECKENBACH, Chairman

The Raymond W. Brink
Selected Mathematical Papers

VOLUME FOUR

SELECTED PAPERS ON
GEOMETRY

Reprinted from the

AMERICAN MATHEMATICAL MONTHLY
(Volumes 1–81)

and from the

MATHEMATICS MAGAZINE
(Volumes 1–47)

Selected and arranged by an editorial committee consisting of

ANN K. STEHNEY, Co-Chairman
Wellesley College

TILLA K. MILNOR, Co-Chairman
Rutgers University

JOSEPH E. D'ATRI
Rutgers University

THOMAS F. BANCHOFF
Brown University

Published and distributed by
THE MATHEMATICAL ASSOCIATION OF AMERICA

© 1979 by
The Mathematical Association of America (Incorporated)
Library of Congress Catalog Card Number 79-65512

Complete Set ISBN 0-88385-200-4
Vol. 4 ISBN 0-88385-204-7

Printed in the United States of America

Current printing (last digit):
10 9 8 7 6 5 4 3 2 1

FOREWORD

The RAYMOND W. BRINK SELECTED MATHEMATICAL PAPERS series of the Mathematical Association of America was established through a generous gift to the Association from Mrs. Carol Ryrie Brink in honor of her late husband, Professor Raymond W. Brink. The series provides a fitting and lasting memorial to Professor Brink, who served the Association in many significant ways including terms as Governor (1934–39 and 1943–48), as Vice-President (1940–41), and as President (1941–42).

Articles for inclusion in the SELECTED PAPERS volumes are selected from past issues of the Association's journals.

In expressing its deep appreciation to Mrs. Brink, the Board of Governors was particularly pleased to note that the gift was made to the Association at a time when our organization has become increasingly dependent on private sources for support. The Board therefore felt grateful to Mrs. Brink not only for her generosity but also for her wish that the gift will stimulate others to contribute to the projects of the Association.

———————

The following RAYMOND W. BRINK SELECTED MATHEMATICAL PAPERS have been published:

Volume 1: SELECTED PAPERS ON PRECALCULUS

Volume 2: SELECTED PAPERS ON CALCULUS

Volume 3: SELECTED PAPERS ON ALGEBRA

Volume 4: SELECTED PAPERS ON GEOMETRY

APOLOGY TO THE READER

These articles and problems on geometry are reprinted from the AMERICAN MATHEMATICAL MONTHLY and the MATHEMATICS MAGAZINE. The selection was made by an editorial committee appointed by E. F. Beckenbach, Chairman of the Committee on Publications of the Mathematical Association of America. Our committee was charged with selecting articles from past issues of those journals which would be most helpful to college students and teachers with an interest in geometry.

The editorial committee read more than 1200 articles. Of the 37 articles finally chosen, many may be used directly in the classroom or assigned as special projects. Others contain background material, provide insight, or discuss matters of controversy. The articles are arranged chronologically to exhibit, as much as possible, the changing emphasis within the geometry curriculum. To aid in locating an article in a particular topic, an Index of Key Words has been provided.

The length of articles in geometry, compared for instance to those in calculus, placed regrettable limitations on the selection process. The editorial committee did not attempt a comprehensive survey of geometry, and the reader will notice that some important topics such as transformation groups and projective geometry are not represented. This situation is alleviated somewhat by the inclusion of a Supplementary Bibliography listing articles which are omitted from the present volume because of their length or because other treatments are widely available. The editorial committee would like to point out that excellent expository articles on geometry are available in other special volumes published by the Mathematical Association. Of particular note is *Studies in Global Geometry and Analysis*, edited by S. S. Chern, from the M.A.A. Studies in Mathematics series.

The editorial committee wrote to all authors of articles selected whose current address was known, soliciting revisions and references to subsequent papers which would serve to update the originals. These and other minor changes, some at the suggestion of the committee, have been incorporated in the text. Typographical errors known to the committee have been corrected.

Editorial notes have been included to acknowledge additional material provided by the author, to give cross-references, and to refer to related articles which are not reprinted here. Notes which clarify the original formulation or indicate current terminology were added only if the committee was unable to contact the author. Some articles are followed by notes to help the reader to place the topic within the development of geometry.

The 42 problems reprinted here were selected for their variety and interesting solutions. In general, problems which are the subject of extensive literature and those which commonly appear in textbooks have been omitted. To encourage the reader to try the problems, their solutions are given separately. Problem enthusiasts should also consult the "Otto Dunkel Memorial Problem Book" which was published as part of Volume 64 (1957) of the AMERICAL MATHEMATICAL MONTHLY. It contains 400 selected problems, 67 in geometry, which appeared in the MONTHLY between 1918 and 1950, but solutions are not included.

The editorial committee is grateful to those who contributed to the production of this volume in various ways. Our thanks to the authors for their cooperation in updating the articles, to E. F. Beckenbach and Raoul Hailpern, Editorial Director of the Association, for their support throughout the selection and revision process, to Eleanor Tutty for her characteristically efficient secretarial services, and to Abigail Thompson, colleague-to-be, for assistance with the indices. The Selected Mathematical Papers series is published by the Association with support from a gift from Carol R. Brink in memory of her late husband, Raymond W. Brink.

THE EDITORIAL COMMITTEE

CONTENTS

FOREWORD   v

PREFACE: Apology to the Reader   vii

ARTICLES

The First Treatise on Non-Euclidean Geometry   GEORGE BRUCE HALSTED   1

Geometric Explanation of a Certain Optical Phenomenon

      WILLIAM H. ROEVER   3

History of the Parallel Postulate   FLORENCE P. LEWIS   6

Some Properties of a Skewsquare   W. H. ECHOLS   13

The Four-Color Problem   H. R. BRAHANA   20

Conformal and Equiareal World Maps   B. H. BROWN   29

Soap Film Experiments with Minimal Surfaces   RICHARD COURANT   40

On the Densest Packing of Spherical Caps   L. FEJES TÓTH   48

Euclidean n-Space Analogue of a Theorem of Galileo

      JOSEPH BARNETT, JR   50

Non-Euclidean Geometry   HERBERT BUSEMANN   51

Area in Hyperbolic Geometry   KENNETH LEISENRING   67

An Elementary Analogue to the Gauss-Bonnet Theorem   G. PÓLYA   74

On Defining Conic Sections   G. B. HUFF   77

Combinatorial Topology of Surfaces   ROBERT C. JAMES   79

Metric Postulates for Plane Geometry   SAUNDERS MAC LANE   114

The Steiner-Lehmus Theorem   G. GILBERT AND D. MACDONNELL   127

A Polygon Problem

      E. R. BERLEKAMP, E. N. GILBERT, AND F. W. SINDEN   128

Generalizations of Theorems about Triangles   CARL B. ALLENDOERFER   136

Higher Curvatures of Curves in Euclidean Space   HERMAN GLUCK   143

Lattice Points and Polygonal Area   IVAN NIVEN AND H. S. ZUCKERMAN   149

A Proof of Minkowski's Inequality for Convex Curves   HARLEY FLANDERS   154

An Area-Width Inequality for Convex Curves   PAUL R. CHERNOFF   167

The Early Development of Algebraic Geometry   SOLOMON LEFSCHETZ   168

The Kiss Precise   W. S. BROWN   177

A Problem in Cartography   JOHN MILNOR   180

Critical Points and Curvature for Embedded Polyhedral Surfaces
                                                    T. F. BANCHOFF   192

Existence of Four Concurrent Normals to a Smooth Closed Curve
                        NARSINGH DEO AND M. S. KLAMKIN   202

After the Deluge   D. A. MORAN   204

Dynamic Proofs of Euclidean Theorems   ROSS L. FINNEY   206

On Fenchel's Theorem   R. A. HORN   214

On a Characterization of the 2-Sphere   KRISHNA AMUR   216

What is a Convex Set?   VICTOR KLEE   218

Some Recent Results on Topological Manifolds   REINHARD SCHULTZ   234

On Involutions of a Circle   W. F. PFEFFER   246

The Historical Development of Algebraic Geometry   J. DIEUDONNÉ   248

An Isoperimetric Inequality for Polyhedra
                        H. H. JOHNSON AND J. OSAKA   288

A Characterization of Curves of Constant Width   G. D. CHAKERIAN   292

SUPPLEMENTARY BIBLIOGRAPHY   295

PROBLEMS AND SOLUTIONS   298

AUTHOR INDEX   333

INDEX OF KEY WORDS   335

ARTICLES

THE FIRST TREATISE ON NON-EUCLIDEAN GEOMETRY*

GEORGE BRUCE HALSTED, University of Virginia

Is it not surprising that a book so remarkable, that it will henceforth forever mark an epoch in human thought, should have been forgotten for more than a century and a half? The first treatise on Non-Euclidean Geometry, a work of extraordinary genius, appeared in 1733, yet so far as I know, not one single copy of this wonderful book is owned on the Western Continent.

That I have had the book for a considerable time in my possession is due to the generosity of a learned prelate of Louvain, R. P. Thirion, S. J., who lent me, across the ocean, his only copy of this rare and inestimable treasure.

I have already made the distinction between Anti-Euclideans, such as Bertrand of Geneva, Legendre, M. Vincent, Ed. T. Dixon, etc., who try to convict Euclid of imperfection by offering short proofs of his celebrated Parallel-Postulate; and the true Non-Euclideans, such as Lambert, Bolyai, Lobatschewsky, Riemann, Helmholtz, Lie, Klein, Clifford, Cayley, etc., all ardent admirers of Euclid, but makers of two companion geometries, called usually Lobatschewsky's and Riemann's.

But a whole century before Lobatschewsky, it must have been by a sort of prophetic instinct that the Italian Saccheri called his book, now ressurrected for immortality, "Euclid vindicated from every fleck." The complete title is as follows: Euclides—ab omni naevo vindicatus:—sive—conatus geometricus—quo stabiliuntur—Prima ipsa universae Geometriae Principia.—Auctore—Hieronymo Saccherio —Societatis Jesu—In Ticinensi Universitate Matheseos Professore.—Opusculum— Exmo Senatin—Mediolanensi—Ab Auctore Dicatum.—Mediolani, MDCCXXXIII.—Ex Typographia Pauli Antonii Montani. Superiorum permissu.

This book is an intensely interesting historical study, not only in science, but in psychology, and ethics; for it is evident to any mathematician reading it with the light we now have, that Saccheri deliberately built up a non-Euclidean geometry, and then covered it with just enough disguise to enable it to pass the ordeal of authorization by Ignatius Vicecomes, Provincial of the Jesuits, who certifies that it had been read by some Theologians of that society and found fit to see the light; and to enable it to receive the "*Imprimatur*" of the Senate, a Cardinal, and the Inquisitor General, Sylvester Martini, by whose order it was carefully read by the Revisor Don Gaspar, Doctor both of law and of Sacred Theology, who declared it to contain nothing against the orthodox faith.

---

* From AMERICAN MATHEMATICAL MONTHLY, vol. 1 (1894), pp. 149–152.

1

Remember DeMorgan's saying: "As to writing another work on geometry, the middle ages would as soon have thought of composing another New Testament...his order of demonstration was thought to be necessary and founded in the nature of our minds;" and remember that Saccheri's book contains not merely "another *work* on geometry," but *another geometry*, a thought so tremendous, so unorthodox, that its discovery in his book by these great Church Dignitaries would have doomed Saccheri to death. Perhaps an after suspicion of the truth *did* doom him to death, for the permission of the Provincial was given August 16th, 1733, and Saccheri was dead October 25th, 1733.

There exists in the library of Modena, Italy, a Manuscript Biography of Saccheri. Dr. Emory McClintock, the able President of the New York Mathematical Society, writes of Saccheri: "He confessed to a distracting heretical tendency on his part in favor of the 'hypothesis anguli acuti,' a tendency against which, however, he kept up a perpetual struggle. After yielding so far as to work out an accurate theory anticipating Lobatschewsky's doctrine of the parallel-angle, he appears to have conquered the internal enemy abruptly, since, to the surprise of his commentator Beltrami, he proceeded to announce dogmatically that the specious 'hypothesis anguli acuti' is positively false." Of course no such confession occurs or could have occurred in the book itself; for with it the book could never have been printed. If such a statement occurs in the Manuscript Biography that must have been the work of a personal friend written after Saccheri's death.

The sudden dogmatic assertion above mentioned does occur, first on page 70 of the work, a quarto. But this, after seventy quarto pages of rigorous logic and elegant demonstration to establish a non-Euclidean geometry, may be looked upon as something like the stucco for the king's inspection with which the immortal architect in Egypt covered the stone bearing his own name.

A whole century later, 1829, Gauss, in the "Lehr-und Lernfreiheit" of a German University, writes to his friend Bessel that he will never publish his researches on this subject, "da ich das Geschrei der Gegner scheue, wenn ich meine Ansicht ganz aussprechen wollte."

Saccheri's marvellous book says all it could have said and existed; apart from any decision of the historical question as to how consciously the Italian Jesuit was in it practising the motto of his order: "The end justifies the means."

The expression on the title-page, "the very First Principles of Universal Geometry," strongly suggests Lobatschewsky's *Pangeometry."*

The book opens with a dedication to the Senate of Milan, followed by a Preface to the Reader.

This, after a powerful eulogy of Euclid, begins by stating that the Parallel-Postulate has ever been a vexed question.

> Though heretofore no one has doubted the truth of the proposition, yet many have maintained that it is not axiomatic. So not a few have attempted to demonstrate it from those propositions of the First Book of Euclid which precede the twenty-ninth, wherein it is first used. But since all these attempts have signally failed, many, more recently, have attacked the matter by setting up a certain new definition of parallels.

Whereas Euclid defines parallels thus "Any straight lines, which are in the same plane, and being produced indefinitely towards both sides meet each other on neither, are parallels;" these, for the latter words substitute "*are always mutually equidistant*." But here arises a new split.

For some, and those certainly the more acute, have endeavored to show the existence of parallel straight lines as thus defined, whence they make the transition of the proposition to be demonstrated under Euclid's own words.

But others (not without a great sin against rigorous logic) assume as given such parallel straight lines, forsooth *equidistant*, that thence they may make the step to what remains to be proved.

Saccheri then goes on to say that he will not at first go into the question of the nature of those lines equidistant at all points from a certain line supposed straight, but will return to it later. "But any one can see that herein is occasion for subjecting to a rigorous examination the very first principles of universal geometry."

This is Saccheri's excuse, his plea, his defense for introducing a new kind of geometry.

"Atque hinc incipit diuturnum praelium adversus hypothesin anguli acuti."

Editorial Note: This is the third in a series of 37 articles, translating and interpreting the work of Saccheri, which Halsted wrote for the first five volumes of this MONTHLY. The early issues of the journal reflect considerable controversy over the "validity" of non-Euclidean geometry, and Halsted did much to expound the view that was bound to prevail. His numerous biographies of famous mathematicians also make excellent reading. (For precise references to the biographies and the rest of the series on non-Euclidean geometry, see the index to volumes 1–56.)

GEOMETRIC EXPLANATION OF A CERTAIN OPTICAL PHENOMENON*†

WILLIAM H. ROEVER

Description of the Phenomenon. In the parcel checking-room of the new Union Station at Kansas City, Missouri, there is a counter covered with brass plates which have, during the course of time, received numerous scratches by the baggage which is moved around upon the counter. The scratches are not very deep and they seem to be of fairly uniform distribution in both density and direction, as one might expect them to be after the cause of their formation has been in operation for some time. The baggage room is lighted by large electric lamps which are not very close together, so that an observer near the counter may regard the illumination in his

* From AMERICAN MATHEMATICAL MONTHLY, vol. 26 (1919), pp. 111–113.
† Presented to the American Mathematical Society (Southwestern Section), December 1, 1917.

immediate neighborhood as being due to a single lamp. Notwithstanding the apparently lawless nature of the manner of formation of these scratches, an observer anywhere near the counter and regardless of the direction of the illuminating electric light, will observe what appears to be a one-parameter family of illuminated ellipses which are approximately concentric and similar (see the accompanying pictures on the opposite page).

The Explanation. A prolate spheroid, *i.e.*, an ellipsoid obtained by revolving an ellipse around its major axis, has the property that each of its points is a brilliant point with respect to its foci. In other words the focal radii to any point of the surface make equal angles with the normal to the surface at that point, and the normal lies between and in the plane of the focal radii. Consequently any reflecting surface or curve* which is tangent to such an ellipsoid of which a point source of light and an observer's eye are the foci, will appear to have at its point of contact with the ellipsoid a luminous point, *i.e.*, a brilliant point. The electric light and the eye of the observer are the foci of a one-parameter family of confocal ellipsoids of revolution. These ellipsoids intersect the plane of the brass-covered counter in a one-parameter family of ellipses (which are neither concentric nor similar in general, but are approximately so for the smaller curves of the family). For different positions of the observer's eye (and of the lamp, which, however, is fixed) there are, of course, different families of ellipses on the counter. Those scratches on the counter which are tangent to the members of this one-parameter family of ellipses, will have brilliant (or luminous) points at their points of contact with these curves. Owing to the fact that a scratch may have some curvature and some width and also because the source of light is not a point, it follows that not merely a point but that a small arc of the scratch becomes illuminated. These small illuminated portions of scratches (though short and disconnected) are well distributed, and even though few may lie along any individual ellipse, they do, in the aggregate, give the general impression of a one-parameter family of illuminated ellipses, *i.e.*, they make visible, so to speak, the geometric ellipses described above in much the same way that iron filings distributed in a magnetic field make visible the lines of force of that field.†

Editorial Note: The reader may be interested in other articles by Roever which examine optical phenomena and suggest geometric models to explain them: The curve of light on a corrugated dome, vol. 20 (1913), 299–303, Optical interpretation in higher geodesy, vol. 21 (1914), 69–77, and Lines of illumination caused by the passage of light through a screen, vol. 29 (1922), 149–56, all in this MONTHLY.

---

* We will regard as a curve the exterior surface of a wire of small cross section, or the gutter-like surface of a scratch.

† This brilliant point phenomenon is different from any of those described by the author in the *Transactions of the American Mathematical Society*, Vol. 9, No. 2, pp. 245–279; *Bulletin of the American Mathematical Society*, Vol. XXII, No. 5, p. 218; THE AMERICAN MATHEMATICAL MONTHLY, Vol. XX, No. 10, pp. 299–303, and Vol. XXI, No. 3, pp. 69–77.

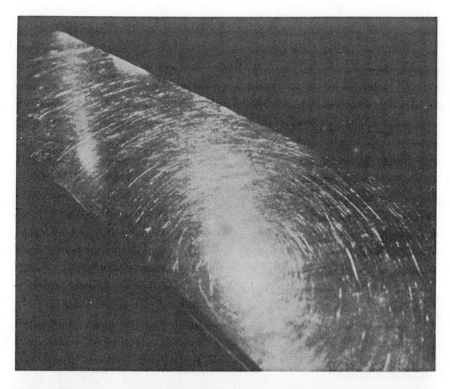

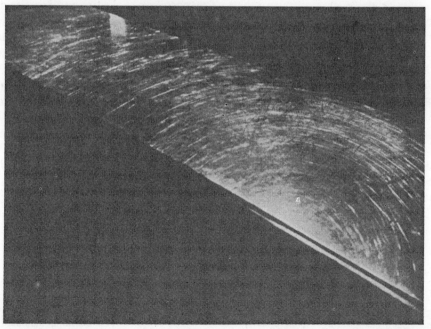

HISTORY OF THE PARALLEL POSTULATE*

FLORENCE P. LEWIS, Goucher College

Like the famous problems of construction, Euclid's postulate concerning parallels is a thought that links the ages. Its history is a long story with dramatic climax and far-reaching influence on modern mathematical and general scientific thought. I wish to recall briefly the salient features of the story, and to state what seem to me its suggestions in regard to the teaching of elementary geometry.

Euclid's fifth postulate (called also the eleventh or twelfth axiom) states: "If a straight line falling on two straight lines makes the interior angles on the same side less than two right angles, the two straight lines if produced indefinitely meet on that side on which are the angles less than two right angles." The earliest commentators found fault with this statement as being not self-evident. Concerning the meaning of *axiom*, Aristotle says: "That which it is necessary for anyone to hold who is to learn anything at all is an axiom;" and "It is ignorance alone that could lead anyone to try to prove the axiom." Without going into the difficult question of the precise distinction to the Greek mind between axiom and postulate, we may take it that the character of being indisputable pertained to each. Postulates stating that a straight line joining any two points can be drawn, that a circle can be drawn with given center and radius, or that all right angles are equal, were accepted, while the postulate of parallels was scrutinized and admitted at best with reluctance.

Proclus, writing in the fifth century A.D., gives some of the reasons for this attitude, and we may surmise others. The postulate makes a positive statement about a region beyond the reach of possible observation or geometrical intuition. Proclus insinuates that those who "suppose they have ground for instantaneous belief" are "yielding to mere plausible imaginings"; the conclusion is "plausible but not necessary."† The converse is proved in Proposition 27, book I of Euclid's *Elements*, and there seems to be no reason why this proposition should be more or less self-evident than its converse. The fact that the two lines continually approach each other‡ was not a convincing argument to the Greek geometer who was acquainted with the relation of the hyperbola to its asymptote. The form of statement of the postulate is long and awkward compared with that of the others, and its obviousness thereby lessened. There is evidence that Euclid himself endeavored to prove the statement before putting it down as a postulate; for in some manuscripts it appears not with the others but only just before Proposition 29, where it is indispensable to the proof. If the order is significant, it indicates that the author did not at first intend to include this among the postulates, and that he finally did so only when he found that he could neither prove it nor proceed without it.

---

* From AMERICAN MATHEMATICAL MONTHLY, vol. 27 (1920), pp. 16–23.

† Cf. Heath's Euclid, Vol. I, and Bonola's *Non-Euclidean Geometry*, Chicago, Open Court, 1912, to which reference is made throughout this paper.

‡ Even the meaning of this phrase requires further elucidation.

Most of the early geometers appear to have attacked the problem. Proclus quotes and criticizes several proofs, and gives one of his own. He instances one writer who even attempted to prove the falsity of the statement, the argument being similar to those used in Zeno's paradoxes. The common opinion, however, seems to have been that the postulate stated a truth, but that it ought to be proved. Euclid had proved two sides of a triangle greater than the third, which is far more obvious than this. If the statement was true it should be proved in order to convince the doubters; if false, it should be removed. In no case should it be retained among the fundamental presuppositions. Sir Henry Savile (1621) and the Italian Saccheri (1733) refer to it as a blot or blemish on a work that is otherwise perfect, and this expresses the common attitude of mathematicians until the first quarter of the nineteenth century.

Early attempts at proof usually took the form of a change in the definition of parallels, or the substitution, conscious or unconscious, of a new assumption. Neither of these methods resulted in satisfaction to any but their inventors; for the definitions usually concealed an assumption, and the new postulates were no more obvious than the old. Posidonius, quoted by Proclus, defines parallels as lines everywhere equidistant. This begs the question; surely such parallels do not meet, but may there not be in the same plane other lines, not equidistant, which also do not meet? The definition involves also the assumption, that the locus of points in a plane at a given distance from a straight line is a straight line, and this was not self-evident.* Ptolemy says that two lines on one side of a transversal are *no more parallel* than their extensions on the other side; hence if the two angles on one side are together less than two right angles, so also are the two angles on the other side, which is impossible since the sum of the four angles is four right angles. This is another way of saying that through a point but one parallel to a given line can be drawn, which is exactly Euclid's postulate. Proclus himself assumes (with some concealment) that if a line cuts one of two parallels it cuts the other, which is again postulate 5. Even as late as the close of the eighteenth century we find this argument advanced by one Thibault, and attributed also to Playfair: Let a line segment with one end A at a vertex of a triangle be rotated through the exterior angle. Translate it along the side until A comes to the next vertex and repeat the process. We finally arrive at the original position and must therefore have rotated through 360°. Hence the sum of the interior angles of a triangle is 180°; and, since Legendre had satisfactorily proved that this proposition entails Euclid's postulate of parallels, the latter is at last demonstrated. The fact that the same process could equally well be carried out with a spherical triangle, in which the angle-sum is not 180°, might have given him pause. The assumption that translation and rotation are independent operations is in fact equivalent to Euclid's postulate. Heath gives

---

* It should be noted that even the meaning of the criterion suggests several questions of logic. If two lines are so placed that perpendiculars to one of them from points on the other are equal, will the same statement hold when the rôles of the two lines are reversed? Will a perpendicular to one of two non-intersecting lines necessarily be perpendicular to the other? Of course the answers to these questions are closely bound up with the very postulate under discussion.—EDITOR.

(l.c.) a long and instructive list of these substitutes. In the course of centuries the minds of those interested became clear on one point: they did not wish merely to know whether it was possible to substitute some other assumption for Euclid's, though this question has its interest; they wished to know primarily whether exactly his form of the postulate was logically deducible from his other postulates and established theorems. To change the postulate was merely to re-state the problem.

After certain Arabs and Persians had had their say in their day, the curtain rises on the Italian Renaissance of the sixteenth century, where the problem was attacked with great vigor. French and British assailants were not lacking. The first modern work devoted entirely to the subject was published by Cataldi in 1603. When the eighteenth century took up the unfinished business of proving the parallel postulate, we find most of the giants of those days attacking the enemy of geometers with an even keener sense that without victory there could be no peace. Yet d'Alembert toward the close of the century could still refer to the state of the theory of parallels as "the scandal of elementary geometry." Klügel in 1763 examined thirty demonstrations of the postulate. He was perhaps the first to express doubt of its demonstrability. Lagrange, according to De Morgan, in about 1800, when in the act of presenting to the French Academy a prepared memoir on parallels, interrupted his reading with the exclamation, "Il faut que j'y songe encore," and withdrew his manuscript.

While the results of these investigations were on the whole negative, certain positive and valuable results were nevertheless obtained. The relation between the parallel postulate and the angle sum of a triangle was clearly brought out. Legendre proved that if in a single triangle the angle sum is two right angles, the postulate holds. Other equivalents are of interest. John Wallis and Laplace wished to assume: There exists a figure of arbitrary size similar to any given figure. Gauss could proceed rigorously provided he could prove the existence of a rectilinear triangle whose area is greater than any previously assigned area. W. Bolyai could have succeeded with the assumption that a circle can be passed through any three points not in a straight line. It must be borne in mind, moreover, that few mathematical questions have served so well as whetstones on which to sharpen the critical powers of mankind.

The work of the Italian priest Saccheri deserves notice because his method is that which finally brought the discussion to a close. Though published in 1733 his results did not become well known until after 1880, and therefore had little influence on other investigations. Legendre's *Réflexions*, published a hundred years later, covered much of the same ground without advancing quite so far. The title of Saccheri's work is *Euclides ab omni Naevo Vindicatus*, Euclid Vindicated of every Flaw. His plan was to prove the postulate by assuming its contradictory and showing that an inconsistency followed. He succeeded in proving that, according as in one triangle the angle sum is greater than, equal to, or less than two right angles, the same holds in every triangle, and that accordingly Euclid's postulate or one of its contradictories will hold. He makes three hypotheses which were recognized later to correspond to the elliptic, Euclidean and hyperbolic geometries. But at the end of his work, in order to exhibit a contradiction when Euclid's postulate is denied, he is forced to make use of a somewhat vague and unacceptable assumption about "the nature of a straight line."

Gauss's activity in connection with the parallel postulate is of especial interest because of its psychologic aspect. It is difficult for us to picture a mathematician hesitating to publish a discovery for fear of the outcry that its publication might produce—perhaps not many would be displeased to awaken an echo; yet this is believed by some to have been the attitude of Gauss. Though he was keenly interested and thought deeply on the subject of parallels, he published nothing; he feared, as he said, "the clamor of the Bœotians." When forced to write a letter on the subject, he begs his correspondent to keep silence as to the information imparted. In 1831 he writes in a letter: "In the last few weeks I have begun to put down a few of my Meditations [on parallels] which are already to some extent forty years old. These I had never put in writing, so that I have been compelled three or four times to go over the whole matter afresh in my head. Also I wished that it should not perish with me." It is only when we call to mind the unrivalled place of honor held by Euclidean geometry among branches of human knowledge—a respect no doubt enhanced by the prominence given it in Kant's *Critique of Pure Reason*—that we realize the uncomfortable position of one who even appeared to attack its validity. Gauss's meditations were leading him through tedious and painstaking labors to the conclusion that Euclid's fifth postulate was not deducible from his other postulates. The minds of those not conversant with the intricacies of the problem might easily rush to the conclusion that Euclid's geometry was therefore untrue, and feel the whole structure of human learning crashing about their ears.

Between 1820 and 1830, the conclusion toward which Gauss tended was finally made sure by the invention of the hyperbolic non-Euclidean geometry by Lobachevsky and Johann Bolyai, working simultaneously and independently. The question, Is Euclid's fifth postulate logically deducible from his other postulates? is answered by showing that the denial of this postulate while all the others are retained leads to a geometry as consistent as Euclid's own. The method, we recall, was that used by Saccheri, whose intellectual conservatism alone prevented his reaching the same result. The famous postulate is only one of three mutually exclusive hypotheses which are logically on the same footing. Thus was Euclid "vindicated" in an unexpected manner. Knowingly or not, the wise Greek had stated the case correctly, and only his followers had been at fault in their efforts for improvement. To quote Heath: "We cannot but admire the genius of the man who concluded that such an hypothesis, which he found necessary to the validity of his whole system, was really indemonstrable."

Thus in some sense the problem of the parallel postulate was laid to rest, but its spirit marches on. If the fifth postulate could without logical error be replaced by its contradictory, could the other postulates be similarly treated? What is the nature of a postulate or axiom? What requirements should a satisfactory system of axioms fulfill? Are we sure that accepted proofs will bear as keen scrutiny as that to which proofs of the postulate have been subjected? The facing of these questions has brought us to the modern critical study of the foundations of geometry. It has been realized that if geometry is to continue to enjoy its reputation for logical perfection, it should at least try to deserve it. The edge of criticism, sharpened on the parallel postulate, is turned against the whole structure. Out of this movement has grown the critical examination of the foundations of algebra, of projective geometry, of mechanics, of logic itself; and the end is not in sight.

One obvious result of this critical study is that geometrical axioms are not necessary truths, but merely presuppositions: they are the hypotheses on which the whole body of theorems rests. It is essential that a system of axioms should be consistent with each other, and desirable that they be non-redundant, and complete. No one has found Euclid's system inconsistent,* and redundancy would be a crime against elegance rather than against logic. But on the score of completeness Euclid is far from giving satisfaction. He not infrequently states conclusions which could be arrived at only by looking at a figure, *i.e.*, by space intuition; but we are all familiar with cases where space intuition misleads (for example in the fallacious proof that all triangles are isosceles), and if we accept it as a guide how can we be sure that our intuitions will always agree? In constructing an equilateral triangle Euclid says, "From point *C* where the two circles meet, draw····." Perhaps they do meet—but not on the basis of anything previously stated. In dropping a perpendicular from a point to a line, he says a certain circle will meet the line twice. Why should it not cut thrice or not at all? In another proof he says that a certain line will lie within a certain angle. I *see* that it does, but I do not see it proved. We are told in the midst of a proof to bisect an angle of a triangle and produce the bisector to meet the opposite side. How do we know it will meet? Because it is not a parallel. And probably it is not parallel because it is inside the triangle. How do we know it is inside; or, being inside, that it must get outside? When have these terms been defined? You may answer: It is not necessary to define them because everyone with common sense knows inside from outside without being told. "Who is so dull as not to perceive····?" says Simson, one of Euclid's apologists. This may be granted. But it must be pointed out that common sense knows that two straight lines cannot enclose space, yet this is given prominence as an axiom; or that a straight line is the shortest distance between two points, yet this is proved as a proposition. To state in words what distinguishes the inside from the outside of a polygon is not easy. The word "between" is likewise difficult of definition. But the modern geometer imbued with the critical spirit feels it necessary to define such terms, and what is more, he finds a way of doing it.

Hilbert's *Grundlagen der Geometrie*, published in 1899, is a classic product of this movement. It presupposes no space concepts at all, but only such general logical terms as "corresponds to," "associated with," "determined by." Contrary to tradition, it does not begin by defining terms. The first sentence is: "I think of three systems of things which I call points, lines and planes." Note the unadorned simplicity of the concept *things*. The axioms serve as definitions. They state, in

---

* It is in fact possible to show that the system is consistent, provided we agree to accept the axioms of arithmetic as consistent. While this is only a transformation of the problem, it is logically important to recognize the possibility of such a transformation.—EDITOR.

non-spatial terms, relations between these "things"; that is to say, the points, lines and planes are such things as have such and such relations. "That is all ye know on earth, and all ye need to know." Twenty-one axioms are found necessary, as against Euclid's meager five. The whole work could be read and comprehended by a being with no space intuitions whatever. We could substitute the names of colors or sounds for points, lines and planes, and get on equally well. The ideal of making a thing "so plain that a blind man could see it" is literally realized. And the age-old ideal of a body of proved propositions, close-knit together by unassailable logic, is immeasurably nearer realization.

Although to the best of my knowledge no one has yet had the hardihood to invite a child of fourteen to consider "three systems of things,"* the modern critical movement is not without bearing on problems of teaching. I wish, with proper humility, to put forward a few ideas on this subject.

If it is true that our traditional formal geometry, taken directly or indirectly from Euclid, is not the logically perfect thing we had imagined, and if its modern perfected descendant is so abstract that not even the most rationalistic of us would venture to force it on beginners, why not acknowledge these facts and bravely face anew the question of how we can best make the study of elementary geometry serve its proclaimed purpose of training the mind? I would suggest two lines along which progress might be made. First, by sacrificing the ideal of nonredundancy in our underlying assumptions we could save time and stimulate interest by arriving more quickly at propositions whose truth is not immediately evident and which could be presented as subjects for investigation. Must we, because Euclid did it, prove that the base angles of an isosceles triangle are equal? A child that has cut the triangle out of paper and folded it over knows as much as any proof can teach him. If to treat the proposition in this way is repugnant to the teacher's logical conscience, let him privately label it "axiom" or "postulate," and proceed, even though this proposition could have been proved. *The place to begin producing arguments is the place where the truth of the proposition is even momentarily in doubt.* One statement which presents itself with a question mark and is found after investigation to be true or to be false is worth ten obviously true statements proved with all the paraphernalia of hypothesis, conclusion, step one and step two, with references. The only apparent reason for proving in the traditional way the theorems on the isosceles triangle and congruent triangles is in order to familiarize the student with the above-mentioned paraphernalia. This brings me to my second suggestion.

Formal geometry has been looked upon as a complete and perfect thing to which the learner can with profit play the sedulous ape. Yet I sometimes think that by emphasizing too early the traditional form of presentation of geometrical argument, and paying too little attention to the psychology of the learner, we may have corrupted some very good minds. "I wish to prove...," says the student; meaning, I wish to prove something stated and accepted as true in advance of argument. Should we not prefer to have our students say, "I wish to examine..., to

---

* It is possible that Halsted's *Rational Geometry* has a somewhat similar purpose.—EDITOR.

understand, to find out whether..., to discover a relation between..., to invent a means of doing..."? What better slogan could prejudice desire than "I wish to prove"? The conscienceless way in which college debaters collect and enumerate arguments regardless of the issues involved is another aspect of the same evil. A student said not long ago, "The study of mathematics would be good fun if we did not have to learn proofs." It had never been brought home to her that mathematical reasoning is not a thing to be acquired, like a knowledge of Latin verbs, but a thing to be participated in like any other form of exercise. Another said, "I cannot apply my geometry because all we did in school was to learn the proofs and pass the examinations."

In the midst of a proof the student hesitates and says, "I am sure this is the next step, but I cannot recall the reason for it." The step and its reason would occur simultaneously to a mind that had faced the proposition as a problem and thought it through. I should like to see in every text an occasional page of exercises to prove or disprove. And if formal proofs must be printed in full, by all means let some of the proofs be wrong.

When the student has thus halted with one foot in the air in this progress from step to step down the printed page, on what does the ability to proceed depend? On the ability to quote something: to quote, usually, a single statement—compact, authoritative, triumphantly produced. Surely it is bad training that leads the mind to *expect* to find support for its surmises in a form so simple; and the temptation to substitute ability to quote in place of the labor of finding out the truth may be a real danger. What wonder if the mind so trained quotes Washington's Farewell Address or the Monroe Doctrine and feels that its work is done?

I do not mean that formal proof should never be given. It has its place as an exercise in literary composition; for it deals with the form in which thought is expressed. We should, however, take every possible precaution to see that the thought is first there to be expressed, lest the form be mistaken for the substance. Just how this is to be brought about I am not prepared to discuss, although I suspect that drawing and measurement in the early stages of study, problems of construction and investigation, and the total absence of complete proofs from the printed page, would help. I wish merely to state my belief that only insofar as we succeed in these aims shall we succeed in making geometry really train the mind.

> It can be done, said the butcher, I *think*;
> It must be done, I am sure.

One point further. Perhaps we are a little too modest about the importance of having our students retain something of the subject-matter of the courses we teach. Evidently it is here that memory, based on understanding, should rightly be used. I sometimes think we might in some way collectively take out insurance against a student's arriving at the junior year in college in the belief that two triangles are similar whenever they have a side in common.

Editorial Note: This article should be compared with Halsted's. Clearly Lewis and Halsted disagreed on the extent to which Saccheri was conscious of the implications of his work.

SOME PROPERTIES OF A SKEWSQUARE*

W. H. ECHOLS, University of Virginia

A plane quadrilateral in which two diagonals are perpendicular and equal to each other has a number of rather interesting properties. Such a figure for convenience of reference in the present paper will be called a *skewsquare*, although I am given to understand it has at some time been called a pseudo-square.† The writer has seen no reference to it in print and it is possible that some or all of the properties enunciated below are not new. A number of the properties admit of easy demonstration by elementary geometry and also as easy exercises in interpreting complex number relations. Furthermore the figures involved in the geometrical constructions furnish excellent exercises in mechanical drawing inasmuch as the constructions check themselves as to accuracy in various ways.

The writer was led to the consideration of this type of quadrilateral in trying to place the complex number

$$w_{ij} \equiv \frac{1}{z_j - z_i} \int_{z_i}^{z_j} f(z)\,dz,$$

with reference to $f(z)$ in such a manner as to locate ζ that $f(\zeta) = w_{ij}$, wherein $f(z)$ is an analytic function. To any pair of points in the z-plane corresponds a point in the w-plane, the mean value of $f(z)$.

To the quadrilateral points z_1, z_2, z_3, z_4 correspond the quadrilateral points $w_{12}, w_{23}, w_{34}, w_{41}$, and in this figure it can be easily shown (by the properties of the integral) that the triangles $w_{12}w_{23}w_{31}$ and $w_{13}w_{34}w_{41}$ are, respectively, similar to $z_3z_1z_2$ and $z_4z_1z_3$. Also triangles $w_{24}w_{32}w_{34}$ and $w_{12}w_{24}w_{41}$ are, respectively, similar to $z_3z_4z_2$ and $z_4z_1z_2$. In particular, if the four z-points are the corners of a square, then the corresponding four w-points are the corners of a skewsquare. In like manner, with n z-points taken around a polygon‡ can be constructed a skew-polygon of n w-points, and in addition the w-points corresponding to the diagonals of the z-polygon give a figure with more complicated but interesting relations.

The following are some properties of skewsquares:

(1) In a skewsquare the two squares constructed on two opposite sides as diagonals have a common vertex. The two points thus determined will be called the *foci* of the skewsquare.

(2) Conversely, if the squares constructed on two opposite sides of a quadrilateral as diagonals have a common vertex, then will the squares on the other two sides as diagonals also have a common vertex, and the figure is a skewsquare. Otherwise,§ if any two isosceles right triangles have the vertex common, their bases are opposite sides of a skewsquare.

---

* From AMERICAN MATHEMATICAL MONTHLY, vol. 30 (1923), pp. 120–127.
† For example, *Mathesis*, 1894, p. 268.—EDITORS.
‡ Regular polygon.—ED.
§ In other words.—ED.

(3) The foci and the mid-points of the diagonals of a skewsquare are the corners of a square (called the focal square). The center of this square will be called the *center* of the skewsquare. The mid-points of the diagonals of a skewsquare will be called the *conjugate foci* of the skewsquare.

(4) The foci lie on the bisectors of the angles between the diagonals of the skewsquare, respectively.

(5) The center of a skewsquare is the centroid of its corners, and also the centroid of the corners of the four squares having for diagonals the sides of the skewsquare.

(6) The segment joining a focus to the midpoint of a side of a skewsquare is perpendicular to and equal to half a side.

(7) The sum of the squares of the opposite sides of a skewsquare are equal.

(8) The four triangles (not right angled) whose bases are the sides of a skewsquare and whose vertices are the foci are equal in area.

(9) Segments joining any vertex of a skewsquare to the two foci make equal angles with the sides at that vertex.

(10) The foci of a skewsquare are the foci of a conic tangent to the sides of the skewsquare. The points of contact are points in which the bisectors of the angles between the diagonals cut the sides. A point of contact divides the side in the ratio of the adjacent segments of the diagonals determined by their intersection.

(11) The midpoints of the sides of a skewsquare form a square (called the midsquare), its side is equal to half the diagonal of the skewsquare. The circle circumscribing the midsquare (called the midcircle) is the auxiliary circle on the transverse axis of the tangent conic. The square of the transverse diameter of the conic is equal to half the square of the diagonal of the skewsquare.

(12) The area of each of the four triangles in (8) is equal to the square of the semi-conjugate diameter of the tangent conic. The foci are both inside, on, or outside the boundary of the skewsquare, the tangent conic is respectively an ellipse, a line-segment, or a hyperbola. Also the product of two opposite sides of a skewsquare and the cosine of the angle between them is equal to the square of the conjugate diameter of the conic.

(13) The sides of all skewsquares, whose diagonals are perpendicular diameters of two fixed equal circles, are tangent to a fixed conic. The power of the center with respect to either circle is equal to the sum of the squares of the semiaxes of the conic.

Otherwise, the locus of the vertices of skewsquares circumscribing a given central conic is two equal circles whose centers are the other corners of a square on the segment joining the foci as diagonal, the radius of the circles is equal to the side of a square inscribed in the auxiliary circle of the conic.

(14) When one side of a skewsquare is fixed and the diagonals become infinite in given directions the conic becomes a parabola tangent to two right-angled sides at the ends of the latus rectum.

(15) The four corners of the four squares having as diagonals the sides of a skewsquare (and which are not foci) are the corners of a second skewsquare, called the conjugate skewsquare (the first being called the primitive skewsquare). The

square of the diagonal of the conjugate is twice that of the primitive skewsquare. The vertices of the conjugate lie on the bisectors of the angles between the diagonals of the primitive, the foci and the conjugate foci of the primitive are respectively the conjugate foci and the foci of the conjugate skewsquare.

Otherwise, the construction of the primitive from the conjugate skewsquare follows. The vertices of four squares whose diagonals are the four segments joining each vertex of a skewsquare to the midpoint of the diagonal not containing that vertex determine only four points which are the corners of another skewsquare (called the primitive skewsquare). Its vertices lie on the bisectors of the angles between the diagonals of the conjugate, the foci and the conjugate foci of the conjugate skewsquare are respectively the conjugate foci and the foci of the primitive.

Thus, associated with any given skewsquare there is a conjugate and a primitive skewsquare which are confocal, their diagonals are collinear and their midpoints coincide, the diagonal of the first being twice that of the second. Associated with any skewsquare there is a series of derived skewsquares, alternate members are confocal and their tangent conics are members of a confocal system.

(16) Segments joining the center to the vertices of a primitive skewsquare are perpendicular to and equal to half the corresponding sides of its conjugate skewsquare.

(17) The diagonals of a skewsquare are parallel and equal to the diagonals of the midsquare of the conjugate skewsquare.

(18) Lines through the midpoints of the sides of a skewsquare parallel to the transverse axis of the tangent conic pass through the corners of its primitive skewsquare. The segments between these midpoints and the corners of the primitive skewsquare are equal to each other and to half the distance between the foci of this conic.

(19) The center of a skewsquare is the radical center of the four circles whose centers are the corners of the skewsquare, the radius of each circle is equal to the segment joining its center to either of the two adjacent corners of the primitive skewsquare.

(20) If Z is any point in the plane of a skewsquare $ABCD$ and ZA, ZB, ZC, ZD be rotated about Z, in the same direction, through one, two, three, four right angles, respectively, to ZA', ZB', ZC', ZD', then will the last four segments be in equilibrium.*

(21) The four points which divide the four sides of a skewsquare in the same given ratio ($m:n$) are the corners of a skewsquare (m and n may be real or complex numbers).

(22) If $ABCD$ and $A'B'C'D'$ are any two skewsquares, then the four points dividing the segments AA', BB', CC', DD' in given ratio ($m:n$ real or complex) are the corners of a skewsquare.

(23) If the corners of any one of a system of confocal skewsquares are the roots

---

* As vectors.—ED.

of a polynomial, the roots of its second derivative and the foci are the corners of a lozenge* composed of two equilateral triangles.

(24) If s_1, s_2, s_3, s_4 are the lengths of the segments into which the intersection of diagonals of a skewsquare divides those diagonals, then the two foci and the intersection of diagonals of the skewsquare (whose corners are z_1, z_2, z_3, z_4) are the roots of the derivative of the function

$$(z - z_1)^{s_1}(z - z_2)^{s_2}(z - z_3)^{s_3}(z - z_4)^{s_4}.$$

(25) If r_1, r_2, r_3, r_4 are the segments joining any point Z to the corners of a skewsquare, and s_1, s_2, s_3, s_4 are defined as in (24), then forces acting at Z whose directions pass through the corners and whose magnitudes are proportional to

$$\frac{s_1}{r_1} : \frac{s_2}{r_2} : \frac{s_3}{r_3} : \frac{s_4}{r_4}$$

are in equilibrium when Z is at either focus of the skewsquare.

The following abbreviated proofs are offered to establish the properties enunciated above. In the accompanying diagram, $ABCD$ is a skewsquare, the diagonals intersect at I, the foci are E and F, M and N are the midpoints of diagonals and K, L, P, Q the midpoints of sides. The figure is drawn for a convex skewsquare, it may however be either reëntrant or two opposite sides may cut internally. $A'B'C'D'$ is the conjugate, and $A_1B_1C_1D_1$ the primitive of $ABCD$.

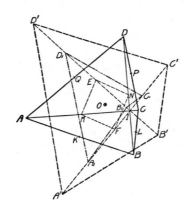

(1) The diagonal AC can be brought to BD by a rotation through a right angle, first A to B and C to D by rotation about E, second A to D and C to B by rotation about F.

(2) In the two right isosceles triangles AEB, CED the rotation AEC about E through a right angle brings A to B and C to D and AC to BD at right angles. There is a second center of rotation F accomplishing the same result, as in (1).

(3) The triangles AEC and BED are congruent, EM turns through a right angle to EN, in like manner FM turns through a right angle to FN. Hence $EMFN$ is a square.

---

* Diamond-shaped figure.—ED.

(4) The altitudes of the congruent triangles AEC, BED are equal, hence E (and in like manner F) is equidistant from the diagonals.

(5) The mid-points of the sides of a skewsquare are the corners of a square called the mid-square. Its sides are obviously parallel and equal to half the diagonals of the skewsquare. The centroid of the corners of the skewsquare is obviously the center of the mid-square, and the centroid of the foci is O the mid-point of MN which is also the centroid of the corners of the skewsquare. The centroid of the corners of the four squares on the sides of the skewsquare as diagonals is the centroid of the corners of the mid-square.

(6) This is obvious when the focus is the vertex of the right isosceles triangle having a side as the base. In any other case such as EQ, since QL and EF bisect each other at O, then QE is parallel and equal to FL which is perpendicular and equal to half BC.

(7) $AB^2 + CD^2 = AI^2 + BI^2 + CI^2 + DI^2 = CB^2 + AD^2$.

(8) The triangles AED and BEC have equal area, for AE is equal and perpendicular to EB, ED is equal and perpendicular to EC. The included angles are therefore equal or supplementary.

(9) $\angle DAB$ is half a right angle plus either $\angle EAD$ or $\angle FAB$ (if E and F are outside use *minus*).

(10) The property in (9) is a well-known fundamental property of a conic having foci E, F and tangent to the sides of the skewsquare. The circumcircle of the square $AEBA'$ passes through I, therefore IA' makes with IA half a right angle and contains F, by (4). FA' cuts AB in T such that

$$\angle KTA' = \angle ETK = \angle FTB.$$

Since the focal radii to the point of contact make equal angles with the tangent, T is the point of contact. Since IT bisects AIB, T divides AB in ratio $AI : BI$.

(11) The sum (difference) of the focal radii to the point of contact is equal to FA'. Also PF is parallel and equal to $EK = KA'$, (6), each being perpendicular and equal to $\frac{1}{2}AB$. Therefore FA' is parallel and equal to KP. Hence the mid-circle is the auxiliary circle of the tangent conic. Also $KP^2 = 2KQ^2 = \frac{1}{2}DB^2$.

(12) The product of the perpendicular from F on AB and $EK = KB$ is equal to the square on the semi-minor axis, and to the area of AFB which is equal to

$$\tfrac{1}{2}AF \cdot FB \sin AFB = \tfrac{1}{4}AD \cdot BC \sin AFB.$$

$\angle AFB + \angle LFQ = 2\pi - \frac{1}{2}\pi = \frac{3}{2}\pi$, $\angle LFQ = \pi - \angle(AD, BC)$. Therefore $\angle AFB = \angle(AD, BC) + \frac{1}{2}\pi$. Therefore $\sin AFB = \cos \angle(AD, BC)$. This establishes the statement. Obviously the eccentricity of the conic is equal to the ratio of the radius of the focal circle to that of the midcircle. In virtue of the triangles in (8) the foci E and F must be inside, on, or outside the boundary of the skewsquare, the areas of the triangles being counted positive for inside foci and negative for outside foci.

(13) The midpoints M and N are fixed and therefore the foci are also. The diagonal being of constant length, the transverse axis is fixed in length and position, and therefore the conic which the sides touch is fixed. The power of O with respect to the circle on AC as diameter is equal to

$$AM^2 - OM^2 = QP^2 - OM^2 = 2OP^2 - OM^2 = OP^2 + (OP^2 - OE^2).$$

We note by holding the midsquare and the focal circle fixed we derive the forms of skewsquares touching congruent concentric conics. If only the midsquare be fixed the skewsquares touch concentric conics having equal tranverse axes. When the foci are on the boundary the skewsquare degenerates into an isosceles right triangle, one focus is the midpoint of the hypothenuse and the other is at the vertex of the right angle and the conic degenerates into the segment joining them. When two opposite sides of the skewsquare cut internally the conic is an hyperbola, since the foci lie on opposite sides of a tangent.

(14) If AB is fixed and the diagonals extend indefinitely intersecting in I, the other two sides whose ends are A and B become parallel to the diagonals and cut at right angles, in say Z. The single focus is E and the bisector IE makes equal angles, $\frac{1}{4}\pi$, with the right-angled sides, and ZE makes this same angle with these sides. This establishes the statement from the known properties of the parabola.

(15) In the figure, the circle on AB as diameter passes through A' and I, therefore AIA' is half the right angle AKA' and A' is on bisector IF, and so for each corner. By (16) FP is parallel and equal to EK, hence KF is parallel and equal to EP or $\frac{1}{2}EC'$. Therefore from similar triangles $A'KF$ and $A'EC'$, we have $A'F=FC'$. Also $A'F$ is parallel and equal to KP.

$$A'C'^2=4KP^2=8KQ^2=2BD^2.$$

In like manner $B'D'^2=2AC^2$, and E is the midpoint of $B'D'$. Hence M,N are the foci of $A'B'C'D'$.

(16) A_1O is parallel and equal to $\frac{1}{2}A'E$, therefore A_1O is perpendicular and equal to $\frac{1}{2}AB$.

(17) A_1C_1 is parallel and equal to $\frac{1}{2}KP$, by (18).

(18) $A'F=BN\sqrt{2}$ and $BN=A_1F\sqrt{2}$. Therefore $A'F=2A_1F$. $A'A_1=A_1F$, $A'K=KE$. Therefore A_1K is parallel and equal to $\frac{1}{2}FE$, and so for the other points.

(19) $AA_1=AD_1$, $BA_1=BB_1$. The circles, centers A and B having radii AA_1, BB_1, respectively, have their radical axis passing through $A_1 \perp AB$ and therefore passing through O, by (16).

(20) The characteristic property defining a quadrilateral $ABCD$ whose corners are the complex numbers z_1,z_2,z_3,z_4, as a skewsquare is

$$z_3-z_1=i(z_2-z_4),$$

which can be written

$$z_1+\omega z_2+\omega^2 z_3+\omega^3 z_4=0, \tag{1}$$

where ω is the principal fourth root of unity, i. Any four numbers which satisfy this equation are the corners of a skewsquare. This then is the equation of a skewsquare and it establishes at once the truth of (20) on multiplying (1) through by i.

(21) If the corners of a skewsquare satisfy equation (1) in (20), then the same points satisfy the equation

$$z_2+\omega z_3+\omega^2 z_4+\omega^3 z_1=0. \tag{2}$$

On multiplying (1) by $m/(m+n)$ and (2) by $n/(m+n)$ and adding, the resulting equation proves the statement in (21). If m,n are real numbers, division of a segment in ratio $m:n$ is conventional and familiar. When m,n are complex numbers then the point w "dividing the segment" z_1 to z_2 in ratio $m:n$ is determined, as before, from $w=(mz_1+nz_2)/(m+n)$, or what is the same thing

$$\frac{w-z_2}{z_1-w}=\frac{m}{n}.$$

The point w is therefore the vertex of a triangle constructed on segment z_1 to z_2 as base similar to the triangle whose sides are m and n. Therefore, if similar triangles be similarly constructed (all outwards or all inwards) on the sides of a skewsquare as bases, then the vertices are the corners of a skewsquare.

(22) Let equation (1) in (20) be any skewsquare and $z_1'+\omega z_2'+\omega^2 z_3'+\omega^3 z_4'=0$ be any other skewsquare. Then on multiplying the first equation by $m/(m+n)$ and the second by $n/(m+n)$ and adding, the statement is verified.

(23) Take the origin at the center. Let ζ and η represent the midpoint of and half of a diagonal respectively.* Then the corners are the roots of

$$(z-\zeta-\eta)(z-\zeta+\eta)(z+\zeta-i\eta)(z+\zeta+i\eta)=0,$$

or

$$(z^2-\zeta^2)^2-4\zeta\eta^2 z-\eta^4=0.$$

The roots of the second derivative are $z=\pm\zeta/\sqrt{3}$.

(24) Let $f(z)$ be the function. Take the logarithm and differentiate. Then

$$f'(z)=f(z)\left(\frac{s_1}{z-z_1}+\frac{s_2}{z-z_2}+\frac{s_3}{z-z_3}+\frac{s_4}{z-z_4}\right).$$

The roots of $f'(z)$, not common to $f(z)$, are gotten by the vanishing of the parenthesis. Also

$$s_1+s_3=s_2+s_4=d,$$

the length of the diagonal. Also

$$\frac{s_1 z_3+s_3 z_1}{s_1+s_3}=\frac{s_2 z_4+s_4 z_2}{s_2+s_4}=I,$$

the intersection of the diagonals. Clearing the equation of fractions and dividing by d, there results $(z-I)[(z-z_1)(z-z_3)+(z-z_2)(z-z_4)]=0$. The quadratic in the square bracket is $z^2-\frac{1}{2}(z_1+z_2+z_3+z_4)z+\frac{1}{2}(z_1 z_3+z_2 z_4)=0$. Transfer the origin to the centroid (center). Then $z_1=\zeta+\eta$, $z_3=\zeta-\eta$, $z_2=-\zeta-i\eta$, $z_4=-\zeta+i\eta$, $\zeta\equiv OM$, $\eta=MA$. Then $z_1 z_3+z_2 z_4=2\eta^2$. The equation then becomes $(z-I)(z+i\zeta)(z-i\zeta)=0$, as required.

(25) In the parentheses in (24) put $z-z_1=r_1(\cos\theta_1+i\sin\theta_1)$, etc. Equating to zero the real and imaginary components shows that the sum of the components along the right-angled axes vanishes, and so also must their resultant, and the segments are in equilibrium, when Z is at either focus.

---

* Here η is the vector from the origin to the midpoint and ζ is a vector parallel to and with half the length of the diagonal. The corners of the skewsquare are $\zeta+\eta$, $\zeta-\eta$, $-\zeta+i\eta$, and $-\zeta-i\eta$.—ED.

THE FOUR-COLOR PROBLEM*

H. R. BRAHANA, University of Illinois

1. Origin. Of the origin of the problem, A. B. Kempe† in 1879 says: "It has been stated somewhere by Professor De Morgan that it has long been known to map-makers as a matter of experience—an experience however probably confined to comparatively simple cases—that four colors will suffice in every case.... Whether this statement was one merely of belief, or whether Professor De Morgan, or anyone else, ever gave a proof of it, or a way of coloring any given map, is, I believe, unknown; at all events, no answer has been given to the query to that effect put by Professor Cayley to the London Mathematical Society on June 13, 1878." A later reference to an earlier date is made by Guthrie‡ in which he says, "Some thirty years ago...my brother," who had been attending Professor De Morgan's class, "showed me the fact that the greatest necessary number of colors...is four." Professor De Morgan was pleased with the result and was in the habit of acknowledging to his classes the source of his information. The brother's proof, however, "did not seem altogether satisfactory to himself."

The problem is still unsolved. It has afforded many mathematicians experience and very little else.

2. Statement of the problem. The problem may be stated as follows: Given on the surface of a sphere any map, each region of which is such that any two points on its boundary can be joined by an arc made up wholly of points of the boundary, is it possible to color the map with four colors so that no two regions which touch along an arc have the same color?§

3. The problem in general. It is obvious that there would be a similar problem for a surface of genus 1, the anchor ring; another for a surface of genus 2, and so on. A natural conclusion to draw off-hand would be that if the problem in the case of the sphere is difficult, the similar problems would become increasingly more difficult as the value of the genus increases. This conclusion is contrary to experience. The problem has been completely solved for each value of the genus p, between 1 and 6, and the way pointed out for $p > 6$. Heawood** solved the problem for $p = 1$, and indicated the method of procedure for each other case. His method is given here because the point of difference between the problem on the sphere and that on a surface of higher genus is interesting. First we need only consider maps on which there are three and only three lines at every vertex—such a vertex is said to be of *degree* 3. For, suppose P (Fig. 1) is a point at which there are n ($n > 3$) lines, we may move one end of a along b to the position indicated by the

---

* From AMERICAN MATHEMATICAL MONTHLY, vol. 30 (1923), pp. 234–243.

† *American Journal of Mathematics*, 2 (1879), p. 193.

‡ *Proceedings of the Royal Society of Edinburgh*, 10 (1880), p. 727.

§ We do not consider the problem of giving the same color to two distinct regions; we consider no countries of more than one region.

***Quarterly Journal of Mathematics*, 24 (1890), p. 332.

FIG. 1

dotted line. This introduces a new vertex of degree 3 and reduces to $n-1$ the degree of P. It is evident that if the second map M_1 can be colored with k colors, the first one M may be given the same colors, for on returning from M_1 to M we introduce no new contacts between regions. Now if M_1 contains a vertex of degree n greater than 3, we may obtain an M_2 that has one more vertex of degree 3 and has a vertex of lower degree in place of the vertex of degree n in M_1. This may be continued until there are no vertices of degree greater than 3. The system of vertices and lines joining pairs of vertices is called a *linear graph*. If the vertices are all of the same degree the graph is *regular*, and the degree of a regular graph is the same as the degree of one of its vertices. We may note that in a regular graph of degree 3, $2E=3V$, where E is the number of lines and V is the number of vertices.

Next Heawood obtains a number that he calls the *average number of contacts* for the map. This is $2E/F$ where F is the number of regions in the map. Since each line is on the boundary of two regions, it is evident that if we count the number of sides* of each region and add all these numbers, the sum will be $2E$. Now, making use of the generalized Euler formula $V-E+F=2-2p$, where p is the genus of the surface, and of the relation $2E=3V$, we get an expression for the average number of contacts in terms of p and F as follows:

$$\frac{2E}{F}=6+12\frac{(p-1)}{F}.$$

For each value of p it is possible to find a value for F that makes $2E/F$ a maximum. The smallest integer n_p greater than this maximum is evidently a maximum for the number of colors required for any map on a surface of genus p. For, consider the map M of k regions where k is a number such that any map of fewer than k regions can be colored with n_p colors; M must contain at least one region of fewer than n_p sides since n_p is greater than the average. Let $a_1{}^2$ be such a region and $a_1{}^1$ one of its sides with ends at $a_1{}^0$ and $a_2{}^0$.† Let $a_2{}^1$ and $a_3{}^1$ be the other two lines with ends at $a_1{}^0$, and $a_4{}^1$ and $a_5{}^1$ the lines with ends at $a_2{}^0$. Then by removing the side $a_1{}^1$ replacing $a_2{}^1a_1{}^0a_3{}^1$ by a line $a_6{}^1$, and $a_4{}^1a_2{}^0a_5{}^1$ by a line $a_7{}^1$ we get a new map whose linear graph is regular and which contains one fewer regions than the original map. This map may therefore be colored with n_p colors; only n_p-1 of them can appear about the region $a_1{}^2$, and so we may return the region $a_1{}^2$ giving it the n_pth color. Obviously $k>n_p$, so our induction is complete.

The following are values of n_p for small values of p. $n_0=6$, $n_1=7$, $n_2=8$, $n_3=9$, $n_4=10$, $n_5=11$, $n_6=12$, $n_7=12$, $n_8=13$, $n_9=13$.

---

* By *side* of a region we mean an arc along which it borders a single region.

† By this notation we designate with a superscript the dimensionality of the element,—thus, $a_i{}^0$ is a point, $a_i{}^1$ is an arc, $a_i{}^2$ is a region.

Next Heawood sought a minimum m_p for the number of colors that would be required that one might be assured that any map on a surface of genus p could be colored with m_p colors. His method is to determine a number m_p which is such that the average number of contacts is $m_p - 1$ (in general, a map for which this is true does not have a regular graph). The proof of the validity of m_p for a minimum is complete only when a map of m_p regions each touching each other region on a surface of genus p is given. He gave the map for $p = 1$.* The value of m_0 is 4, $m_1 = 7$, $m_2 = 8$, and he showed that this number m_p is the same as n_p for $p > 0$. The sphere is the only surface for which the two do not agree.

4. Kempe's solution. Not only did Heawood treat rather successfully the problem for surfaces of higher genus, but he pointed out an error in the solution of the problem for the sphere that had been given by A. B. Kempe (*cf.* §1) and that had been generally accepted as correct. Kempe showed that we need consider only maps whose graphs are regular and of degree 3. He introduced the notion of *chains* which has probably come in for consideration by everybody who has worked on the problem since.

The work which follows uses the same considerations as Kempe used but is in a different form. We will use a method of induction. Four colors suffice for any map of four regions. Then either four colors suffice for any map, or there exists a map of k regions that cannot be colored with four colors, where k is such a number that any map of fewer than k regions can be colored. We investigate this hypothetical map M of k regions.

Now M contains no two-sided region, also no triangular region, for if it does contain one of either kind, we remove one side and combine it with one of the surrounding regions; the resulting map can be colored. No more than 3 colors are used around the region. We introduce it again giving it the fourth color and so get M colored in four colors.

Likewise, it can have no four-sided region. Here we make use of the notion of *chains*. Suppose M has a four-sided region. We remove one side of it combining it with one of the surrounding regions. The resulting map can be colored. The regions surrounding the four-sided one may be colored in order A, B, A, and C, in which case we give the fourth color D to the four-sided region. Or they may be colored A, B, C and D (Fig. 2). In this case let us consider the group of regions, each region having the color A or the color C, that can be reached starting with region a_1^2 of the figure and adding to it every C region which touches it, then adding each A region that touches one of these C regions, and so on. This group of regions Kempe called an *AC chain*. Two *AC* chains can have no vertices in common, and no regions in common. No *BD* chain can cross an *AC* chain.

Now either the *AC* chain beginning at a_1^2 contains region a_3^2 or it does not. If it does, then the *BD* chain beginning at a_2^2 does not contain a_4^2, and if we

---

* Heffter gave the maps for $p = 1, 2, \cdots 6$. *Mathematische Annalen*, 38 (1891), p. 477. Another set of "maps of verification" is given for $p = 1, 2, 3$, by the author in a paper soon to appear.

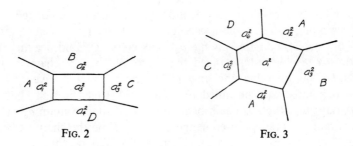

FIG. 2   FIG. 3

interchange B and D on that chain we have but three colors about a_5^2 and may give it the color B. In the opposite case we interchange colors A and C on the chain starting at a_1^2 and give a_5^2 the color A.

Next, as Kempe showed, there must be five-sided regions in M. Let a_i be the number of i-sided regions in the map, and suppose there is no region with more than l sides. Then $\Sigma_{i=2}^{l} a_i$ is the number F. Also, $2E = \Sigma_{i=2}^{l} i a_i$, for the latter is the sum of the sides of countries, which gives every line twice. Also $2E = 3V$. Using the Euler formula, we get

$$4a_2 + 3a_3 + 2a_4 + a_5 = 12 + \sum_{i=6}^{l} (i-6)a_i.$$

We have already shown that $a_2 = a_3 = a_4 = 0$, so a_5 is at least 12.*

Let us make a short digression to observe that five colors are sufficient for any map on a sphere. It is evident that the minimum map which cannot be colored in five colors must contain a five-sided region. Let one such region a_1^2 be removed as above and let the map be colored. The only possibility that need engage our attention is the case where the regions $a_2^2 \cdots a_6^2$ about a_1^2 have the colors A, B, C, D, and E respectively. The AC chain beginning at a_2^2 contains a_4^2 or does not. In the first case the BD chain beginning at a_3^2 does not contain a_5^2, and after permuting B and D on this chain we may give the color B to a_1^2; in the opposite case we interchange A and C on the AC chain beginning at a_2^2 and give a_1^2 the color A.

Returning to our problem, let a_1^2 be the five-sided region (Fig. 3) surrounded by $a_2^2 \cdots a_6^2$. The only case to consider is the one where $a_2^2 \cdots a_6^2$ take the colors A, B, A, C, and D respectively.

First let us notice that the BC chain beginning at a_3^2 must include a_5^2, otherwise we may interchange colors B and C on this chain and then give the color B to a_1^2. Also, the BD chain beginning at a_3^2 must include a_6^2, otherwise we may interchange B and D on it and give a_1^2 the color B. As a result of the first chain it follows that the AD chain which contains a_4^2 does not contain a_6^2 and that by interchanging A and D on it we have two D countries and only one A country

---

* By considering the number of lines, Wernicke (*Mathematische Annalen*, 58 (1901), p. 413) shows that there must be a pentagon adjacent to another pentagon, or else to a hexagon.

touching a_1^2. Or, as a result of the second chain we may interchange A and C on the chain beginning at a_2^2, without changing the color of a_5^2. Kempe interchanged the color on the AD chain, giving a_4^2 the color D, and leaving the others unchanged, *and then* he interchanged the colors on the AC chain, giving a_2^2 the color C and leaving the other unchanged. Thus he was able to give a_1^2 the color A. As Heawood pointed out, he failed to notice that after the first change it is no longer necessary that the BD chain beginning at a_3^1 shall contain a_6^2. If this chain does not contain a_6^2, then an interchange of A and C on the chain beginning at a_2^2 would put C at a_2^2 and A at a_5^2 and there would still be four colors about a_1^2. Heawood not only pointed out this fault in the argument but he presented a map where the difficulty suggested above was actually realized. This map did not, however, disprove the theorem that four colors suffice, for he gave a way of coloring it.

5. The results obtained by use of chains. In a paper entitled *du Coloriage des Cartes et de quelques Questions d'Analysis Situs** Errera carries to its conclusion another method of attack. The conclusion is, as often happens with this problem, an example which shows that the method will not solve the problem. His method is as follows: Make the first interchange of colors as in the paragraph above. This gives the color D to a_4^2. Then we may suppose that there is an AC chain containing both a_2^2 and a_5^2. Interchange B and C on the chain beginning at a_2^2. He continues in this manner; each step, once the first change has been made, is determined. If at any time any of these chains joining certain pairs of regions are missing, the map can be colored. Either that must happen or else the process continues indefinitely. Errera hoped to prove that the latter was impossible.

There are but a finite number of ways of putting four colors on a finite number of regions, so the operation he describes must lead at some stage to the original map. He notes that only after the twentieth operation do the same colors occupy the same positions around a_1^2 as at the beginning. This does not mean that the whole map has taken its original coloring, but rather that the number of operations required to return the map to its original state is a multiple of 20. He hoped to show that no map could have twenty substitutions as the whole or part of its period, and thus that the process would always lead to a map that could be colored. He ended by constructing a map of 17 regions which took on its original coloring after 20 operations. Also he gave a way of coloring the map, thus leaving the problem as much a problem as ever.

In a paper "On the Reducibility of Maps"[†] Birkhoff uses the notion of chains. He defines a *ring* as a set of k regions $a_1 a_2 \cdots a_k$ "such that each of these regions has a boundary line in common with the one preceding and following it in cyclical order, but with no other region of the set. A ring R of regions of this kind divides the map into two sets of regions M_1 and M_2 which together with R make up all the regions of the map." He then attempts to color the partial maps $M_1 + R$ and

---

† *American Journal of Mathematics*, 35 (1913), p. 115.

* Paris, Gauthier-Villars et Cie. (1921).

$M_2 + R$ in such a way that the regions of R will have the same colors in both cases or in such a way that by the use of the Kempe chains the regions of R can be given the same colors in both cases. Then putting the partial maps together, one could get the original map colored. He is able to show that the minimum map M (§4) contains no rings of four regions; no ring of five regions except about a single region; no ring of six regions surrounding only five-sided regions where the number of such regions is greater than 3; and no ring of n regions within which are n 5-sided regions surrounding a n-sided region; and no ring of $4n$ regions within which are $2n$ 6-sided regions about a single region.

6. Other forms of the problem. There are other interesting forms of the problem. We are considering only the problem of regular maps—maps whose linear graphs are regular and of the third degree, and which contain no 1-, 2-, 3-, or 4-sided regions.

One way to treat the problem is to consider only the linear graph of the map. Tait* was the first to notice that the problem was identical with the problem of assigning numbers 1, 2, and 3 to the lines of the graph in such a way that a line bearing each number would be at each vertex.

To see that the two problems are equivalent suppose first that a given map has been colored in four colors and consider the countries at any vertex. If we number the lines according to the following scheme,

$$1 = \frac{A}{B} = \frac{C}{D}$$

$$2 = \frac{A}{C} = \frac{B}{D}$$

$$3 = \frac{A}{D} = \frac{B}{C},$$

by which we mean that a 1-line is between an A country and a B country or between a C country and a D country, and so on, we see that every vertex will have a 1-line, a 2-line and a 3-line.

Conversely, suppose we have the numbers 1, 2, and 3 attached to the lines in the above-described manner. We may assign the color A to any region a_i^2 and the color of every other region is determined.

To see that no two adjacent regions are given the same color, consider two consecutive sides of the region a_i^2 which was given the color A. One of those sides must be either a 1-side or a 2-side. Such a 1-line or a 2-line will be included in some $(1-2)$ circuit—i.e., a set of lines $l_1 l_2 \cdots l_n$ where l_1 and l_2 have a vertex in common, l_2 and l_3, and so on to l_n and l_1 and where l_1 is a 1-line, l_2 is a 2-line, l_3 is a 1-line, and so on. Such a circuit must contain an even number of lines and an even number of vertices; there must be a 3-line at every vertex of the circuit. All of the 1-lines and 2-lines constitute a set of circuits of the above type. A $(1-2)$ circuit divides the sphere into two regions, one of which may be called the inside of the

---

* *Proceedings of the Royal Society of Edinburgh*, 10 (1880), p. 501.

circuit and the other the outside. There may or may not be $(1-2)$ circuits inside of a given $(1-2)$ circuit C; if there is a single $(1-2)$ circuit C_1 inside of C, then C_1 is connected to C by an even number of 3-lines. For C_1 has an even number of vertices and an even number (which may be zero) of those are joined in pairs by 3-lines. The remaining vertices, of which there are an even number, must be joined to C by 3-lines. If C is a circuit with n circuits $C_1, C_2, \cdots, C_n$ inside, no one of which contains more than one of the others inside it, then C is joined to the system by an even number of 3-lines; for each C_i $(1 \leqslant i \leqslant n)$ has an even number of vertices joined by 3-lines to some circuit outside it, and so there will be an even number for the whole system, and each 3-line joining two of them reduces by 2 the number of 3-lines joining the system to C. Thus we see that there are an even number of 3-lines going outward from C.

Now let the circuit which touches the region a_i^2 be C. The region a_i^2 is given the color A. Then color every region that touches the circuit C and that can be reached from a_i^2 by crossing 3-lines. If this method ever leads us back to a_i^2 we must have crossed an even number of 3-lines and so a_i^2 would not be given a color different from A. If this set of regions which has just been colored touches any other $(1-2)$ circuit, we may continue crossing 3-lines and coloring regions A and D without ever coming to a contradiction. When this has been carried as far as possible we will have obtained a Kempe chain. Now cross a 1-line or a 2-line from any region of this chain. According to the table above, this leads to a B region or a C region. From this continue crossing 3-lines as before and assigning colors according to the table. This gives a BC chain. Continuing in this way as long as there are any uncolored regions, we finally arrive at a coloring of the map. Thus we have shown that the two problems are equivalent.

We will now introduce the following definitions:

The number of vertices of a linear graph is called the *order* of the graph.

A regular graph is said to be *factorable* if it is possible to select a set of lines of the graph that constitutes a regular graph of the same order but of lower degree; the set of lines selected is one *factor*, and the remaining set of lines is the other.

A *leaf* is a part of the graph that is connected with the remainder by a single line provided it contains no such part itself.

An affirmative to the map problem is equivalent to the following theorem:

Any regular graph of degree 3 which contains no leaves and which can be put on a sphere can be factored into three first degree factors.

Tait, accepting Kempe's solution and believing that he had solved the problem in several ways himself, stated this theorem as a corollary. (*Cf.* reference above.)

The most that has been accomplished in this direction is the statement and proof by Petersen[*] of the following theorem:

Any linear graph containing less than three leaves can be factored into a first degree factor and a second degree factor.

It follows *a fortiori* that the linear graph of a regular map on a sphere can be factored into a linear and quadratic factor. This second degree factor is made up of

---

[*] *Acta Mathematica*, 15 (1891), p. 193. A second proof was given by the author, *Annals of Mathematics*, (2) 19 (1917), p. 59; and later a proof was given by Errera (*loc. cit.*).

a number of circuits, which may, however, contain an odd number of vertices and lines.

We may go a step further in refinement. The problem of attaching numbers 1, 2, and 3 to the lines of the linear graph so that one of each kind is at every vertex is equivalent to the problem of attaching $+1$ or -1 to each vertex so that the sum of the numbers on the vertices of any circuit is equal to zero (mod 3). Heawood[†] was the first to put the problem in this form. Veblen[‡] applied modular equations to the problem and so translated it into a problem in finite geometry.

Let the number of vertices be α_0, the number of lines be α_1, and the number of regions be α_2. Suppose we were to attach a number to each region, each number to represent a single color and no two distinct numbers to represent the same color. The condition that two adjacent regions have the same color could be expressed by the modular equation

$$y_a + y_b = 0 \quad (\text{mod } 2), \tag{1}$$

where y_a and y_b are the numbers attached to the regions a and b which touch along a particular line. For every line we may write an equation of the above type. The problem may then be considered as the problem of finding a set of α_2 values $(y_1, y_2, \cdots y_{\alpha_2})$, where y_i may have any one of four values, which does not satisfy any of the α_1 equations (1).

He then takes the field $GF(2)$ consisting of 0 and 1 combined modulo 2 and extends it by the Galois imaginaries[§] satisfying the relation $i^2 + i + 1 = 0$.[||] The extended field $GF(2^2)$ has four elements 0, 1, i and $i+1$. A set of values $(y_1, y_2, \cdots y_{\alpha_1})$ may be considered as a point on a finite projective space of $\alpha_2 - 1$ dimensions provided we exclude the set $(0, 0, \cdots 0)$ and consider $(ky_1, ky_2, \cdots ky_{\alpha_2})$ to be the same point as $(y_1, y_2, \cdots y_{\alpha_2})$. If the variables y_i range over the $GF(2)$ there will be $2^{\alpha_2} - 1$ points in the space, if they range over the $GF(2^2)$ there will be $(4^{\alpha_2} - 1)/3$ points. The first space is included in the second and the points of the second not included in the first may be regarded as *imaginary* with respect to the first space.

Each of equations (1) represents an $(\alpha_2 - 2)$ space. *Any point which does not lie on any one of these $(\alpha_2 - 2)$-spaces represents a solution of the problem.* In general, no real point can represent a solution because then the map would require but two colors which is impossible when the map contains a vertex at which there are an odd number of regions. Every imaginary point is on one and only one real line. But if $(y_1 + iy'_1, y_2 + iy'_2, \cdots y_{\alpha_2} + iy'_{\alpha_2})$ satisfies one of the above equations, so must $(y_1, y_2, \cdots y_{\alpha_2})$ and $(y'_1, y'_2, \cdots y'_{\alpha_2})$, and conversely. Hence a solution is given by *each real line which does not lie on any of the $(\alpha_2 - 2)$-spaces which are represented by equations* (1).

He also states the problem in terms of modular equations in a form equivalent to the theorem on linear graphs, and still again in a form equivalent to the problem

[†] *Quarterly Journal of Mathematics*, 29 (1898), p. 270.

[‡] *Annals of Mathematics*, 25, 14 (1912), p. 86.

[§] For a discussion of Galois imaginaries see Dickson: *Introduction to the Theory of Algebraic Equations*, p. 42 ff.

[||] i is no longer used as an index.

of attaching numbers $+1$ and -1 to the vertices. The only question remaining is whether or not the systems of equations have solutions. Similarly, Birkhoff* gives a formula for $P_n(\lambda)$, the number of ways in which a given map of n regions may be colored in λ colors. This, however, does not settle the question, for we have yet to find out whether or not $P_n(4)$ is positive for all values of n.

Finally, Franklin† arrives at the rather disheartening conclusion that the most we can say about the minimum uncolorable regular map, taking account of all the reductions that have been made and many more which he makes himself, is that it contains at least 26 regions. He gives a map of 42 regions to which none of the reductions apply. He establishes by a method similar to that used by Birkhoff (see §5) that the minimum uncolorable regular map does not contain:

(1) A side of a hexagon surrounded by the hexagon and three pentagons;

(2) A pentagon in contact with three pentagons and a hexagon;

(3) A pentagon surrounded by two pentagons and three hexagons;

(4) An even sided region completely surrounded by hexagons and (some or no) pairs of pentagons, the two of each pair being adjacent;

(5) A pentagon surrounded by four pentagons and two hexagons;

(6) A $2n$-sided region surrounded by $2n-2$ pentagons and two other adjacent regions;

(7) A $(2n-1)$-sided region surrounded by $2n-2$ pentagons and one other region.

By a method similar to that used by Wernicke (see footnote, §4), he shows that, as a result of the above and other like restrictions on the minimum uncolorable map, it contains one of the three arrangements which follow:

(1) A pentagon adjacent to two other pentagons;

(2) A pentagon adjacent to a pentagon and a hexagon;

or

(3) A pentagon adjacent to two hexagons.

Editorial Note: The problem introduced in this article fascinated mathematicians for many years. The activity it generated is shown in [3] and [4]. An affirmative resolution of the conjecture was announced in 1976, see [1] and [2].

References

1. K. Appel and W. Haken, Every planar map is four colorable, Bull. Amer. Math. Soc., 82 (1976), 491–2.

2. G. B. Kolata, The four-color conjecture: a computer-aided proof, Science, 193 (1976), 564–5.

3. Roy B. Levow, Relative colorings and the four-color conjecture, Amer. Math. Monthly, 81 (1974), 491–2.

4. Thomas L. Saaty, Thirteen colorful variations of Guthrie's four-color conjecture, Amer. Math. Monthly, 79 (1972), 2–43.

---

* *Annals of Mathematics*, 25, 14 (1912), p. 42.

† *American Journal of Mathematics*, 24 (1922), p. 225.

CONFORMAL AND EQUIAREAL WORLD MAPS*

B. H. BROWN, Dartmouth College

1. *Introduction.* The surface of the earth is not developable; that is, it can not be spread out in a plane without stretching. If we consider even an infinitesimal region of the earth, the map of this region must, in general, be distorted either in area or in shape; it may be distorted in both. But if for every infinitesimal region there is no distortion in shape the map is said to be *conformal*: if there is no distortion in area, *equiareal*. An equiareal map is equiareal over all. Conformality, however, is a property of an infinitesimal region only; it is wrong to say that a conformal plane map of Greenland has the shape of Greenland, for Greenland does not possess a plane shape. A conformal map is desirable whenever the accurate representation of directions is required as in navigation, meteorology, and in artillery fire. An equiareal map is desirable for political, agricultural, and general statistical purposes.

On our earth the system of lines of latitude and longitude has a significance which makes this system more important, geographically, than any other possible co-ordinate system. A map of the earth is effected as soon as the system or network of curves in the plane corresponding to the lines of latitude and longitude has been formulated and drawn. Other things being equal, it is desirable that this plane system consist of curves which are easily constructed. Such curves are the line, the circle, the conic sections. It is then very natural to propose the *Problem: to find all the conformal, and all the equiareal maps which carry the lines of latitude and longitude into plane systems composed of lines, circles, or conics.*

As will appear in paragraphs 2 and 3, two basic mappings, the conformal map of Mercator, and the cylindrical equiareal map of Lambert, reduce the problem proposed to the study of transformations in the plane. It is then sufficient to give the results of four investigations each concerned with the problem of finding all the plane transformations which carry the rectangular network:

I conformally into a system of lines and circles, Lagrange,[†]
II conformally into a system of lines and conics, Von der Mühll,[‡]
III equiareally into a system of lines and circles, Gravé,[§]
IV equiareally into a system of lines and conics, the author.[**]

A detailed bibliography of the projections mentioned in this paper would run to excessive length. The reader interested in knowing more of the subject is advised to begin with the article "Map" in the Encyclopaedia Britannica,

---

* From AMERICAN MATHEMATICAL MONTHLY, vol. 42 (1935), pp. 212–223.

† Lagrange, *Sur la construction des Cartes géographiques*, Oeuvres, t. 4, pp. 635–692.

‡ Von der Mühll, *Ueber die Abbildung von Ebenen auf Ebenen*, Crelle, vol. 69, pp. 264–285.

§ Gravé, *Sur la construction des Cartes géographiques*, Liouville, 5 ser., t. 2, pp. 317–361.

**Brown, *Equiareal maps with conic meridians and parallels*, presented to Am. Math. Society. 1933.

then turn to the section on Kartographie by Bourgeois and Furtwanger in volume 6 of the Encyklopädie der Mathematischen Wissenschaften.

Finally we include illustrations of all the maps which can be constructed under I, II, III, and a few under IV. Mention is made of all known cases where cartographers have constructed such maps. The remainder of the maps (as constructions) are new. The formulation of IV is new. In all of the maps we have purposely included as much of the earth's surface as was conveniently possible, so that the amount of distortion is greater than the average user of maps is accustomed to see. This should be kept in mind when considering the usefulness of a map for limited portions of the earth's surface. The maps bear numbers corresponding to the equations which define them.

2. *The Basic Conformal Map.* The well-known Mercator projection (Fig. 1)

(C)
$$\begin{cases} u = \lambda \\ v = \log \tan (\pi/4 + \phi/2), \end{cases}$$

where ϕ and λ are latitude and longitude respectively, maps the sphere on the

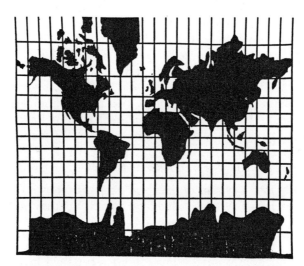

FIG. 1

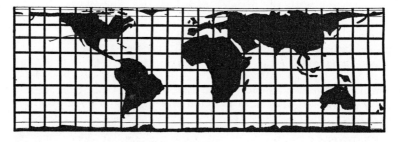

FIG 8

plane conformally, and carries the lines of latitude and longitude into a rectangular co-ordinate system (u, v). The extension here, and in paragraph 3, to an ellipsoid of revolution is immediate, and can be effected in terms of elementary functions.

3. *The Basic Equiareal Map.* The Lambert cylindrical equiareal projection (Fig. 8)

(E)
$$\begin{cases} u = \lambda \\ v = \sin \phi, \end{cases}$$

carries the lines of latitude and longitude of a unit sphere into a rectangular co-ordinate system and is equiareal. The projection may be effected by wrapping a tangent cylinder about the equator, and projecting the sphere onto this cylinder by rays through the axis which are parallel to the equatorial plane. This mapping was first formulated by Lambert, but the principle which allows us to conclude that the resulting map is equiareal is due to Archimedes.

4. *The Problem of Lagrange.* Lagrange showed that all the conformal plane transformations which carry the rectangular network (u, v) into a system of lines or circles fall into 4 types. We may with no essential loss in generality factor out translations, rotations, expansions, and reflections, and enumerate these types as follows:

(1)
$$\begin{cases} x = u \\ y = v, \end{cases}$$

(2)
$$\begin{cases} x = e^v \cos u \\ y = - e^v \sin u, \end{cases}$$

(3)
$$\begin{cases} x^2 + y^2 - x/u = 0 \\ x^2 + y^2 + y/v = 0, \end{cases}$$

(4)
$$\begin{cases} x^2 + y^2 - 2x \tan u = 1 \\ x^2 + y^2 + 2y \coth v = - 1. \end{cases}$$

Geometrically the networks thus obtained are:

(1) the identical rectangular network,
(2) a polar co-ordinate network,
(3) a family of circles tangent to the x-axis at the origin, and a family of circles tangent to the y-axis at the origin,
(4) a family of circles through the two real points $(0, \pm 1)$, and a family of circles through the two imaginary points $(\pm i, 0)$.

The identity (1) combined with (C) gives of course the Mercator projection. The transformation (2) similarly gives a stereographic projection from the North Pole on a plane tangent at the South Pole. Substitution of $-u$ for u and $-v$ for v, reverses the poles. Substitution of cu for u and cv for v gives the general case, the Lambert conformal conical projection (Fig. 2). The sector in which the map is confined is arbitrary and depends only on c. The so-called Lagrange projection is a special case of (4), see Fig. 4.

If in the transformation (2) we replace u by v, and v by $-u$, we obtain an entirely different appearing map (Fig. 2'), based on the same network, but with the rôles of the lines of latitude and longitude reversed. And in general, whenever the system of curves in the plane consists of two essentially different types of families, two distinct types of maps are possible. This is true for (2) and (4), but not for (1) and (3). We thus obtain six fundamental maps.

5. *Von der Mühll's Extension of Lagrange's Problem.* Generalizing Lagrange's problem to include conics, Von der Mühll showed that in addition to transformations (1) to (4), the only conformal transformations involving systems of conics are:

$$(5) \qquad \begin{cases} x^2/\cos^2 u - y^2/\sin^2 u = 1 \\ x^2/\cosh^2 v + y^2/\sinh^2 v = 1, \end{cases}$$

$$(6) \qquad \begin{cases} y^2 + 4u^2 x = 4u^4 \\ y^2 - 4v^2 y = 4v^4, \end{cases}$$

$$(7) \qquad \begin{cases} x^2 - y^2 = u \\ 2xy = v. \end{cases}$$

Geometrically the networks thus obtained are:

(5) confocal ellipses and hyperbolas,
(6) confocal parabolas,
(7) equilateral hyperbolas, the axes of either family being asymptotes of the other family.

Transformation (5) thus yields two new maps, (6) and (7) yield each a single map.

From the standpoint of functions of a complex variable it is interesting to note that the transformations (1) to (7) are, without loss in generality, given by

(1)	$Z = z$,	(5)	$Z = \cos z$,
(2)	$Z = e^z$,	(6)	$Z = z^2$,
(3)	$Z = 1/z$,	(7)	$Z = \sqrt{z}$,
(4)	$Z = \tan z$,		

where $Z = x + iy$, and $z = u + iv$.

6. *The Problem of Gravé.* Gravé showed that all the equiareal plane transformations which carry the rectangular network into a system of lines and circles, are of 6 types, as follows:

$$(8) \qquad \begin{cases} x = c_1 u + c_2 v \\ y = c_3 u + c_4 v, \text{ where } c_1 c_4 - c_2 c_3 = 1, \end{cases}$$

$$(9) \quad \begin{cases} y/x = (c_3v + c_4)/(c_1v + c_2) \\ c_3x - c_1y = \sqrt{2u}, \text{ where } c_2c_3 - c_1c_4 = 1, \end{cases}$$

$$(10) \quad \begin{cases} x = c_1u \\ x^2 + (y - v/c_1)^2 = c_2^2, \end{cases}$$

$$(11) \quad \begin{cases} x^2/(2u) + y^2/(2u) + 2c_1x/\sqrt{2u} + 2c_2y/\sqrt{2u} + c_3 = 0 \\ y = V(v)x, \text{ where } (1 + V^2)/V' + 2(c_1 + c_2V)/\sqrt{V'} + c_3 = 0, \end{cases}$$

$$(12) \quad \begin{cases} x \cos u/c_1 + y \sin u/c_1 = c_2 \\ x^2 + y^2 = c_2^2 + 2c_1v, \end{cases}$$

$$(13) \quad \begin{cases} x^2 + y^2 = 2c_1v + c_2 \\ \{x - c_3 \cos (u/c_1)\}^2 + \{y - c_3 \sin (u/c_1)\}^2 = c_4^2. \end{cases}$$

Geometrically, the networks thus obtained are:

	u-curves*	v-curves
(8)	Parallel lines,	Parallel lines,
(9)	Concurrent lines,	Parallel lines,
(10)	Displaced congruent circles tangent to 2 v-lines,	Parallel lines,
(11)	Concurrent lines,	Circles tangent to 2 u-lines,
(12)	Concentric circles,	Lines tangent to a u-circle,
(13)	Concentric circles,	Congruent circles with centers on a u-circle.

The Lambert projection is of course the identity under (8) combined with (E), see Fig. 8. The Collignon projection is a special case of (9). Certain very special forms of (11) were given by Lambert, and maps based on these bear his name, as the Lambert zenithal equiareal map, and the Lambert conical equiareal map. The Albers projection is also a special case of (11). The general transformation (8) applied to the Lambert projection is shown in Fig. 8A. Each of the transformations (9) to (13) yields two distinct types of maps as illustrated.

7. *The Generalization of Gravé's Problem.* The author has recently determined all the equiareal transformations in the plane which carry the co-ordinate network into a system of lines and conics. Since the transformation (8) will carry a system of conics into a system of conics with preservation of areas, it is possible to simplify materially the formulation *if we assume that any transformation is subject further to* (8). It should be pointed out that with proper use of (imaginary) constants in (8) it may be possible to carry a family of ellipses

---

* Here the u-curves are those along which u varies, i.e., the curves $v \equiv$ constant.—ED.

into a family of hyperbolas. For convenience we define:

$$f_1(x) = c_1x^2 + c_2x + c_3, \qquad\qquad g_1(x) = c_4x^2 + c_5x + c_6,$$

$$f_2(x) = c_1x + c_2 + c_3/(x + c_4), \qquad g_2(x) = c_5x + c_6 + c_7/(x + c_4),$$

$$f_3(x) = c_1x + c_2 + c_3\sqrt{c_4x^2 + c_5x + c_6}, \quad g_3(x) = c_7x + c_8 + c_9\sqrt{c_4x^2 + c_5x + c_6}.$$

The symbol $P_2(S, T)$ shall stand for the most general quadratic polynomial in S and T with constant coefficients. With these conventions, the complete formulation is given by means of the following fourteen types:

(14)†
$$\begin{cases} x = c_1u + c_2v \\ y = c_3u + c_4v, \text{ where } c_1c_4 - c_2c_3 = 1, \end{cases}$$

(15)
$$\begin{cases} x = (c_1v + c_2)\sqrt{2u} \\ y = (c_3v + c_4)\sqrt{2u}, \text{ where } c_2c_3 - c_1c_4 = 1, \end{cases}$$

(16)
$$\begin{cases} \int \int f_i(x)dx = u \\ y = vf_i(x) + g_i(x), \end{cases}$$

(17)‡
$$\begin{cases} y = Vx; \ P_2(1/\sqrt{V'}, \ V/\sqrt{V'}) = 0 \\ P_2(x/\sqrt{u}, \ y/\sqrt{u}) = 0, \end{cases}$$

(18)
$$\begin{cases} x \sin(v/c_1) + y \cos(v/c_1) = c_2 \\ x^2 + y^2 = c_2^2 + 2c_1u, \end{cases}$$

(19)
$$\begin{cases} y = (c_3 + c_1v^{1/3})x + c_2v^{2/3}/2, \text{ where } u = (v^{-1/3}x)^2c_1/6 + (v^{-1/3}x)c_2/3 \\ y = (c_1/f + c_2/f^2)x^2 + c_3x \text{ and } v^{-1/3}x = f(u), \end{cases}$$

(20)
$$\begin{cases} y = 2c_1vx + c_3x + c_1c_2v^2 \\ y = -x^2c_1/c_2 + u/c_2 + c_3x, \end{cases}$$

(21)
$$\begin{cases} xy = v \\ P_2(xe^{-u}, ye^u) = 0, \end{cases}$$

(22)
$$\begin{cases} y = 3Vx^2, \text{ where } P_2(1/V'^{1/3}, 3V/V'^{2/3}) = 0 \\ P_2(x/u^{1/3}, y/u^{2/3}) = 0, \end{cases}$$

(23)
$$\begin{cases} x = c_1u + c_2v \\ y = c_3u + c_4v + f_i(x), \text{ where } c_1c_4 - c_2c_3 = 1, \end{cases}$$

(24)
$$\begin{cases} x = (c_1v + c_2)\sqrt{2u} \\ y = (c_3v + c_4)\sqrt{2u} + f_i(x), \text{ where } c_2c_3 - c_1c_4 = 1, \end{cases}$$

(25)
$$\begin{cases} y = x^2 + v \\ y = f_i(x - u) + x^2 - (x - u)^2, \end{cases}$$

(26)
$$\begin{cases} y = \{c_3 - (3/2)c_1v^{-2}\}x^2 + (c_4 - 2c_2v^{-1})x \\ y = c_3x^2 + c_4x + (-3/2)c_1f^2 - 2c_2f, \text{ where } u = c_1(v^{-1}x)^3 + c_2(v^{-1}x)^2 \end{cases}$$

---

† This is of course type (8), obtained by Gravé.—Ed.

‡ In (17), (22), and (27), V denotes a differentiable function of v alone.—Ed.

and $f(u) = v^{-1}x$,

(27) $\begin{cases} y = x^2 + 2Vx \\ y = x^2 + c_1x^2/\sqrt{u} + c_2x + c_3\sqrt{u}, \text{ where } 2V/\sqrt{V'} = c_1/V' + c_2/\sqrt{V'} + c_3. \end{cases}$

This formulation, which is new, is given without proof that it is complete, the proof being extremely long and tedious. The nature of the families of curves in these transformations is not obvious by inspection, as was the case in each of the previous problems, and is not without interest. Every family, as a family, is completely determined by its envelope, including points (finite or infinite). In fact, it can be shown that in general, in *any* equiareal map, the envelope of one family is entirely composed of curves of the other family and/or the envelope of the other family. The parameter of distribution within the family is, in some cases, difficult of expression, the most complicated case being (22) where the determination of the function $V(v)$ requires the integration of

$$c_1V'^{2/3} + 2c_2VV'^{1/3} + c_3V^2 + 2c_4V' + 2c_5VV'^{2/3} + c_6V'^{4/3} = 0.$$

Despite its apparent complexity, this differential equation is integrable in elementary functions, but the resulting equation is transcendental in V, and it is not possible to solve for V explicitly in terms of v. The same difficulty arises in (16) and in (17), but in no case is there an integration which can not be effected in elementary functions. Further, since our formulation is primarily analytic, some of these, notably (16), contain a large number of sub-cases which are special analytically, but limiting geometrically. The sharp distinction between central conics and parabolas is also noteworthy. A brief characterization of the families follows:

	u-curves	*v-curves*
(14)	Parallel lines,	Parallel lines,
(15)	Concurrent lines,	Parallel lines,
(16)	Conics through 2 points and tangent to 2 *v*-lines,	Parallel lines,
(17)	Concurrent lines,	Similar conics tangent to 2 *u*-lines,
(18)	Lines tangent to a (central) *v*-conic,	Central conics, similar and similarly placed,
(19)	Lines tangent to a *v*-parabola,	Parabolas tangent at a common finite and at a common infinite point,
(20)	Lines tangent to a *v*-parabola,	Parabolas with 4-point contact at infinity,
(21)	Central conics, similar and similarly placed,	Conics tangent to 4 (central) *u*-conics,
(22)	Parabolas tangent at a common finite and at a common infinite point,	Conics tangent to 4 *u*-parabolas,
(23)	Congruent conics tangent to 2 parallel lines,	Congruent conics tangent to the same 2 lines,

(24)   Congruent conics tangent to 2 parallel lines,

Conics tangent to the same 2 lines, passing through 2 points on a line parallel to those lines,

(25)   Conics tangent to 4 v-parabolas,

Parabolas with 4-point contact at infinity,

(26)   Parabolas with 4-point contact at infinity,

Parabolas with same infinite point as u-parabolas, and tangent to 2 u-parabolas,

(27)   Parabolas with same infinite point as v-parabolas, and tangent to 2 v-parabolas,

Parabolas with 3-point contact at infinity and passing through a common finite point.

In addition to the maps noted under paragraph 6, the Mollweide projection is a special case of (16), and the Goode homalographic projections (interrupted) for ocean units and land units are simple rearrangements of the Mollweide projection. Another special case of (16) has recently been employed by Deetz and Adams. The number of possible maps under these transformations is too large (well over 100) to permit of complete illustration. We content ourselves with showing a Mollweide projection (16), and an example of (21), the hyperbolas of the formulation being sheared to a family of concentric circles.

For convenience of reference, the illustrations may be classified as follows, the numbers referring to the equations in the text which define the maps:

Conformal (circles), Lagrange,
    Fig. 1, 2, 2′, 3, 4, 4′.
Conformal (conics), Von der Mühll,
    Fig. 5, 5′, 6, 7.
Equiareal (circles), Gravé,
    Fig. 8, 8A, 9, 9′, 10, 10′, 11, 11′, 12, 12′, 13, 13′.
Equiareal (conics), the author,
    Fig. 16, 21.

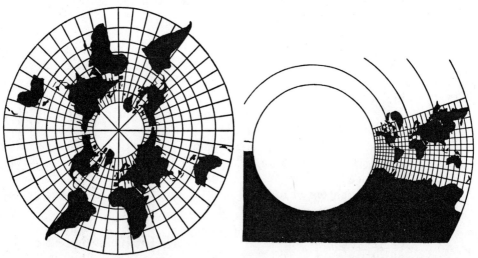

FIG. 2   FIG. 2′

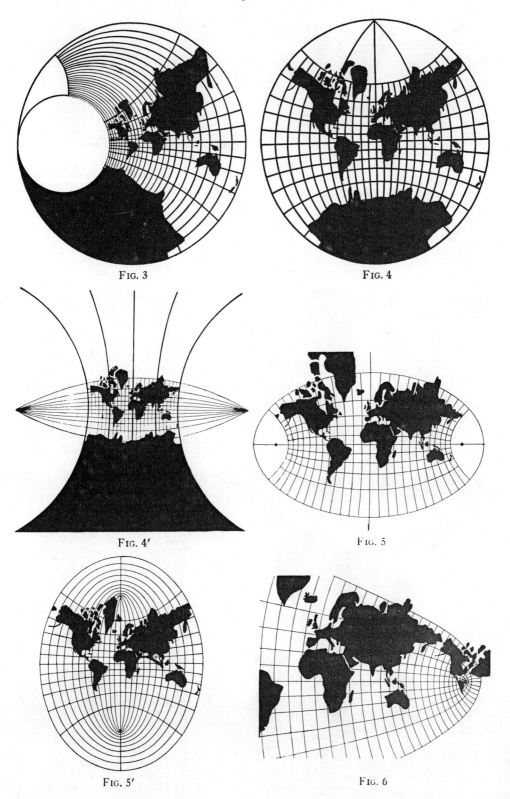

FIG. 3

FIG. 4

FIG. 4′

FIG. 5

FIG. 5′

FIG. 6

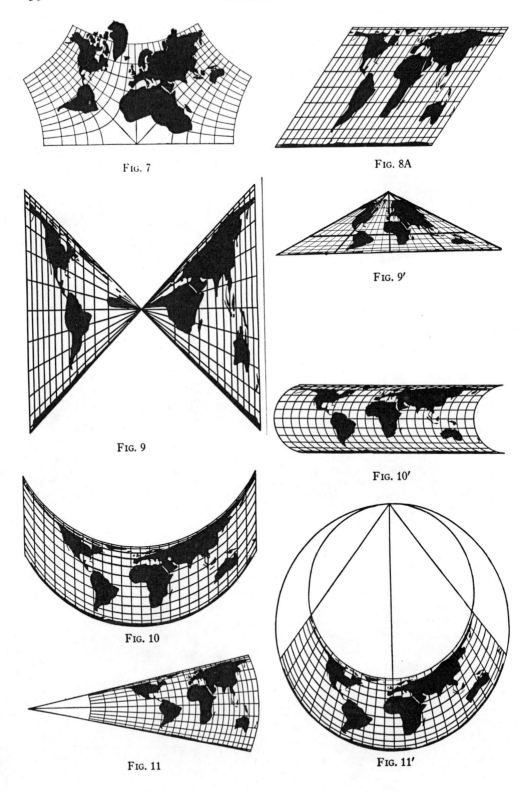

FIG. 7

FIG. 8A

FIG. 9

FIG. 9′

FIG. 10′

FIG. 10

FIG. 11

FIG. 11′

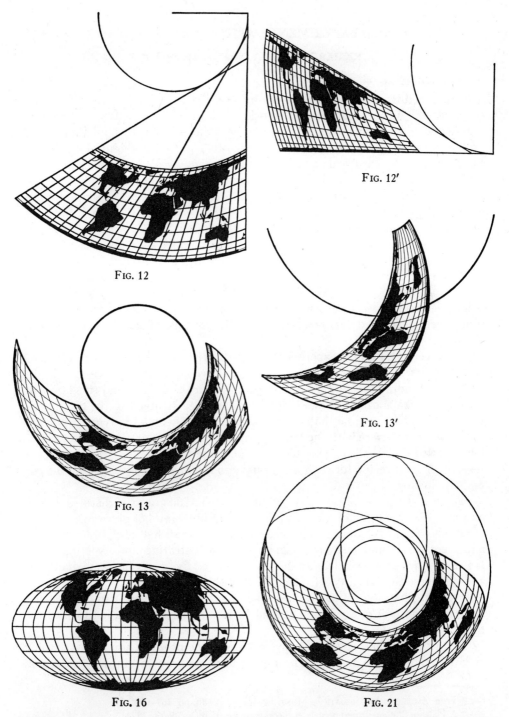

FIG. 12

FIG. 12'

FIG. 13

FIG. 13'

FIG. 16

FIG. 21

Editorial Note: At the time this paper was written, conic sections represented the best-known class of constructible curves. These days computer constructions make almost any curve available to map makers.

SOAP FILM EXPERIMENTS WITH MINIMAL SURFACES*

RICHARD COURANT, New York University

In the development of mathematics there are several instances of valuable help rendered to mathematical theory by physical experiments, *e.g.*, imagined physical experiments with two-dimensional flows of electricity have greatly illuminated geometrical function theory, and the solution of boundary value problems for the Laplace equation and for other partial differential equations can be realized by electrical and also by statistical devices. But perhaps the most striking example of the theoretical mathematical value of simple physical experiments is the demonstration of minimal surfaces by means of soap film experiments as described in detail by the Belgian physicist Plateau,[†] and also used by mathematicians, particularly H. A. Schwarz.[‡]

The problem of finding the smallest surface bounded by a given closed contour was proposed by Euler early in the 18th century. It is easily reduced to a boundary value problem of a non-linear character for a system of partial differential equations. Solutions were given for special types of contours by Riemann, Schwarz, Weierstrass, and others in the 19th century, while general existence proofs were developed only recently by Radó, Douglas, McShane, Garnier, and Courant.

However, the Plateau experiments immediately yield solutions for very general contours. If one dips a wire forming any closed contour into a viscous liquid and then withdraws it, a film suspended in the wire will form, and this film will assume the shape of a minimal surface of least area spanned in the given contour. Thereby we assume that we may neglect the effects of gravity and other causes which interfere with the tendency of the film to attain the smallest possible area and thus the least possible value of the potential energy due to surface tension. A good recipe for such a viscous liquid, which can be easily produced, is the following:

Dissolve 10 grams of pure dry sodium oleate in 500 grams of distilled water, and mix 15 cubic units of the solution with 11 cubic units of glycerin.

Films obtained with this solution and with frames of half hard brass wire are relatively stable and last long enough for demonstrating and experimenting. However, the size of such frames should not exceed five or six inches in diameter, and of course the stability with smaller frames is much higher.

By using solutions of lacquer, instead of soap solution, one can obtain solid and to a certain degree permanent models, because lacquer, having formed the film, dries and solidifies.

With this method it is very easy to "solve" the problem simply by shaping the wire frame in the desired form. Beautiful models are obtained by polygonal

---

* From AMERICAN MATHEMATICAL MONTHLY, vol. 47 (1940), pp. 167–174.

† J. Plateau, Statique expérimentale et théorétique des Liquides, Paris, 1873.

‡ Schwarz's Collected Papers, vol. I, *passim.*

wire frames formed by a sequence of edges of a regular polyhedron. In particular it is interesting to dip the whole frame of a cube into such a solution. The result is first a system of different surfaces meeting each other along lines of intersection. (If the cube is withdrawn very carefully, the result is a symmetric system of planes.) Then we have to pierce and to destroy enough of these different surfaces so that only one surface bounded by a closed polygon remains, which can be done in various ways leading to different surfaces.

But the scope and the informative value of soap film experiments with minimal surfaces is wider than these original demonstrations by Plateau. In recent years the problem of minimal surfaces has been studied when not only one but any number of contours is prescribed and when, in addition, the topological type might be more complicated, when for example, the surface is one-sided or is of a genus different from zero. Also the problem with free boundaries has been discussed. That is the problem where boundaries, or parts of boundaries, are not fixed curves, but are left free on given surfaces. Also these more general problems, which produce an amazing variety of geometrical phenomena, can be illuminated by soap film experiments. In this connection it was very useful to modify the simple experiments by making the system of wire frames or the prescribed boundary surfaces not rigid but flexible, and by studying the effect of deformations of the prescribed boundaries on the solution. Such deformations can be easily effected by attaching handles to the wire frames, or by other obvious mechanical devices.

There are essentially three general questions which should be studied in connection with the problem:

1. The existence of a solution of a given type.

2. The uniqueness of the solution.

3. The dependence of the solution on the prescribed data, particularly the question whether or not the solution depends continuously on the prescribed boundaries.

Of these questions only the first has so far been theoretically solved, while with respect to the second and third much remains to be done. However, physical experiments throw light on all three of them, and in addition suggest other mathematical questions and even their answers. This may be explained by a few typical examples.

1. If the contour is a circle we naturally obtain a plane circular disc. If we continuously deform the boundary circle we might expect that the minimal surfaces would always keep the topological character of a disc. However, this is not the case. If the boundary is deformed into a shape as indicated by Figure 1, the physical experiment yields a minimal surface which is no longer simply connected, but is a one-sided (non-orientable) Moebius strip. We may now start with this frame and with a soap film in the shape of a Moebius strip. By distorting the wire frame, pulling handles suitably soldered to it (Fig. 2), we reach a certain moment where suddenly the topological character of the film changes so that the surface is again of the type of a simply connected disc (Fig. 3). Reversing this deformation process, one again obtains a Moebius strip.

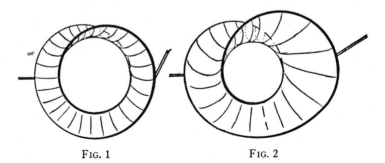

FIG. 1   FIG. 2

However, one observes with this alternating deformation process that, in the reverse process the mutation of the simply connected surface into the Moebius strip takes place at a later stage. This shows that there must be a range of shapes of the contour for which both the Moebius strip and the two-sided simply connected surface are stable (as indicated in Figures 2 and 3), one of which, of course, will furnish only a relative minimum. As a matter of fact, there always exists a two-sided simply connected surface which furnishes a relative minimum. But when the Moebius strip has a much smaller area than the two-sided surface, the latter is too unstable to be really formed.

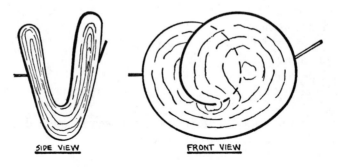

SIDE VIEW   FRONT VIEW

FIG. 3

2. The minimal surface of revolution spanned between two circles is very easily realized. In the actual experiment it happens that first, after the withdrawal of the wire frames, we find not one simple surface but a structure of three surfaces, one of which is a simple circular disc parallel to the prescribed boundary circles (Fig. 4). In this case, only by destroying this intermediate surface is the classical catenoid produced. Then by pulling the two boundary circles apart one can easily demonstrate that there is a moment when the doubly connected minimal surface becomes unstable. At this moment the catenoid jumps discontinuously into the two separated boundary circles, the "Goldschmidt solution." This process is, of course, not reversible.

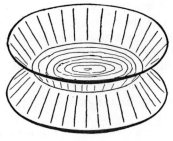

FIG. 4

3. A very instructive model is indicated by Figures 5 and 6. It is obtained from two parallel circles, such as would produce a catenoid, by cutting out two parallel segments and joining them by straight lines. Thus we have a simple closed contour. The interesting fact is that this frame produces essentially two different simply connected minimal surfaces. One is related by a slight deformation to the surface consisting of the two circular discs formed by the boundary

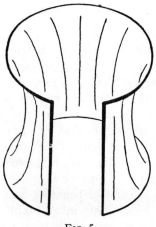

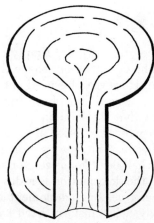

FIG. 5   FIG. 6

circles plus the connecting small strip between the two connecting straight segments. The other is a surface obtained by deformation from the cylinder between the boundary circles, after a strip of the cylinder between the two straight lines is removed. In other words, the first corresponds to the Goldschmidt solution plus the connecting strip; the second to the catenoid minus the connecting strip. The dimensions of this frame can be chosen so that the two different minimal surfaces described have approximately the same area; but even if the frame differs slightly from such a shape, both solutions are relatively stable positions of equilibrium. Therefore, we have here a case where the problem has two different simply connected solutions.

If we deform the frame, for example by pulling the two connecting lines apart or by pressing the opposite circular parts nearer together, we can make one of the solutions so much less stable than the other that at a certain moment of this deformation it becomes unstable and jumps into the other solution. By reverting the deformation of the frame, the transition from one solution to the other is reversed; one can repeat this alternating process very often without breaking the film. This is an instructive example of a problem for which the solution is not uniquely determined, and where the solution does not depend continuously on the data of the problem, namely, the frame.

4. Another even more significant example is provided by the frame of the Figures 7, 8, and 9. This frame again represents a simple curve. It permits the three different minimal surfaces as indicated in these diagrams. All of them are bounded by the same continuous curve; one of them (Fig. 9) has the genus 1,

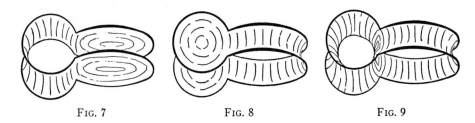

Fig. 7   Fig. 8   Fig. 9

while the other two are simply connected and in a way symmetrical to each other. The two latter ones have the same area if the contour is completely symmetrical. However, this need not be the case, and then only one of them gives the absolute minimum of the area while the other will give a relative minimum, provided that the minimum is sought among simply connected surfaces. The possibility of the solution of genus 1 depends on the fact that by admitting surfaces of genus 1 one may obtain a smaller minimum than by requiring that the surface shall be of genus zero or simply connected. By deforming our frame we must, if the deformation is radical enough, come to a point where the latter is no longer true. In this moment the surface of genus 1 becomes unstable and seemingly will transform itself by a discontinuous jump into a stable solution represented by one of the diagrams (Figs. 7 and 8) and simply connected. If we start with one of these simply connected solutions, as in Figure 7, we may deform it in such a way that the other simply connected solution in Figure 8 becomes much more stable. The consequence is that at a certain moment a discontinuous transition from type 7 into type 8 will take place. By reversing this deformation slowly we return to the initial position of the frame, but now with the solution of type 8 in it. Of course we can repeat the process in the other direction, and in this way swing back and forth by discontinuous transitions between the types in these two diagrams. By careful handling one also is able to transform discontinuously either one of the simply connected solutions into that of genus 1. For this purpose, starting, *e.g.*, with the surface 7, we have to bring the disc-like parts very close to each other so that the surface of type 9 becomes markedly more stable. Sometimes in this process intermediate pieces of film appear first and have to be destroyed before the surface of genus 1 is obtained.

This example shows not only the possibility of different solutions of the same topological type, but also of another and different type in one and the same frame; moreover, it again illustrates the possibility of discontinuous transitions from one solution to another while the conditions of the problem are changed continuously.

It is very easy to construct more complicated models in a similar way and to study their behavior experimentally. This may be left to the reader.

5. A phenomenon of general interest occurs when the simply connected minimal surfaces in a single contour is necessarily self-intersecting. Such self-intersecting surfaces are always unstable because, as is easily seen, they can be replaced by other surfaces with smaller area and the same contour. Therefore in corresponding experiments the surfaces of a simple topological character will not form and instead we automatically obtain surfaces of a higher topological structure. This is the case, for example, if the given contour is knotted; a beautiful one-sided minimal surface is obtained by a frame in the form of a cloverleaf, as in Figure 10.

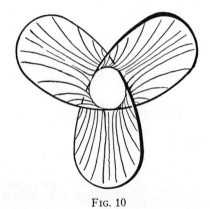

Fig. 10

Closely related to this phenomenon is the appearance of minimal surfaces bounded by two or more interlocking closed curves; for example, two circles. If we first consider the two simply connected minimal surfaces bounded by the two curves respectively, we can immediately see that they must intersect each other. Such an intersecting pair of surfaces does not give the smallest possible area. Therefore we obtain for interlocking curves doubly connected solutions bounded by the two curves. They can easily be realized by the experiment, while it may be hard to visualize them *a priori*. Also in this experiment intermediate films will have to be pierced. Incidentally, it is recommended to move the two interlocking curves into different positions, and to study the changes and discontinuous transitions occurring. Figures 11 and 12 show two different surfaces for the same two interlocking circles in the same relative position; the different orientation of the circles with respect to the surfaces should be noticed.

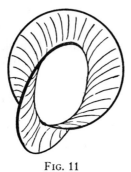

FIG. 11   FIG. 12

6. Interesting minimal surfaces are obtained by experiments in which parts of the boundary are left free to move on a given surface S. As an example, it is recommended to use the frame indicated in Figures 13, 14, and 15. A curve joining two points on a surface S (in our diagram S is not indicated; it is supposed to be approximately a horizontal plane) is the fixed boundary, while the free boundary is on the surface. If the fixed curve stays far enough away from the surface, we obtain a simply connected minimal surface (Fig. 13). If, however, the flat part of this fixed curve comes down near enough to the sur-

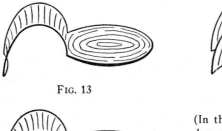

FIG. 13   FIG. 14

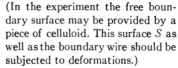

(In the experiment the free boundary surface may be provided by a piece of celluloid. This surface S as well as the boundary wire should be subjected to deformations.)

FIG. 15

face, the stable solution will be a doubly connected minimal surface which has one closed boundary line on the prescribed boundary surface besides the second boundary line on the surface, which completes the fixed boundary curve to a closed curve (Fig. 15). Both solutions can easily be shifted discontinuously into each other. There is a second possibility of a simply connected surface (Fig. 14). Note the relationship of Figures 13, 14, and 15 to Figures 7, 8, and 9, respectively.

All these examples and many others, which the reader will easily construct himself after having some experience, illustrate a general principle which was proved in the recent theoretical investigations concerning minimal surfaces; namely, a solution of a given topological type always exists if for such a type we have greater stability (that means a smaller minimum of the area) than for lower types. For example, the Moebius strip in Example 1 exists as soon as the Moebius strip type gives a smaller area than the disc type. As regards uniqueness and continuity, no general theory has yet been developed. But it is obvious that discontinuous dependence and non-uniqueness are connected with each other.

Finally, it may be remarked that soap film experiments can serve to illustrate not only the theory of minimal surfaces but also that of other minimal problems. The spherical soap bubble, for example, is nothing but the physical realization of the isoperimetric character of the sphere. This isoperimetric problem can be generalized in the following way:

Among all surfaces through a given contour, which together with a prescribed surface in this contour include a given volume, that with least area is sought. The solution must be a surface of constant mean curvature (the minimal surfaces have mean curvature zero). Mathematically this problem is non-linear to a higher degree than that of the minimal surfaces. The theoretical solution has not yet been completed. But soap film experiments very easily "solve" such problems. The given volume simply has to be furnished by blowing air into soap bubbles which are forced into a prescribed boundary. For example, if our prescribed boundary is a square, and if the surface of the square is the prescribed boundary surface, then we may start with our cube. Dip it into the soap solution and blow up a soap bubble from inside until it fills the whole interior of the cube; then by continuing we obtain six surfaces of constant mean curvature as desired.

Interesting phenomena can be observed if we combine the original Plateau soap film experiments with the isoperimetric principle, *e.g.*, we may consider the problem: To find a minimal surface of least area bounded by k given fixed curves and one variable curve of prescribed length. In the corresponding experiments the undetermined curve may be represented by a silk thread; if we pierce the film inside of this curve, the remaining minimal surfaces must form the solution, while the silk thread takes on the optimal position. If the given boundary is plane then this position is, of course, a circle.

The analysis of the corresponding mathematical problems, as well as of problems referring to the phenomena of intermediate films, is still an open question.

Editorial Note: For further reading, see J. C. C. Nitsche, Plateau's problems and their modern ramifications, this MONTHLY, vol. 81 (1974), 945–68.

ON THE DENSEST PACKING OF SPHERICAL CAPS*

L. FEJES TÓTH, Budapest, Hungary

In the present note we shall give a new proof of the following result.

THEOREM [1]. *From $n > 2$ given points of the surface of the unit sphere there always can be found two, having a spherical distance*

$$(1) \qquad d \leqq \arccos \frac{\cot^2 \omega - 1}{2}, \qquad \omega = \frac{n}{n-2}\frac{\pi}{6}.$$

The inequality (1) can not be improved for $n = 3, 4, 6, 12$ [2] and it gives an exact asymptotic estimate for large values of n.

The following inequality, equivalent to (1), improves certain results of A. Thue [3] concerning the densest packing of circles in the plane:

Consider $n > 2$ congruent spherical caps of the unit sphere such that no two of them overlap. If f denotes the area of a cap then

$$\frac{nf}{4\pi} - \frac{n}{2}\left(1 - \frac{1}{2}\sin^{-1}\omega\right) < \frac{\sqrt{3}\,\pi}{6}.$$

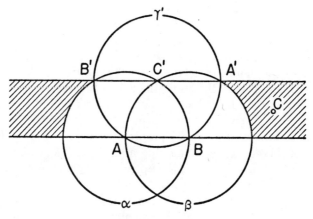

The proof of statement (1) rests on the following lemma.

LEMMA. *If the area of a spherical triangle ABC is less than the area of the equilateral spherical triangle ABC' drawn upon the shortest side AB of ABC then the spherical radius of the circle circumscribed to ABC is greater than AB.*

For suppose that C' lies on the same side of the great circle AB as C. Consider the congruent circles α, β, γ' of radius AB having the center, A, B, C', respectively. Let A' be the point of intersection of β and γ' different from A, and B' the point of intersection of α and γ' different from B.

---

* From AMERICAN MATHEMATICAL MONTHLY, vol. 56 (1949), pp. 330–331.

Since the triangles ABA', ABB', ABC' have the same area,† the arc of the circle through A', B', C' lying "above" the great circle AB is the locus of the vertices C^* of the triangles ABC^* the area of which remains invariant.

C lies, by supposition, in the domain bordered by the above Lexell-circle $A'B'C'$ and the great circle AB. Since, on the other hand, C lies outside α and β, the point C does lie without γ'. This proves the lemma.

Let us now turn to the proof of the inequality (1).

The case $n=3$ being trivial, we can restrict ourself to the case $n \geqq 4$. Obviously we may suppose that the points $P_1, P_2, \cdots, P_n$ lie not all on a hemisphere and thus the convex hull Π of the points contains the center O of the sphere S. Furthermore we may suppose that the polyhedron Π has only triangular faces since otherwise we could decompose them into triangles. The number of the faces of Π is then $2n-4$.

Consider the spherical net N arising by projection of the edges of Π from O upon S. Suppose that—contrary to (1)—

$$P_i P_j > b; \qquad i, j = 1, 2, \cdots, n; \; i \neq j,$$

where we denote by b the bound on the right in (1). Since b equals—as a simple computation shows—the length of the side of an equilateral spherical triangle of area $4\pi/(2n-4)$, the triangle $P_i P_j P_k$ having the least area among the $2n-4$ triangles determined by N satisfies the condition of our lemma.

Let us denote the circle circumscribed to $P_i P_j P_k$ by γ and the (spherical) center of γ by P_{n+1}. Since the spherical cap bordered by γ contains no point of $P_1, \cdots, P_n$, and the radius of γ is by our lemma greater than b, we can complete the point system $P_1, \cdots, P_n$ by P_{n+1} without loss of the property that any two points of the system have a spherical distance greater than b.

This proceeding can be continued. But since the number ν of the points of a system having the above property is obviously bounded [4] we arrive in a finite number of steps to a contradiction. This completes the proof.

Comments

1. See my paper Über eine Abschätzung des kürzesten Abstandes zweier Punkte eines auf einer Kugelfläche liegenden Punktsystems, Jahresbericht d. D.M.V., vol. 53 (1943), pp. 66–68. According to the kind information of Professor H. Hadwiger, he also found the inequality (1) and presented it in a lecture at the Mathematical Seminary of the University of Bern in the winter semester 1942–43.

2. The extremal distribution of the points is determined, for these values of n, by the vertices of an equilateral triangle inscribed in a great circle, of a regular tetrahedron, of an octahedron or of an icosahedron, respectively.

3. Cf., for instance, the paper of B. Segre and K. Mahler, On the densest packing of circles, this MONTHLY, vol. 51 (1944), pp. 261–270.

4. Consider the spherical caps of radius $b/2$ having the top points $P_1, P_2, \cdots, P_\nu$. From the fact that the sum of the areas of these caps is less than 4π, we obtain, for instance, the bound $\nu < \sin \mu^2 (b/4)$.

---

† This follows from the fact that triangles ABC', $AB'C'$, and $A'BC'$ are congruent, hence equiareal.—Ed.

EUCLIDEAN n-SPACE ANALOGUE OF A THEOREM OF GALILEO*

JOSEPH BARNETT, JR., Oklahoma Agricultural and Mechanical College

1. Introduction. The purpose of this paper is to establish an n-space analogue of the following theorem due† to Galileo: *The area of a circle is the mean proportional between the areas of any two similar polygons, of which one is circumscribed about the circle, and the other is isoperimetric with the circle.*

2. Theorem. We shall establish the following result.

THEOREM: *Let V_1 denote the n-dimensional content of a sphere in Euclidean n-space, V the n-dimensional content of a circumscribed polytope, and v that of a similar polytope whose $(n-1)$-dimensional content is the same as that of the sphere. Then $Vv^{n-1} = V_1^n$.*

3. Proof. Let r denote the radius of the sphere, and S_1, S, and s respective $(n-1)$-dimensional contents, so that $s = S_1$ by hypothesis. Since the circumscribed polytope is composed of pyramids of altitude r, having a common apex at the center of the sphere, we have $V = Sr/n$, and similarly $V_1 = S_1 r/n$. Since the two polytopes are similar, it follows that

$$\left(\frac{V}{v}\right)^{1/n} = \left(\frac{S}{s}\right)^{1/n-1}.$$

Solving for v, we get $v = V(s/S)^{n/n-1}$. Therefore we have

$$Vv^{n-1} = \left(\frac{Sr}{n}\right)\left(\frac{Sr}{n}\right)^{n-1}\left(\frac{S_1}{S}\right)^n = \left(\frac{S_1 r}{n}\right)^n = V_1^n.$$

4. Conclusion. Hence we have a direct generalization‡ of Galileo's theorem: *The volume of a sphere in Euclidean n-space is the geometric mean of the volume of a circumscribed polytope, and that of a similar polytope whose $(n-1)$-dimensional content is the same as that of the sphere, with weights 1 and $n-1$, respectively.*

---

* From AMERICAN MATHEMATICAL MONTHLY, vol. 57 (1950), p. 677.

† The writer found the above theorem entitled *Galileo's Theorem* as Ex. 135, pp. 377, in "Plane and Solid Geometry," 1899, by William J. Milne.

‡ The writer is greatly indebted to Professors E. F. Beckenbach and H. S. M. Coxeter for their aid in the preparation of this paper.

NON-EUCLIDEAN GEOMETRY*

HERBERT BUSEMANN, University of Southern California

Introduction. The root of non-euclidean geometry is the problem whether our ordinary geometry† is the only possible one. For centuries mathematicians were convinced that the answer must be positive and tried to prove this. Since all postulates of Euclid seemed self evident, except for the *parallel axiom,‡* the efforts were directed towards establishing this axiom as a consequence of the others. *The failure of all these efforts led Gauss, Bolya and Lobachewsky during the first third of the 19th century to the conclusion that the parallel axiom is not a logical necessity but that consistent so-called non-euclidean geometries exist, in which this axiom does not hold.*

Since then mathematical research has enlarged our concepts regarding geometry and space to such an extent that the non-euclidean geometries appear now as very special examples of general geometric structures which are not euclidean (so that the terms non-euclidean and not euclidean are not synonymous). From the modern point of view the euclidean and the non-euclidean geometries resemble each other very closely. They can be derived jointly from a principle which was discovered by the physicist Helmholtz.§

In an implicit form the principle appears in Euclid's ''proof'' of the first congruence theorem: the triangles ABC and $A'B'C'$ are congruent if $AB = A'B'$, $AC = A'C'$ and $<BAC = <B'A'C'$. Euclid *moves* $<B'A'C'$ until it falls on $<BAC$. Then B' must fall on B and C' on C, etc. Clearly this argument is not the same type of proof as Euclid's other deductions. In fact the presupposed mobility of figures is equivalent to the congruence theorem and should therefore have been stated as an axiom. Helmholtz emphasized that *the mobility of rigid objects, the fact that we recognize an object independently of its position in space, is one of our basic physical experiences.* He therefore put, and partially solved, the mathematical problem of finding those spaces or geometries in which figures can be moved freely

The answer to the problem is that the euclidean and the non-euclidean geometries have that property and that no other geometry does. A complete proof of this fundamental fact is beyond the scope of the present article, but one typical argument will indicate the procedure by which the result is obtained.

---

* From MATHEMATICS MAGAZINE, vol. 24 (1950), pp. 19–34.

† The ordinary geometry, with which the reader was first acquainted in high school, is the main subject of Euclid's Elements. This book remained the standard textbook on geometry until the end of the last century. Therefore ordinary geometry is also called euclidean geometry.

‡ The parallel axiom states: if a line g and a point P not on g are given, then the plane determined by P and g contains exactly one line through P which does not intersect g. The formulation of the axiom in Euclid's Elements is slightly different, but equivalent.

§ There are other unifying principles; for instance, projective geometry furnishes such a principle.

2. *Spherical geometry*. The reader is familiar with one non-euclidean geometry without realizing it, namely with the geometry on a sphere. Let Σ be (the surface of) a sphere with radius ρ and center O in the ordinary or euclidean space. Distances on Σ will be measured not in the space but along Σ. Airplane travel has made everyone familiar with the fact that shortest connections on Σ are arcs of great circles. A great circle is a circle in which a plane through O intersects Σ. From the point of view of spherical geometry going along a great circle means choosing the most direct or straight route from one point to another. The great circles play therefore on Σ the role of straight lines and we shall frequently call them so.

If A,B are points on Σ and α is the radian measure of $\sphericalangle AOB$ then the length of the great circle arc (or the segment) from A to B is $\rho\alpha$, the spherical distance of A and B.

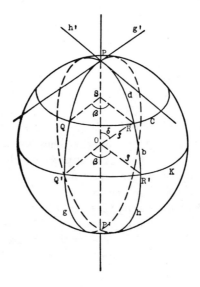

Figure 1

Let the two great circles g and h intersect at P and P'. These points are the intersections with Σ of the line which is common to the planes in which g and h lie. P' and P are diametrically opposite points of Σ or antipodes.

The circle C with radius d and center P on Σ is by definition the locus of those points which have spherical distance d from P. Let $R(Q)$ be one of the two intersections of $h(g)$ with C. If $d = \rho.\delta$ then $\delta = \sphericalangle QOP = \sphericalangle ROP$. Now C is also the euclidean circle whose radius is the perpendicular distance RS of R from the line PP'. The figure shows that $RS/RO = RS/\rho = \sin \delta$ or $RS = \rho \sin \delta$ so that C has length

$$(1) \qquad\qquad L = 2\pi \rho \sin \delta = 2\pi \rho \sin \frac{d}{\rho}$$

This is the first of a series of non-euclidean formulas to be derived here. It solves the problem: given the radius d of a circle, find its length. The euclidean answer is $2\pi d$, the spherical answer is $2\pi\rho \sin d/\rho$. The constant ρ is typical for the specific spherical geometry under consideration. *It is a length and can be determined by purely geometric means* (for instance, the distance of any two antipodes is $\pi\rho$), *whereas in euclidean geometry no definite length can be determined by geometry alone.* We must use physical aids like the standard meter at Paris. The existence of absolute lengths is one of the main differences between the euclidean and the two non-euclidean geometries.

The relation $\sin x/x \to 1$ for $x \to 0$, which is familiar from calculus, and (1) show that

$$2\pi \rho \sin \frac{d}{\rho}/2\pi d = \sin \frac{d}{\rho}/\frac{d}{\rho} \to 1 \text{ for } d \to 0.$$

This means that for d which are small compared to ρ the spherical length of a circle is approximately equal to the euclidean length $2\pi d$. We recognize here a special case of a procedure which we use all the time, namely, *the application of euclidean geometry to small domains on the earth.*

Among the circles about P there is a spherical straight line k, the equator for P and P' as poles, which corresponds to the value $\delta = \pi/2$. Viewed from k the previously considered circle C appears as the locus of those points which have the same distance $b = QQ' = RR' = \rho(\pi/2 - \delta)$ from k (see Figure 1). Hence, *in spherical geometry the locus of the points which are equidistant to a straight line is not a straight line.* Also the lengths of the corresponding arcs $Q'R'$ and QR of k and C are not equal. for if β is the radian measure of the angle $\sphericalangle Q'OR'$ and $\sphericalangle QSR$ then $Q'R'$ has length $s = \rho\beta$ and the arc QR of C has length $t = \beta\rho \sin \delta$ so that

$$(2) \qquad\qquad t = s \sin \delta = s \cos(\pi/2 - \delta) = s \cos b/\rho$$

That t is proportional to s was obvious beforehand, (2) determines the factor of proportionality.

Hitherto only distances or lengths of curves were considered. In euclidean geometry angle and area are two other fundamental concepts. The same goes for spherical geometry. There is no question regarding angles, because locally, or close to a fixed point P, the spherical geometry is almost euclidean, so that angles with vertices at P can be measured in terms of the local euclidean geometry. This amounts to the following: instead of operating on Σ we operate on the tangent plane π to Σ at P. The two great circles g and h have tangents g' and h' at P which lie in π. As measure of the angle between g and h we take the measure of the euclidean angle of g' and h' in π (any unit can be used. Here we always use radian measure). Then the previously used β is also the measure of one of the angles formed by g and h, and we find

(3)
$$C = \beta\rho \, \sin \frac{d}{\rho}$$

as spherical formula for the length of a circular arc with radius d which subtends the spherical angle β at its center P.

Everyone knows the importance of *Pythagoras' Theorem* in euclidean geometry. Unfortunately it is frequently formulated as a relation between areas. The reader, who remembers the applications of the theorem (for instance in trigonometry) will realize that its importance is due to the fact that *it expresses one side of a right triangle in terms of the other two sides*. The corresponding problem is just as important for spherical geometry. To solve it consider on Σ the triangle ABC with a right angle at C. Denote the (spherical) lengths of the sides AB, BC, CA by c, a, b respectively (all sides are assumed to be shortest connections of their endpoints). Let the tangent plane π of Σ at A (see Figure 2) intersect the line OB in B' and OC in C'. The plane is perpendicular to the radius OA and therefore to the plane OAC. The plane OBC is also perpendicular to OAC because the angle at

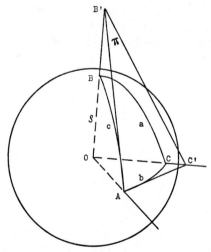

Figure 2

C is a right angle. Therefore the line $B'C'$ is perpendicular to the plane OAC, hence

$$\angle B'C'A = \angle B'C'O = \angle B'AO = \angle C'AO = \pi/2$$

Since $\angle AOB' = c/\rho$, $\angle B'OC' = a/\rho$, $\angle C'OA = b/\rho$ we find

$$OA/OB' = \cos c/\rho, \quad OC'/OB' = \cos a/\rho, \quad OA/OC' = \cos b/\rho$$

and because of $OA/OB' = (OC'/OB') \cdot (OA/OC')$

(4) $\cos c/\rho = \cos a/\rho \cdot \cos b/\rho$ (*Spherical Theorem of Pythagoras*)

Those who are familiar with the expansion of $\cos x$ see that

$$1-\frac{c^2}{2\rho^2}+\ldots=(1-\frac{a^2}{2\rho^2}+\ldots)(1-\frac{b^2}{2\rho^2}+\ldots)=1-\frac{a^2}{2\rho^2}-\frac{b^2}{2\rho^2}+\frac{a^2b^2}{4\rho^4}+\ldots$$

or

$$c^2 = a^2 + b^2 + \ldots,$$

where the omitted terms contain at least ρ^2 in the denominator. This confirms again that the *euclidean geometry holds for triangles whose sides are small compared to ρ.*

The angle at A of the spherical triangle ABC equals (by our agreement) the euclidean angle $\angle B'AC'$. Therefore

(5) $\qquad \cos A = \dfrac{AC'}{AB'} = \dfrac{AC'/OA}{AB'/OA} = \dfrac{\tan b/\rho}{\tan c/\rho} = \dfrac{\sin b/\rho}{\sin c/\rho} \cdot \dfrac{\cos c/\rho}{\cos b/\rho}$

By (4) the last factor on the right equals $\cos a/\rho$. To evaluate the preceding factor observe that

(6) $\qquad \sin A = \dfrac{B'C'}{AB'} = \dfrac{B'C'/B'O}{AB'/B'O} = \dfrac{\sin a/\rho}{\sin c/\rho}$

Since A and B play the same role in the triangle ABC, the relation

(7) $\qquad \sin B = \dfrac{\sin b/\rho}{\sin c/\rho}$

must also hold. (5) and (7) yield

(8) $\qquad \cos A = \sin B \cdot \cos a/\rho$

Contrary to all our euclidean experiences (8) reveals that *in spherical geometry the angles of a triangle determine its sides.* A theory of similar figures does therefore not exist. *There is no such concept as shape independent of size.* This is, of course, closely related to the previously discussed existence of an absolute length.

Finally we turn our attention to area. A two-gon on Σ is a figure bounded by two semi great circles h, k with diametrically opposite points P,P' as vertices. Obviously the area A of a two-gon is proportional to the angle α at its vertices:

$$A = \omega \cdot \alpha$$

where ω is the factor of proportionality. Consider a triangle PQR, where Q and R lie on h and k. Let β and γ be the angles at Q and R of the spherical triangle PQR and Q',R' the antipodes to Q and R. Clearly any triangle is congruent to the triangle formed by its antipodes and has therefore the same area so that, for instance, $PQR = P'Q'R'$ and $PQR' = P'Q'R$.

Now $PQR + PQR'$ is a two-gon with angle γ, hence

(9) $PQR + PQR' = \omega\gamma$

Similarly

(10) $PQR + PRQ' = \omega \cdot \beta$

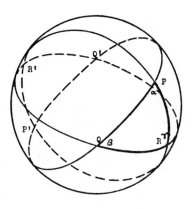

Figure 3

But $PQR' = P'Q'R$ and $PRQ' + P'Q'R$ is a two-gon with angle $\pi - \alpha$ so that

(11) $PQR' + PRQ' = P'Q'R + PRQ' = \omega(\pi - \alpha)$

Adding (9) and (10) and subtracting (11) yields

(12) $2\,PQR = \omega(\alpha + \beta + \gamma - \pi)$

The area of a triangle is therefore proportional to the excess of the angle sum over π. This confirms that the size of a triangle is determined by its angles. The constant ω depends on the unit of area used. The usual agreement is to make the area of a spherical triangle as object of spherical geometry equal to the area of the same triangle as curved surface in euclidean space. This amounts to putting $\omega = 2\rho^2$.

A noteworthy consequence of (12) is that *the sum of the angles of a spherical triangle is always greater than* π.

Spherical geometry was, of course, well known prior to 1800, in fact, it was known to Ptolemy and the Arabs. Why then, did the discovery of non-euclidean geometry come so late? Here is the reason. Spherical geometry does not satisfy some of the euclidean axioms. Two straight lines (or great circles) intersect twice. There is however a geometry, closely related to the spherical, the so-called elliptic geometry, in which straight lines intersect only once. But in elliptic geometry only figures which are not too large can be moved freely. Moreover, neither in spherical nor in elliptic geometry can we continue walking straight ahead without eventually retracing our steps. Euclid assumes that this does not happen. The pioneers looked only for geometries in which the straight lines have the euclidean property of infinite extension in both directions. Thus they discovered the other non-euclidean geometry which is called hyperbolic. Since this geometry cannot be interpreted as geometry on a surface in ordinary space it is much more recondite.

After hyperbolic geometry had been fully developed its formulas proved amazingly similar to spherical geometry and unification became natural.

3. *The hyperbolic functions.* The analogy between spherical and hyperbolic geometries becomes apparent when the so-called hyperbolic trigonometric functions are introduced.

The hyperbolic sine and cosine of x are defined by the formulas:

$$(13) \qquad \sinh x = \frac{e^x - e^{-x}}{2} \qquad \cosh x = \frac{e^x + e^{-x}}{2}$$

Clearly

$$(14) \qquad \frac{d \sinh x}{dx} = \cosh x \text{ and } \frac{d \cosh x}{dx} = \sinh x$$

Since $e^x > 0$ for all x, the function $\cosh x$ is positive, therefore $\sinh x$ increases monotonically over the whole x-range. Because $\sinh 0 = 0$ it follows that

$$\sinh x < 0 \text{ for } x < 0 \text{ and } \sinh x > 0 \text{ for } x > 0$$

Therefore (14) and the elementary theory of maxima and minima yield that cosh x reaches its minimum 1 for $x = 0$. From $e^{-x} \to 0$ for $x \to \infty$ and (13) we see that sinh $x \to \infty$ for $x \to \infty$.

Most formulas for the ordinary trigonometric functions hold without change for the hyperbolic functions. For instance

$$(15) \quad \cosh(x+y) + \cosh(x-y) = (\tfrac{1}{2})(e^{x+y} + e^{-x-y}) + (\tfrac{1}{2})(e^{x-y} + e^{-x+y})$$

$$= (\tfrac{1}{2})(e^x e^y + e^{-x}e^{-y} + e^x e^{-y} + e^{-x}e^y) = (\tfrac{1}{2})(e^x + e^{-x})(e^y + e^{-y})$$

$$= 2 \cosh x \cosh y.$$

But the reader is warned that

$$\cosh^2 x - \sinh^2 x = 1.$$

The formulas of hyperbolic geometry can be obtained from those of spherical geometry by a simple formal procedure: The angles are in both cases determined by the local euclidean geometry and are therefore treated alike. This means that *all functions of angles are left as ordinary trigonometric functions. But the ordinary trigonometric functions of the sides are replaced by the corresponding hyperbolic functions.* For instance the hyperbolic Theorem of Pythagoras is

$$\cosh c/\rho = \cosh a/\rho \cosh b/\rho,$$

whereas (8) becomes

$$\cos A = \sin B \cosh a/\rho.$$

Thus the angles of a triangle determine also in hyperbolic geometry its sides. ρ is again a positive constant typical for the specific hyperbolic geometry under consideration, but it does not have such a simple euclidean interpretation as in spherical geometry.

That *straight lines in hyperbolic geometry intersect at most once* can be seen as follows: Let the lines g and h form the angle α at P.

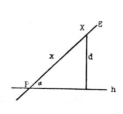

On g lay off from P the distance x to the point X and let d be the perpendicular distance of X from h. The hyperbolic formula which is analogous to (7) shows that $\sin \alpha = \dfrac{\sinh d/\rho}{\sinh x/\rho}$ or sinh $d/\rho = \sin \alpha \cdot \sinh x/\rho$. It was proved above that sinh x/ρ increases monotonically from 0 to ∞ as x goes from

Figure 4

0 to ∞. Therefore sinh d/ρ tends monotonically to ∞ and so does d because of the monotonicity of sinh x. This implies of course that X will never lie on h when $x > 0$. (In the spherical case X lies on h for $x = \rho\pi$, because $\sin \rho\pi/\rho = 0$.)

The preceding discussion is nothing but a purely formal procedure for deriving hyperbolic facts from spherical ones. No geometric argument was advanced for justifying this procedure. The next section will indicate how the mobility principle leads to such arguments.

4. *Mobility of figures. A motion of the euclidean plane* Π *or of the sphere* Σ *is a mapping f of* Π *(or* Σ*) on itself which preserves distances*, which means that the images $f(P)$ and $f(Q)$ of any two points P,Q of Π (or Σ) have the same distance as P and Q:

(16) $$f(P)f(Q) = PQ$$

As example we consider *rotations*. On Π or Σ choose an arbitrary point R and an oriented angle α. A rotation f about R through α is

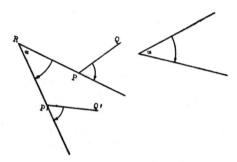

Figure 5

defined as follows: 1) $f(R) = R$ 2) For any point $P \neq R$ we lay off the angle $\sphericalangle PRP' = \alpha$ with RP as initial and RP' as terminal side and choose $P' = f(P)$ such that $RP' = RP$. The reader will easily prove that (16) holds for rotations of Π or Σ.

Reflections in a line give a second example of motions. Let g be a straight line on Π or Σ. For any R on g we put $f(R) = R$. If P is

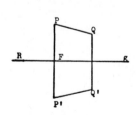

not on g let F be a point of g closest to P (the point F is uniquely determined unless (on Σ) P happens to be a pole to g as equator). On the prolongation of PF beyond F we choose $P' = f(P)$ such that $P'F = PF$. The reader will again easily confirm (16). The present definition of rotations and reflections carry over to hyperbolic geometry.

Figure 6   *The most general motion can be composed from rotations and reflections.* Let ABC and $A'B'C'$ be any two congruent triangles in Π or Σ. If R is any point which has the same distance from A and A', then a rotation f_1 about R through a suitable angle will carry A' into A or $f_1(A') = A$.

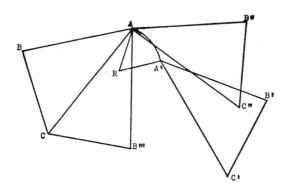

Figure 7

Let $f_1(B') = B''$, $f_1(C') = C''$. Then the triangle $AB''C''$ is congruent to $A'B'C'$ and therefore to ABC. Since $AC = AC''$ a rotation f_2 about A (through $C''AC$) will carry C'' into C or $f_2(C'') = C$. Let $f_2(B'') = B'''$. Since ACB''' is congruent to ACB, the point B''' either coincides with B or (as in the figure) B can be obtained from B''' by a reflection in the line AC. In any case we can move the triangle $A'B'C'$ into the triangle ABC by combining at most two rotations and a reflection.

The spherical case exhibits a phenomenon which is foreign to the other two geometries: a rotation f about R also moves a straight line h, the equator to R as pole, into itself. Or in the standard terminology: f is a *translation* of Σ along h. In the other two geometries translations can be combined from rotations (or reflections) but they are not themselves rotations. They are defined as follows: Let an oriented **straight** line h and a positive number x be given.

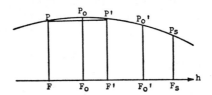

Figure 8

To find the image $P' = f(P)$ of a given point P drop the perpendicular PF from P on h ($P = F$, if P lies on h). From F lay off x in the positive direction on h to the point F'. On h erect at F' a perpendicular $F'P'$ on the same side as P and such that $F'P' = FP$. Then $P' = f(P)$. If x changes $f(P)$ varies on an equidistant curve k to h because every point of k has distance PF from h.

Only in the euclidean case is k a straight line, f is then also a translation along k. In the hyperbolic and spherical cases f cannot be interpreted as translation along any other line but h.

The mobility principle is the requirement that congruent figures can be moved into each other. Here we restrict ourselves to the two-dimensional case. It then suffices to require mobility for triangles. A rigorous form of the principle is this:

(17) $$If \ AB = A'B', \ BC = B'C', \ CA = C'A'$$

then a motion f exists such that $f(A') = A$, $f(B') = B$, $f(C') = C$. Since a combination of motions is again a motion the preceding considerations show that euclidean and spherical geometries satisfy the mobility principle. We want to understand why hyperbolic geometry is the only other geometry in which this principle holds.

The problem presupposes a set of general geometries from which the mobility principle selects the three geometries. How can "general geometries" be defined? (17) shows that the concept of distance is essential. We therefore assume that our space is such that for any two points x,y a distance xy is defined and has the following three properties whose content will be clear without further comment:

1) $$xy = 0 \text{ means } x = y$$

2) $$\text{If } x \neq y, \text{ then } xy = yx > 0$$

3) $$xy + yz \geqslant xz$$

These conditions are, of course, still much too weak as a foundation. For instance, take any set of points and define $xy = yx = 1$ for $x \neq y$ and $xx = 0$. Then properties 1), 2), 3), and (17) are satisfied (to see the latter we define f at the points A', B', C' as in (17) and $f(P) = P$ for $P \neq A'$, B', C', A, B, C, finally $f(A) = A'$, $f(B) = B'$, $f(C) = C'$). It is clear what is wrong with the example: it is impossible to go continuously from one point to another. For our problem *the distance of x and y must be the length of the shortest route from x to y.* In exact language we require

4) If $x = y$ then a curve $z(t)$, $0 \leq t \leq xy$ from x to y (that is $z(0) = x$, $z(xy) = y$) exists such that $z(t_1)z(t_2) = |t_1 - t_2|$

This implies that for $t_1 < t_2 < t_3$

$$z(t_1)z(t_2) + z(t_2)z(t_3) = |t_1 - t_2| + |t_2 - t_3|$$

$$t_2 - t_1 + t_3 - t_2 = t_3 - t_1 = |t_1 - t_3| = z(t_1)z(t_3)$$

so that distances along a shortest route add up (instead of only 3)). Requirement 4) excludes the above example, but it also excludes Σ, if the distances of ordinary space are used instead of the spherical distances along great circles (going through the earth instead of along its surface).

4) guarantees only that y can be reached from s by a curve of length xy. It is also necessary to know that we can go beyond y, at least if y is not too far from x. (On Σ the antipode P' to a given P is a point beyond which it is not possible to go if we start at P). Moreover we need to know that there is only one shortest route from x to y, again only if y is not too far from x. (There are infinitely many different shortest routes from P to P' on Σ). All this can be easily expressed by conditions similar to 4). But 1) to 4) will suffice to give the reader an idea how the problem of defining ''general geometries'' is approached in modern mathematics.

Without going into further details we assume that *every point P is contained in a plane piece in which shortest connections are unique and can be prolonged.* If P is on Σ, a hemisphere with P as pole would be such a piece. It is true that a hemisphere is not a piece of a plane in the usual sense, but it can be bent and stretched into one if we imagine it to be of rubber. After this bending process we define the distance of two atoms x and y as equal to the distance of the same atoms in their original position on Σ. Thus we obtain a plane piece with a spherical geometry. Intrinsically the hemisphere does not differ from the plane piece. A difference appears only if we try to imbed the piece in ordinary space, that is derive its geometry from the geometry of the surrounding space as we did in section 1. In the plane form it cannot be imbedded. The reader will wee that the following argument is entirely intrinsic.

Consider a straight line h and an equidistant curve k to h obtained by translation of a point P along h (see Figure 8). If the feet F, F' of P and P' are such that they have the same distance (and orientation) as the feet F_0, F_0' of P_0 and P_0', then the figures $PFF'P'$ and $P_0 F_0 F_0' P_0'$ consisting of three straight segments and an arc of k each are congruent, because one figure can be obtained from the other by translations along h. Therefore the arcs PP' and $P_0 P_0'$ of k have equal lengths. We see that equal arcs of k correspond to equal distances of their feet on h. Therefore, if F_s is the point on h obtained by laying off the distance s from F_0 in the positive direction and P_s is the corresponding point of k, then the length t of the arc $P_0 P_s$ must be proportional to s; the factor of proportionality depends on the distance $b = PF = P_0 F_0 = P_s F_s$, or in a formula,

$$(18) \qquad t = \phi(b) \cdot s.$$

We have to determine the function $\phi(b)$. For that purpose construct the equidistant curves to h (all on the same side of h) at distances x, $x + y$, and $x - y$ from h where $x > y > 0$. Divide the segment from F_0 to F_s into n equal parts, where n is any even positive integer.

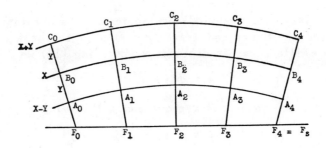

Figure 9

Denote the points of division by F_0, F_1, $F_2, \ldots, F_n = F_s$, and the points corresponding to F_i on the equidistant curves by A_i, B_i, C_i as indicated in the figure which corresponds to $n = 4$. The relation (18) yields for the length of the arc of the equidistant curve

(19) $\qquad$ length $A_0 A_n = \phi(x - y) \cdot s$, length $B_0 B_n = \phi(x) \cdot s$,

$$\text{length } C_0 C_n = \phi(x + y) \cdot s$$

The quadrangles $A_i C_i C_{i+1} A_{i+1}$ bounded by the arcs $A_i A_{i+1}$ and $C_i C_{i+1}$ of equidistant curves and the straight segments $A_i C_i$ and $A_{i+1} C_{i+1}$ are all congruent. By using the mobility principle we put these quadrangles together in a new way, namely we put every second one upside down, so that A_1 coincides with C_1, etc. as shown in Figure 10.

In the new figure the points $B_0, \ldots, B_n$ lie on a straight line. The length of the middle arc is the same as before, that is $z = \phi(x) \cdot s$ because it consists of the same arcs. The upper and lower arcs consist each of half the previous upper and lower arcs and have therefore

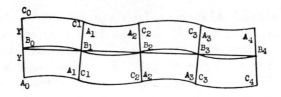

Figure 10

length (compare (19))

$$w = \tfrac{1}{2}(\text{length } A_0 A_n + C_0 C_n) = \tfrac{1}{2}[\phi(x - y) + \phi(x + y)] \cdot s$$

The segments $C_i A_i$ are still perpendicular to the new middle curve.

Therefore as n increases the middle curve tends to a straight segment of length z and the upper and lower curves tend to the equidistant curves to the middle curve at distance y. By using (18) we find

$$w = \phi(y) \cdot z,$$

and substituting the values of w and z and multiplying by 2

(20) $$\phi(x + y) + \phi(x - y) = 2\phi(x) \cdot \phi(y).$$

We have reduced the problem to solving this so-called functional equation for ϕ. Everyone notices at once the trivial solution $\phi(x) \equiv 0$, but this is excluded because $\phi(0) = 1$, which expresses that for $b = 0$ or $x = 0$ the points P_s and F_s coincide so that $t = s$. Another obvious solution is $\phi(x) \equiv 1$. This leads to the euclidean geometry in which P_s moves on a straight line and $t = s$.

From trigonometry the reader is familiar with the identity

$$\cos (x + y) + \cos (x - y) = 2 \cos x \cos y$$

Therefore $\cos x$ and, more generally, $\cos ax$, where a is any constant, is a solution of (20). Formula (15) shows that $\cosh ax$ is also a solution of (20).

It is quite within the reach of a sophomore to see that 0, 1, $\cos ax$, and $\cosh ax$ are the *only continuous solutions* of (20), but the proof will be omitted due to lack of space. If we replace a by $1/\rho$ we can say (compare (2)):

If the mobility principle holds, then the length t of the arc of an equidistant curve to a straight line at distance b has one of the following three forms:

$$t = s, \quad t = \cos b/\rho \cdot s, \quad t = \cosh b/\rho \cdot s$$

The third expression originates from the second by the formal procedure explained in Section 2. But now it was obtained by geometric arguments. In the same way, not only all the hyperbolic formulas can be derived, but the reader sees that the euclidean and spherical formulas appear at the same time and that *no other formulas are compatible with the mobility principle.*

5. *Hyperbolic geometry.* After the procedure of Section 2 has been justified hyperbolic geometry can be legitimately based on it. We describe here, without proofs, some of its properties.

An equidistant curve k to a straight line h is a straight line in euclidean geometry. In spherical geometry it is a circle whose center is on the other side of k from h. Therefore k turns its convex side towards h. In hyperbolic geometry k is neither a circle nor a straight line, but a new kind of curve which turns its concave side towards

h as in Figure 8. The segment $P_s F_s$ forms with k (and h) a right angle. Therefore the straight quadrangle $PP'F'F$ has at P and P' angles which are (equal and) smaller than $\pi/2$, so that the angle sum in $PP'F'F$ is less than 2π. This leads to the fact that

In hyperbolic geometry the sum of the angles of a triangle is less than π.

The area of a hyperbolic triangle with angle α, β, γ *is proportional to $\pi - \alpha - \beta - \gamma$, and remains therefore below a fixed number although triangles with arbitrary large sides exist* (for instance, equilateral ones).

Consider a straight line g in the hyperbolic plane and a point P not on g. Drop the perpendicular PF from P on g and let X traverse one of the rays of g determined by F. The line PX turns monotonically

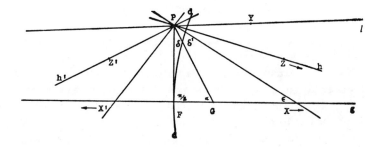

Figure 11

and tends to a limit position h. If G is a fixed point on the same ray as X we find, with the notation of Figure 11,

$$\pi/2-(\delta+\delta')>\pi/2-(\delta+\delta'+\epsilon)=\pi-(\alpha+\delta+\pi/2)+\pi-(\delta'+\epsilon+\pi-\alpha).$$

The parentheses on the right are the angle sums in the triangles PFG and PGX respectively. Since these are smaller than π

$$\pi/2 - (\delta + \delta') > \pi - (\alpha + \delta + \pi/2) > 0.$$

Because P, F, G do not depend on X we have

$$\lim \sphericalangle FPX = \lim(\delta + \delta') = \sphericalangle FPZ < \pi/2$$

By reflecting this figure in the line PF we see: if X' traverses the other ray of g then

$$\sphericalangle FPZ' = \lim \sphericalangle PFX' = \lim \sphericalangle PFX < \pi/2$$

Therefore $\sphericalangle ZPZ' < \pi$, so that the limit position h' of PX' is different from h. The lines through P inside $\sphericalangle ZPZ'$ intersect g, all the other lines through P do not intersect g. *The parallel axiom does not hold.*

A line l inside the supplementary angle to $\sphericalangle ZPZ'$ is usually called a hyper parallel to g. The distance of a variable point y on l from g reaches a minimum at some point and tends to ∞ as y tends in either direction to ∞.

If Z traverses h in the indicated direction its distance from g tends to 0. The lines h and h' are called parallels to g. The angle $\beta(d) = \sphericalangle FPZ = \sphericalangle FPZ'$ which depends only on the distance $d = PF$, is called the *parallel angle* belonging to d. Its value is given by the relation

$$\tan \tfrac{1}{2}\beta(d) = e^{-d/\rho}$$

In spherical geometry a straight line is at the same time a circle (about the pole to the line as equator). In euclidean geometry a straight line is not a circle but it is the limit of circles whose radii tend to ∞. *In hyperbolic geometry a circle with center X through F (Figure 11) does not tend to a straight line as X tends on g to ∞, but to an important new kind of curve d which is orthogonal to all parallels to the orientation of g in which X recedes.*

These facts explain why the discovery of hyperbolic geometry had such a deep influence on mathematics. Because of its many new phenomena it freed us from the shackles of euclidean tradition. *It is somewhat richer than the other two geometries.* The following fact elucidates this contention: In three dimensional hyperbolic space the geometry on a sphere is spherical (but its hyperbolic radius is not equal to the ρ belonging to the same sphere if imbedded in euclidean space). As the hyperbolic sphere becomes large it does not tend to a plane, but to a curved surface, whose geometry is euclidean. Thus both the other geometries can be derived from hyperbolic geometry as the spherical from euclidean. On the other hand, three dimensional spherical geometry does not contain models of either of the other two geometries, a result which is made plausible by the finite extent of spherical geometry.

We conclude with a few hints for the reader who wants to see the approach through mobility carried out rigorously in detail.[*]

The very ingenious idea leading to (20), whose originator I was never able to trace, is probably quite old, from the pre-critical period, because it takes for granted that the limit of the lengths of a converging sequence of curves equals the length of the limit curve. This is, in general, false, e.g., the curve $y = (\cos n^2 x)/n$, $0 \leqslant x \leqslant \pi$ in the (x,y)-plane tends uniformly to $y = 0$, its length to ∞ instead of π. In the present case the assertion is correct, but at the stage where it enters, the proof, although not hard, is quite cumbersome.

---

[*] Prof. Busemann added these conclusions and the references for the present volume.—ED.

Therefore other approaches, in all of which ϕ plays an important part, have been devised. A place where all details can be found is [1, Section 47].

Besides [1] we mention below three of the very many books on non-Euclidean geometry, which will guide the reader to the widely different possible treatments of the subject.

References

1. H. Busemann, The Geometry of Geodesics, New York 1955.
2. H. Busemann and P. J. Kelly, Projective Geometry and Projective Metrics, New York 1953.
3. H. S. M. Coxeter, Non-Euclidean Geometry, 2nd edition, Toronto 1947.
4. H. E. Wolfe, Introduction to Non-Euclidean Geometry, New York 1948. (This is by far the easiest book on the subject.)

AREA IN HYPERBOLIC GEOMETRY*

KENNETH LEISENRING, University of Michigan

1. Introduction. Because of its origin in connection with the problem of parallelism, hyperbolic geometry, the non-Euclidean geometry of Lobatchewsky and Bolyai, has from the beginning been treated in as close analogy as possible with Euclidean; particularly in synthetic discussions. However, it has long been known that hyperbolic geometry is perfectly realized on a sphere of imaginary radius. The writer believes that the key to a quick intuitive grasp of this geometry is to minimize the analogy with the Euclidean plane, and to stress instead the analogy with the real sphere.

On the sphere (or in the "elliptic plane," which may be regarded as the sphere with opposite points abstractly identified) any two "lines" both intersect and have a unique common perpendicular. In the hyperbolic plane line pairs either intersect or have a unique common perpendicular, "parallels" being the singular, intermediate case. On the sphere the common normal of two lines is greater than any other distance to one from a point of the other, while in the hyperbolic plane the common normal of two "ultra-parallels" is the shortest distance between them, and away from it in either direction the lines diverge indefinitely. Parallels are asymptotic.

The angular excess of a spherical triangle is replaced, in the hyperbolic plane, by angular deficiency. The theorem that the defect of the sum of two polygons is the sum of their defects is proved precisely as in the case of the excess. As regards area, however, there is an important difference, for the hyperbolic plane is infinite

---

* From AMERICAN MATHEMATICAL MONTHLY, vol. 58 (1951), pp. 315–322. The original title was 'Area in Non-Euclidean Geometry".

in extent. This means that the fundamental theorem, that area is proportional to deficiency, must be handled differently.

The present paper makes use of a direct axiomatic approach. It also puts to use the analogy with the elliptic case, this being our chief motivation. There is a further point of interest in the use that is made of a class of polygons most characteristic of the non-Euclidean planes, but frequently neglected, the orthogons—polygons all of whose angles are right. In the elliptic plane ortho-n-gons exist for $n=2$ and $n=3$; in the hyperbolic plane for $n \geqslant 5$. The hyperbolic plane can be tessellated with duplicates of an arbitrary ortho-n-gon, $n \geqslant 5$, without the requirement of regularity.

What makes the area problem so simple on the sphere is the finiteness of the total area and the existence of a triangle one angle of which is proportional to its area, the triangle with two right angles. The figure corresponding to this in the hyperbolic plane is not a triangle but a quadrilateral with three right angles. The area is not proportional to the remaining acute angle, but is proportional to its complement. This is the key theorem in the argument which follows; the fundamental theorem for the triangle is a simple consequence.

2. The area function. By the term "area" we shall mean here any function μ satisfying the following postulates:
1. The function μ is defined for every simple closed polygon.
2. The function μ is a real number, positive and finite.
3. If the simple closed polygons s and t are congruent, then $\mu(s) = \mu(t)$.
4. If the simple closed polygon w is the sum of two others, s and t, then $\mu(w) = \mu(s) + \mu(t)$.

For such a function we shall use the conventional terminology. We shall say that a polygon has "an area," one polygon has "greater area" than another, *etc.*

In the Euclidean plane there exists no intrinsic function μ. Euclid, in effect, assumed the existence of such functions and then showed that any two can differ only by a factor of proportionality. In the non-Euclidean geometries the situation is different: angular excess in the elliptic plane, and defect in the hyperbolic, satisfy these postulates—the argument being familiar. What remains to be shown, in each case, is that *any* function satisfying them is proportional to the excess or defect. We are concerned with the hyperbolic plane only. To be completely explicit: writing $d(s)$ for the angular defect of s, we have to show that for any function μ, we can write $\mu(s) = k^2 d(s)$. Besides the existence of the angular defect, we shall use the axiom of Archimedes, and the basic theorems on lines, such as the uniqueness of the common normal of two ultra-parallels, and the theorem that away from their common normal ultra-parallels diverge.

3. The area theorems.

THEOREM 1. *If two trirectangles have equal acute angles, then they have equal area.*

Proof: Let the two trirectangles be $t = ABCD$ and $t' = KLMN$, the equal acute angles being A and K, and let us suppose $\mu(t) \geqq \mu(t')$.

Let the figures be superimposed so that KL falls along AB and KN along AD. (See Figure 1.) If L now coincides with B, then M must coincide with C, for CD is the unique common normal of the ultra-parallels AD and BC. The trirectangles are thus congruent and have the same area by Postulate 3. If L falls between A and B, then D will lie between A and N. For, the only alternative is that N lie between A and D; but then M would lie within $ABCD$, and the hexagon $LBCDNM$ would have no angular defect.

Let us suppose that L does fall between A and B, and let LB and CD intersect in E. We see that in this case t and t' are simultaneously decomposed into new quadrilaterals: a birectangular part in which they overlap, and new trirectangles, BE and EN, which have equal acute angles at E.

These new trirectangles can be superimposed in the same way and the procedure can be carried on indefinitely, unless brought to a stop after a finite number of steps by the coincidence of the last pair of trirectangles. If it is thus brought to a stop we have $\mu(t) = \mu(t')$ by Postulates 3 and 4.

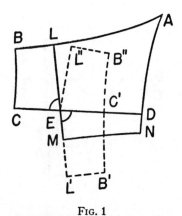

Fig. 1

To obtain a proof when the process is infinite, we must be more precise about the manner in which the successive superpositions are carried out. We shall arrange it in such a way that the process falls into distinct stages, each of which contains a finite number of superpositions. At the end of each stage it will be possible to show that the trirectangles which remain are each less than half the trirectangles with which that stage began.

Suppose in the initial superposition $EC < ED$ (Figure 1), and let the trirectangle EB be rotated through a straight angle about E to the position EB'. Let EB' be reflected in ED, giving rise to the trirectangle $EL''B''C'$. This reflection lies wholly within $ALED$. Otherwise EL'' or $C'B''$ must have a point in common with AL, which is impossible. $EL'' = EL$, and the perpendicular is shorter than any oblique segment from a point to a line. Also, by the divergence of ultraparallels, mentioned above, $C'B'' = CB$ is shorter than any other perpendicular to CD from a point of AL.

By Postulate 4, $EL''B''C'$ has a smaller area than $ALED$. And since, by assumption, $\mu(EB) \geqq \mu(EN)$, we have $\mu(BE) < 1/2 \cdot \mu(t)$ and $\mu(EN) < 1/2 \cdot \mu(t')$. This is the goal of the first stage. We now let $BE = t_1$ and $EN = t_1'$, and proceed in the same way with the second stage.

Suppose, on the other hand, $EC > ED$ in the initial superposition. In this case, the first stage must contain further superpositions, as indicated in Figure 2. Upon rotation of BE about E, the vertex E_1 of the new acute angles will fall on AN.

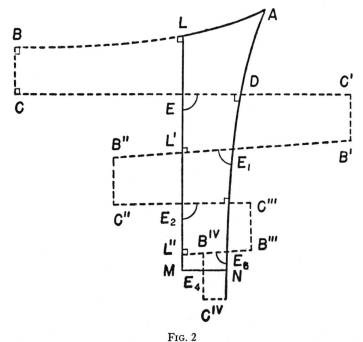

FIG. 2

We proceed to rotate E_1C' about E_1, and compare E_1B' with E_1L'. If $E_1B' > E_1L'$, as in the figure shown, we rotate again about E_2, etc. At each step in this process the sides of the successive trirectangles which are adjacent to BC, $B'C'$, etc., are diminished by an amount greater than MN. It follows from the axiom of Archimedes that the sides will become in a finite number of steps less than MN. If we consider the various possible resulting configurations and argue from the angular defect as in the first paragraph of this proof, we can rule out all alternatives but that, for some k we shall find E_{k+1} on MN. By the same reasoning as before it will follow that $\mu(E_kC^k) < \mu(E_kE_{k-1}) < \mu(E_kL)$. And since $\mu(E_kM) < \mu(E_kC^k)$ we have again reached our goal. We let $E_kC^k = t_1$, and $E_kM = t_1'$. After n stages, we can write for the difference of the original areas:

$$\mu(t) - \mu(t') = \mu(t_n) - \mu(t_n') < \mu(t_n) < 1/2 \cdot \mu(t_{n-1}) < 1/2^n \cdot \mu(t).$$

The fixed numbers $\mu(t)$ and $\mu(t')$ therefore differ by as little as we please and must be equal.

For later reference, we state the following obvious corollary:

COROLLARY 1a. *In a variable trirectangle of fixed acute angle, either side adjacent to the acute angle can be arbitrarily increased without altering the area.*

(Note that the side in question cannot be arbitrarily diminished, since the adjacent sides must remain ultra-parallel.)

THEOREM 2. *If the acute angle of one trirectangle is greater than that of another, then its area is less.*

Proof: Let the trirectangle with the greater acute angle be $ABCD$, the acute angle being at A. Consider the side $A\overset{\frown}{B}$ and one of the sides adjacent to the acute angle in the other trirectangle. Let them be made equal by extending the shorter one by Corollary 1a, and let these equal sides be superimposed as in Figure 3, where the second trirectangle is represented by $ABRS$. Segment BR

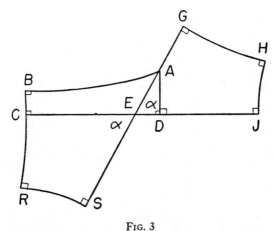

FIG. 3

will then be longer than BC. Let AS meet CD at E and call the vertical acute angles α. On EA let segment EG be marked off such that $EG > EA$ and also greater than the segment whose angle of parallelism is α. Let GH be perpendicular to EG. The line GH is then ultra-parallel to CD; let the common normal be HJ. Segment EJ is then definitely longer than ED. By Theorem 1 the trirectangles $ECRS$ and $EGHJ$ have the same area. Since triangle EAD is a proper part of $EGHJ$, the theorem follows.

The following theorem is now evidently true:

THEOREM 3. *If two trirectangles have the same area they have the same acute angle.*

Thus the area of a trirectangle is a monotonic decreasing function of its acute angle, and a monotonic increasing function of its angular defect, which is just the complement of the acute angle. It remains to show that the latter function is a proportionality.

THEOREM 4. *All orthopentagons have the same area.*

Proof: If two orthopentagons have a side of one equal to a side of the other they can be superimposed, and the conclusion then follows by Theorem 1. But any side of an orthopentagon can be arbitrarily increased without changing the area; hence, the superposition can always be carried out.

Remark: For the area of an orthopentagon we shall use the letter U.

COROLLARY 4a. *The area of an ortho-n-gon is $(n-4)U$.*

Proof: If s_1, s_2, s_3, s_4 are four consecutive sides of any ortho-n-gon, then s_1 and s_4 are ultra-parallel. Their common normal lies inside the figure, and is the side opposite s_2 and s_3 of an orthopentagon. Removing from the figure this orthopentagon leaves an ortho-$(n-1)$-gon and diminishes the area by U.

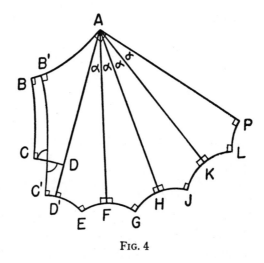

FIG. 4

THEOREM 5. *The area of a trirectangle is proportional to its angular defect.*

Proof: Since the area function has been shown to be a monotonic function of the defect, it is sufficient to show that if the ratio of the defects of two given trirectangles is m/n then this is also the ratio of the areas.

Let the trirectangles be t and t', and let $d(t)/d(t')=m/n$. Let $t=ABCD$, with the acute angle at A. Let AP be perpendicular to AB, as in Figure 4. The defect of t is then angle DAP. Let that angle be divided into m equal parts by the lines AF, AH, AK, etc., the equal angles having the value α. Now if side AD is not greater than the segment whose angle of parallelism is α, let it be extended by Corollary 1a, so that $C'D'$ is ultra-parallel to AF. Let the common normal be EF; EF is then ultra-parallel to AH, by the equality of the angles α, and similarly GH to AK, etc. The resulting figure, represented in our figure by $AB'C'EGJLP$, is an ortho-$(m+4)$-gon whose area, by Corollary 4a, is mU.

If we write μ_α for the area of one of the trirectangles with acute angle α, we have $\mu(t) = mU - m\mu_\alpha = m(U - \mu_\alpha)$. Now consider the trirectangle t'; let the complement of the acute angle be divided into n equal parts—by hypothesis these equal angles are again α. Proceeding as above, we obtain $\mu(t') = n(U - \mu_\alpha)$, which proves the theorem. For the trirectangle t we may write $\mu(t) = k^2 d(t)$.

COROLLARY 5a. *The area unit U can be written $k^2 \cdot \pi/2$.*

Proof: It is easily seen that an orthopentagon is the sum of two trirectangles whose total defect is $\pi/2$.

THEOREM 6. *The area of a triangle is proportional to its angular defect.*

Proof. Let ABC be any (proper) triangle and let the sides be extended, AB through B, BC through C, and CA through A. (See Figure 5). On these exten-

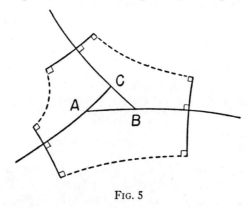

FIG. 5

sions let perpendiculars be erected, sufficiently far out to be pair-wise ultra-parallel. With the common normals, these then make an orthohexagon. Each vertex of the triangle is then the vertex of a pentagon with four right angles, a figure which is the sum of two trirectangles whose total defect is just the corresponding angle of the triangle. The area of the hexagon is $k^2\pi$. Subtracting the three pentagons, we have $\mu(\triangle ABC) = k^2(\pi - A - B - C)$.

4. The factor of proportionality. In the hyperbolic plane the factor of proportionality is strictly arbitrary; but it is natural, and customary, to make it agree with the otherwise locally Euclidean character of the plane. In a hyperbolic plane of parameter ρ the factor is then ρ^2. This can be shown in various ways; the following argument is simpler than some.

Consider an isosceles trirectangle t, the length of the sides opposite the acute angle being a. Let the complement of the acute angle be θ. By trigonometry we have then

$$\sin \theta = \sinh^2 \frac{a}{\rho} \cdot$$

The area of t is then

$$A = k^2 \arcsin\left(\sinh^2\frac{a}{\rho}\right).$$

We want it to be true that

$$\lim_{a\to 0} A/a^2 = \lim_{a\to 0} 1/a^2 \cdot k^2 \arcsin\left(\sinh^2\frac{a}{\rho}\right) = 1.$$

Since the sine and the hyperbolic sine both differ from their arguments by terms of the third degree and higher, we have at once $k^2 = \rho^2$.

Remark. For another treatment of area distinct from the classical, see H. S. M. Coxeter's book *Non-Euclidean Geometry* (Toronto, 1942), pages 241–251.

AN ELEMENTARY ANALOGUE TO THE GAUSS-BONNET THEOREM*

G. PÓLYA, Stanford University

1. We consider a polyhedral surface P of the topological type of a circular disk, bounded by a (skew) closed polygon R (the rim) without double points. We shall appropriately define S, the *area of the spherical image* of P, and T, *the total change of direction* of R on P, and prove that

$$S + T = 2\pi.$$

This is an elementary analogue to, or an elementary limiting case of, the Gauss-Bonnet theorem which itself can be considered as a limiting case of our elementary theorem (when the faces become infinitely small and their number infinitely large).

Let F, E, and V denote the number of faces, edges, and vertices of P, respectively. Since P is a (simply connected) open polyhedral surface, Euler's well-known theorem assumes here the form

$$F - E + V = 1.$$

Euler's theorem is closely connected with our analogue of the Gauss-Bonnet theorem. In fact, we shall kill two birds with one stone, and derive both theorems together.

2. Our proof needs two well-known lemmas and the introduction of appropriate notation.

---

* From AMERICAN MATHEMATICAL MONTHLY, vol. 61 (1954), pp. 601–603.

LEMMA I. *The sum of the angles in a plane polygon with n sides is $(n-2)\pi$.*

LEMMA II. *We consider the polyhedral angle H and its polar polyhedral angle H'. We call L the sum of the face angles of H, and A' the measure of the solid angle included by H'. Then*

$$A' = 2\pi - L.$$

According to the well-known definition of the polar polyhedral angles, each edge of H' corresponds, and is perpendicular to, a face of H. If we assume for a moment that H and H' have the same vertex and describe a unit sphere about this common vertex as center, each polyhedral angle intersects the sphere in a spherical polygon: the spherical polygon associated with H has the perimeter L and that associated with H' the area A'.

Let F_3, F_4, F_5, $\cdots$ denote the number of the triangular, quadrilateral, pentagonal, $\cdots$ faces of P, respectively. Obviously

(1) $$F_3 + F_4 + F_5 + \cdots = F.$$

Some of the edges of P do not, and others do, belong to the boundary R of P. Let E_i denote the number of the former (interior) edges, and E_b that of the latter (boundary) edges. Obviously

(2) $$E_i + E_b = E.$$

Some of the vertices of P do not, and others do, belong to the boundary R. Let V_i denote the number of the former, V_b that of the latter. Obviously

(3) $$V_i + V_b = V.$$

Let α stand for any angle of any of the F faces of P, $\sum \alpha$ for the sum of all such angles, $\sum_i \alpha$ for the sum of those angles whose vertex does not belong to R, and $\sum_b \alpha$ for the sum of those whose vertex does belong to R. Obviously

(4) $$\sum_i \alpha + \sum_b \alpha = \sum \alpha.$$

It is easily seen that

(5) $$3F_3 + 4F_4 + 5F_5 + \cdots = 2E_i + E_b,$$

(6) $$E_b = V_b,$$

(7) $$\sum \alpha = \pi F_3 + 2\pi F_4 + 3\pi F_5 + \cdots.$$

In deriving (7), we use Lemma I.

3. The total change of direction T of the boundary R of P is the sum of the changes of direction of R at its several vertices. If one of these vertices lies on the boundary of just one of the F faces of P, and the angle of that face at that vertex is α, the change of direction is $\pi - \alpha$. If, however, one of the vertices of R belongs to several contiguous faces of P and these faces have at that vertex the angles α', α'', α''', $\cdots$ respectively, the change of direction is $\pi - \alpha' - \alpha''$

$-\alpha''' - \cdots$. Therefore, the total change of direction is

(8) $$T = \pi V_b - \sum_b \alpha.$$

4. In taking the next step, we restrict ourselves to the most intuitive case. We assume that we started from a convex polyhedron C, drew the not self-intersecting closed polygonal line R on its surface, and obtained P as one of the two portions into which R divides the surface of C. We draw to each of the faces of P a normal (an *outer* normal to C). Then we take a unit sphere about a fixed point O of space, and draw to each normal a parallel radius. These radii intersect the surface of the unit sphere in F points; we shall use these F points in constructing the spherical image of P. In fact, let H be the polyhedral angle (interior to C) included by those faces of P that meet in one of the V_i interior vertices of P. Let us draw H', the polar polyhedral angle to H, with O as vertex. The vertices of the spherical polygon associated with H' are among the F points that we have just constructed. The full spherical image of P consists of such spherical polygons, each corresponding to an interior vertex of P. Computing the area of each of these V_i polygons on the basis of Lemma II, we find that the total area of the spherical image is

(9) $$S = 2\pi V_i - \sum_i \alpha.$$

5. We use first (9), (8), and (4), and we pass to the following lines by using (7), then (5) and (1), and finally (6):

$$\begin{aligned} S + T &= 2\pi V_i + \pi V_b - \sum \alpha \\ &= \pi[2V_i + V_b - (F_3 + 2F_4 + 3F_5 + \cdots)] \\ &= \pi[2V_i + V_b - (2E_i + E_b - 2F)] \\ &= \pi[2F - 2E_i - 2E_b + 2V_i + 2V_b]. \end{aligned}$$

Using (2) and (3) we obtain:

(10) $$S + T = 2\pi[F - E + V].$$

Now let us observe an important point. There is a "flat" polyhedral surface P_f that has exactly F faces, E edges, and V vertices, all contained in the same plane. (If we have obtained P from a convex polyhedron C by dividing the surface of C into two portions by the closed line R, let us choose any point p on the "other side" of R; that is, p belongs to the surface of C, but does not belong to P or R. Now, P is "fully visible" from p; that is, we can project P from p as center of projection onto a plane in one-one fashion: the projection so obtained can be taken as P_f, which is, in fact, the division of a plane polygon with E_b sides into F polygons. Obviously, we could flatten P into P_f by and by, that is, pass from one to the other by a continuous deformation.) Our foregoing derivation and its result (10) are valid not only for P, but also for P_f. Yet, obviously, the spherical image of P_f reduces to a point, its area S_f is nil, and T_f

the total change of direction of the boundary of P_f, is 2π (by Lemma I), and so the left hand side of (10) reduces to $0+2\pi$: yet the right hand side is the same for P_f as for P. Therefore, we obtain at one stroke

$$S + T = 2\pi, \qquad F - E + V = 1.$$

6. In order to apply the variant just proved of the Gauss-Bonnet theorem, let us consider each of the F faces of P as a rigid plate; these plates are joined along the E_i interior edges as with hinges. If some of the V_i interior vertices have more than three edges, the polyhedral surface P, although consisting of rigid plates, may be flexible, of variable shape. When P varies, also the shape of its spherical image may vary. Yet T, the total change of direction along the rim of P, remains unchanged, and so must remain also the *area* $S = 2\pi - T$ of the spherical image. This is the elementary counterpart of the *theorema egregium* of Gauss according to which the (Gaussian) curvature of a surface remains unchanged when the surface is bent.[*]

The scope of the theorem proved would be considerably widened by a clear discussion of the validity of Lemma II for non-convex polyhedral angles.

ON DEFINING CONIC SECTIONS[†]

G. B. HUFF, The University of Georgia

The conventional American text in analytic geometry begins the study of conic sections with a paragraph describing these curves as sections of a right circular cone by suitably chosen planes. It is customary to remark that this definition of a conic section is not adapted to the methods and objectives of the course, and each one of the parabola, the ellipse, and the hyperbola is then given its own definition as a locus in a later paragraph. Ordinarily there is no attempt to give even an intuitive proof of the equivalence of the two definitions.

Apparently it is not well known that this may be done quickly and simply in the case of the ellipse and the hyperbola. In *Anschauliche Geometrie* (Hilbert, Cohn-Vossen; Dover; 1944; pp. 7, 8), a figure is given from which a simple intuitive argument may be made that the two definitions for the ellipse (or the hyperbola) are equivalent. This note is written to call these figures to the attention of analytic geometry teachers and to give the construction of a third figure which leads to a demonstration of the equivalence of the two definitions of the parabola.

Let the parabola be defined as the intersection of a right circular cone Γ

---

[*] *Cf.* D. Hilbert and S. Cohn-Vossen, Anschauliche Geometrie, pp. 172–173.

[†] From AMERICAN MATHEMATICAL MONTHLY, vol. 62 (1955), pp. 250–251.

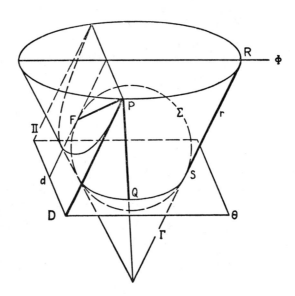

and a plane Π parallel to a ruling r of Γ and perpendicular to the plane on r and the axis of Γ. Construct a sphere Σ internally tangent to Γ and tangent to the plane Π. (The existence and unicity of Σ is made clear by thinking of fitting a small balloon in Γ and inflating it until it meets Π.) Let Θ be the plane determined by the intersection of Σ and Γ. Let F be the common point of Π and Σ, and let d be the common line of Π and Θ. I claim that F is the focus and d is the directrix of the parabola defined by Γ and Π.

Let P be any point of the parabola and let Φ be the plane through P and perpendicular to the axis of the cone Γ. Let D be the foot of the perpendicular from P to d; let Q be the intersection of the ruling on P with the plane Θ, and let R, S be the points on r cut out by Φ and Θ. A simple argument shows that RS is parallel to PD.

Since PF and PQ are tangents to the sphere, then $PF = PQ$.

Since PQ and RS are segments of rulings of Γ cut out by the parallel planes Θ and Φ, then $PQ = RS$.

Since PD is parallel to RS, these segments are cut out on parallel lines by parallel planes and $RS = PD$. It follows that $PF = PD$, as claimed.

It is easy to generalize the figure and the argument to definition of a conic of eccentricity e in terms of its directrix and focus. (Zwikker, *Advanced Plane Geometry*, Amsterdam, 1950, p. 108.)

COMBINATORIAL TOPOLOGY OF SURFACES*

ROBERT C. JAMES†

The first section of this paper will be devoted to the study of two particular surfaces. The methods used to treat these surfaces will then be extended to develop a classification of surfaces, each class consisting of combinatorially equivalent surfaces. Applications will be made to the study of the topological nature of covering surfaces and of Riemann surfaces in particular. It will be assumed that the reader is familiar with the material of Chapter V, pages 235–244 and 256–264, in the book *What Is Mathematics?* by Courant and Robbins [8]. Other than for references to this book, no previous knowledge of topology will be assumed.

1. Two Examples.

Example 1.1. Consider the surface of Figure 1. It can be described as a sphere with three appendages, a "handle", a "crosscap", and a "cuff". Cut the surface along the path which consists of path x followed by paths y and z in the directions indicated in Figure 2. After this cut has been made, the surface consists of four pieces, three containing the appendages and the fourth the remaining portion of the sphere. These pieces will be studied individually.

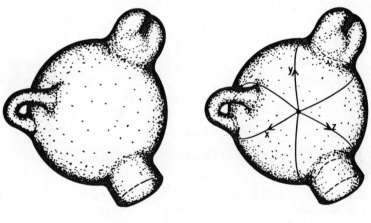

FIG. 1   FIG. 2

The appendage cut off by the cut x is called a *handle*. It can be continuously deformed into a torus with a hole, which is shown in Figure 3. After making the cuts a and b shown in Figure 4, the surface can be unfolded as shown in Figure 5.

* From MATHEMATICS MAGAZINE, vol. 29 (1955), pp. 1–39. This paper is based on lectures given by Professor A. W. Tucker of Princeton University while a Phillips visitor at Haverford College in the fall semester 1953–54.

† Prof. James is now at Claremont Graduate School.

With suitable stretching and shrinking, it can then be spread out to form the plane region of Figure 6.

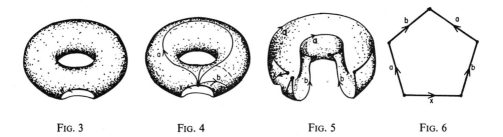

FIG. 3   FIG. 4   FIG. 5   FIG. 6

The appendage cut off by the cut y is called a *crosscap*. It can be regarded as a hemisphere which has first had its "cap" cut off, as shown in Figure 7. The cap is then distorted by pulling the right side of the back through the right side of the front as the front is pushed back. This causes the cap to cross itself, as indicated in Figure 8. In order to be able to put the cap back on the base, the top of the base is pinched together to join the points B_1 and B_2 as the single point B. After the pieces are rejoined, the surface

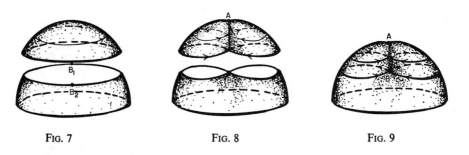

FIG. 7   FIG. 8   FIG. 9

is the crosscap of Figure 9. It crosses itself along AB. The points A and B are single points, but each point on the line between A and B is regarded as a double point—one on each of the portions of the surface crossing along AB. If one travels along the "figure-eight" path through Q, he would go as indicated by the arrows in Figure 9. When moving from left to right through Q, he would not be aware of the other point at Q, which is on the section of this path which goes through Q from right to left. The crosscap can also be described as follows. Make a cut in a hemisphere and spread the cut apart as shown in Figure 10. Originally, the lines A_1B_1 and A_2B_1 were identified, and the lines A_1B_2 and A_2B_2 were identified (two lines being identified means that their corresponding points are regarded as being identical). Now identify the line A_1B_1 with the line A_2B_2 and the line A_2B_1 with the line A_1B_2. This forces A_1 and A_2 to be identified and B_1 and B_2 to be identified, again giving the crosscap of Figure 9. Now cut the crosscap along the path c and pull it through itself, as shown in Figure 11. The resulting surface can then be laid

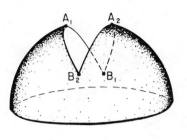

FIG. 10

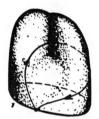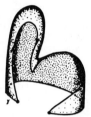

FIG. 11

out as the triangle of Figure 12. A crosscap is topologically the same as a Moebius band in the sense that the points of the two surfaces can be put in a one-to-one correspondence that is continuous in both directions. In fact, if the Moebius band of Figure 13 is cut along the path c as indicated, it can be deformed into the

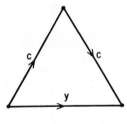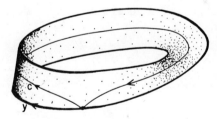

FIG. 12   FIG. 13

triangle of Figure 12 and hence into a crosscap by reversing the process of Figures 11-12 (see Courant-Robbins [8], pages 260–262). The surface shown in Figure 14 is also topologically the same as a Moebius band (see Tuckerman [22]). It does not cross through itself and has a triangular boundary ABC. If this surface is cut along CD, DG, and DEF, it can be spread out as shown in Figure 15. This figure can be used as a pattern for constructing a model, with flaps as indicated to aid in rejoining corresponding edges. All folds are to be made upward, except for the flaps on DE and EF.

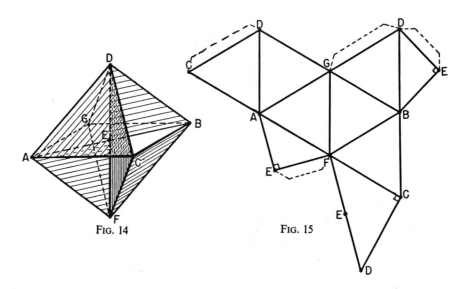

FIG. 14   FIG. 15

A *nonorientable surface* is one for which it is not possible to define orientation for circles in the surface in such a way that this orientation is preserved as a small circle is moved about in the surface. The Moebius band is a nonorientable surface. Figure 16 illustrates the fact that an oriented circle which is moved around a Moebius band returns to its original position with its orientation reversed. A nonorientable surface is also called a one-sided surface (Courant-Robbins [8], pages 259–264). However, there is a technical objection to considering orientability and two-sidedness as synonomous. (Orientability is an intrinsic property of the surface, but whether a surface is two-sided depends on the space in which it is embedded; *e.g.*, a torus can be embedded in a certain 3-manifold so as to be one-sided.)

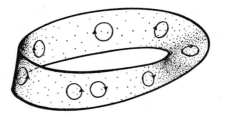

FIG. 16

The third appendage on the sphere of Figure 1 is the *cuff*. If the free edge of the cuff is labeled as *e* and a cut *d* is made from the path *z* to the path *e*, the cuff of Figure 17 can be laid out as shown in Figure 18.

The remainder of the sphere can now be stretched out as the triangle of Figure 19.

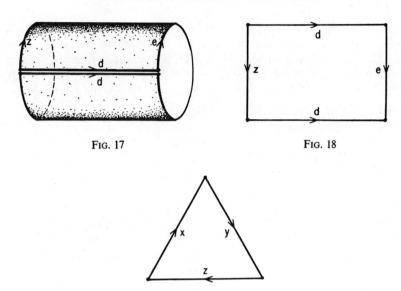

FIG. 17   FIG. 18

FIG. 19

The surface of Figure 1 has been cut into four pieces, each of which can be represented as a plane polygonal region. This piecewise description of the surface is conceptually very important because it is purely two-dimensional. The embedding of the surface in three-dimensional space has been eliminated. But the cutting into pieces is reversible. The pieces can be rejoined by pasting together edges labeled with the same symbol in such a way that directions correspond. Any two surfaces obtained in this way from the same pieces will be said to be combinatorially equivalent (see Definition 3.1). For example, the surfaces of Figures 9, 13, and 14 are the results of three geometrically different ways of pasting together (identifying) the two edges labeled c in Figure 12, but these three surfaces are combinatorially equivalent.

Each of the pieces of the surface of Figure 1 (Figures 6, 12, 18, 19) can be represented symbolically by describing the order of the paths and the orientation of each path. Thus by traveling clockwise around each piece and using the $^{-1}$ sign to indicate that a path is being traveled in the opposite direction to the arrow on that particular path, one can obtain the following "presentation" of the four parts of the surface:

Handle: $aba^{-1}b^{-1}x^{-1}=1,$

Crosscap: $ccy^{-1}=1,$

Cuff: $ded^{-1}z^{-1}=1,$

Remainder: $xyz=1.$

These four relations are merely symbolic descriptions of the four pieces into which the surface of Figure 1 has been cut. It is clear that the order one traces

ROBERT C. JAMES

around a piece and the point at which one starts are immaterial; both $zde^{-1}d^{-1}=1$ and $ed^{-1}z^{-1}d=1$ describe the same "cuff" as the relation $ded^{-1}z^{-1}=1$ given above. The 1 in the right member of a relation is a symbolic formality that suggests the possibility of other forms of the same relation. For example, by multiplying on the right by z, the cuff relation becomes $ded^{-1}=z$. Also, the relations describing the four pieces of Figure 1 can be written as:

Handle:	$x=aba^{-1}b^{-1}$,
Crosscap:	$y=cc$,
Cuff:	$z=ded^{-1}$,
Remainder:	$xyz=1$.

Each of these relations represents a plane polygonal region, it being understood that whenever one symbol occurs twice the corresponding edges are to be identified (or pasted together) with the directions on the edges matching. Thus the four pieces of Figures 6, 12, 18, and 19 can be pasted together to form Figure 20. After the x,y,z edges have been pasted together, they can be eliminated. This produces the

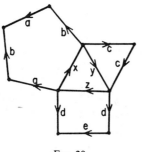
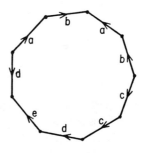

FIG. 20   FIG. 21

polygonal region of Figure 21, which can be described by the single relation

$$aba^{-1}b^{-1}ccded^{-1}=1.$$

This region represents the surface obtained by stretching and twisting the region, or even passing it through itself, in such a way that each pair of edges labeled with the same symbol (or a symbol and its inverse) are joined together with the directions on the edges matching. This surface will be said to be "equivalent in the sense of Combinatorial Topology", or *combinatorially equivalent*, to the surface of Figure 1. Since all cuts have been pasted together in the same way they were joined before cutting, and all other deformations are merely stretchings and shrinkings, it follows that the points of these two surfaces can be put in a one-to-one correspondence that is continuous in both directions (see Courant-Robbins [8], pages 241–243). In such *point-set* terms, the two surfaces are said to be homeomorphic (see Chapter 3 of [16]).

Example 1.2. The above procedures will now be used to determine the topological nature of a particular interwoven covering surface of a sphere. This surface consists of two spheres (as indicated in Figure 22) which have been cut and rejoined so that the outer sheet of the Western Hemisphere joins the inner sheet of the Eastern Hemisphere along b_1 and d_1, and the inner sheet of the Western Hemisphere joins the outer sheet of the Eastern Hemisphere along b_2 and d_2. The resulting surface crosses itself along the curves A_1B_1 and A_2B_2. The points A_1, B_1, A_2, B_2 are single points, but the curve A_1B_1 represents the two distinct superimposed paths b_1 and b_2, and the curve A_2B_2 represents the two distinct superimposed paths d_1 and d_2.

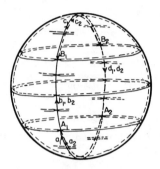

FIG. 22

Now cut the surface along the paths a_1, a_2; b_1, b_2; c_1, c_2; d_1, d_2. After the cutting, the surface will be in four pieces, which can be represented by the four polygonal regions of Figure 23. The first two of these were originally Western Hemispheres and the other two Eastern Hemispheres.

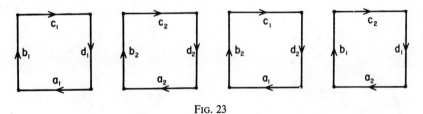

FIG. 23

Eliminate d_1 and d_2 by pasting together the edges labeled d_1 and pasting together the edges labeled d_2. This produces the two polygonal regions of Figure 24.

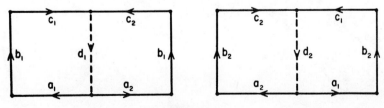

FIG. 24

Next paste these two regions together along $a_1{}^{-1}a_2$. After doing this, let $b_1{}^{-1}b_2 = \alpha$ and $c_2c_1{}^{-1} = \beta$. This produces the polygonal regions of Figures 25 and 26. Figure 27 shows that this polygonal region represents a torus. Thus the original interwoven covering surface of a sphere has been shown to be "equivalent" to a torus.

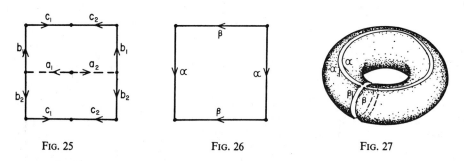

FIG. 25   FIG. 26   FIG. 27

The sequence of steps shown in Figure 28 illustrates how a torus and the surface of Example 1.2 can be continuously deformed into each other. The relation used to define the torus of Figure 28 is

$$(a_1a_2)(b_1b_2)(a_1a_2)^{-1}(b_1b_2)^{-1} = 1.$$

Care should be taken to imagine paths labeled with the same symbol as identified so that directions match and to note that the surface crosses itself along the curves A_1B_1 and A_2B_2.

The surface discussed in Example 1.2 is actually the Riemann surface which uniformizes the following function (also discussed by Curtiss [9, page 167]):

$$w^2 = (1 - z^2)(1 - k^2z^2).$$

Each complex value of z except $z = \pm 1$ and $z = \pm 1/k$ determines two values of w.

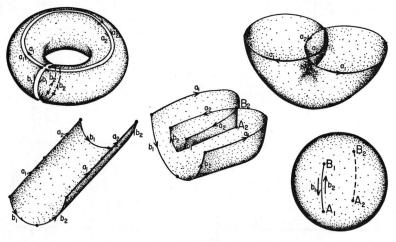

FIG. 28

In order to get a continuous single-valued function of z, one can introduce a pair of z-planes (or sheets) and regard each value of z as determining exactly one value of w. These two sheets are connected, since for $z = \pm 1$ and $z = \pm 1/k$ there is only one value for w, namely $w = 0$. These four points are called *branch points*. The surface can be formed as indicated in Figure 29, as two sheets covering the z-plane and passing through each other along the dashed lines in the same way the crosscap of Figure 9 crosses itself (also see Example 5.2). The complex plane can be mapped onto a sphere by stereographic projection (Hilbert and Cohn-Vossen [12], pages 248–259). The Riemann surface then becomes the interwoven surface illustrated in Figure 22.

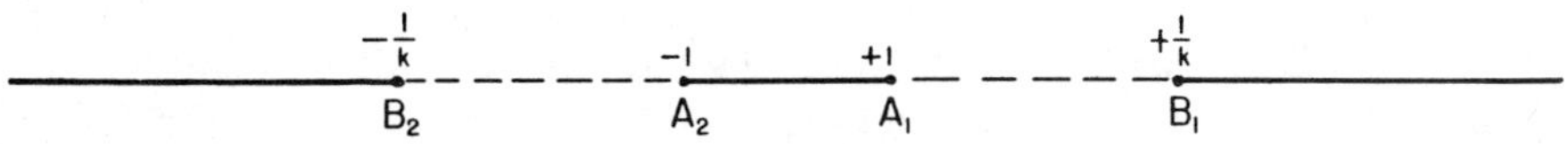

FIG. 29

The surface of Figure 22 was cut into the four pieces represented by the polygonal regions of Figure 23. These regions can be described symbolically by the relations given in the following table.

	Western Hemisphere	*Eastern Hemisphere*
Outer Sheet	$a_1 b_1 c_1 d_1 = 1$	$a_1 b_2 c_1 d_2 = 1$
Inner Sheet	$a_2 b_2 c_2 d_2 = 1$	$a_2 b_1 c_2 d_1 = 1$

The pasting operations which were used to assemble these pieces to form a torus correspond to symbolic operations which will now be used to obtain the relation $\alpha \beta \alpha^{-1} \beta^{-1} = 1$, which describes a torus. First, write the relations for the two pieces covering the Eastern Hemisphere as

$$d_2 = c_1^{-1} b_2^{-1} a_1^{-1} \quad \text{and} \quad d_1 = c_2^{-1} b_1^{-1} a_2^{-1}.$$

Use these relations to replace d_1 and d_2 in the relations for the two pieces covering the Western Hemisphere, to get

$$a_1 b_1 c_1 c_2^{-1} b_1^{-1} a_2^{-1} = 1 \quad \text{and} \quad a_2 b_2 c_2 c_1^{-1} b_2^{-1} a_1^{-1} = 1.$$

Now rewrite these two relations as

$$b_1 c_1 c_2^{-1} b_1^{-1} = a_1^{-1} a_2 \quad \text{and} \quad b_2 c_2 c_1^{-1} b_2^{-1} a_1^{-1} a_2 = 1.$$

Substitute the expression for $a_1^{-1} a_2$ into the second relation, giving

$$b_2 c_2 c_1^{-1} b_2^{-1} b_1 c_1 c_2^{-1} b_1^{-1} = 1 \quad \text{or} \quad b_1^{-1} b_2 c_2 c_1^{-1} \left(b_1^{-1} b_2 \right)^{-1} \left(c_2 c_1^{-1} \right)^{-1} = 1.$$

With the substitutions $b_1^{-1} b_2 = \alpha$ and $c_2 c_1^{-1} = \beta$, this relation becomes

$$\alpha \beta \alpha^{-1} \beta^{-1} = 1,$$

which represents the torus of Figure 27.

2. Surfaces defined by use of systems of relations.

Our examples suggest that any surface might be represented symbolically by a finite system of relations in a finite number of symbols for which each symbol a has an associated symbol a^{-1}, called the *inverse* of a. These relations are statements equating two expressions, each of which is either 1 (1 is not one of the symbols), or an indicated product of symbols, inverses of symbols, and 1's. They are called *relations* to emphasize the difference between them and formal identities such as $(a^{-1})^{-1} = a$ or $a \cdot a^{-1} = 1$, which are used to manipulate relations (see Definitions 2.1 and 3.1). There are certain operations on a relation which do not in any way change the polygonal region which the relations can be interpreted as describing. These are given in the following definition.

Definition 2.1. Two systems of relations are *equivalent* if the relations of one system can be changed into the relations of the other by use of the following operations.

1) If x is a new symbol, then any symbol a can be replaced by x and a^{-1} by x^{-1}, provided this is done wherever a or a^{-1} appears.

2) Wherever it appears, any one of the expressions a, $1 \cdot a$, $a \cdot 1$, or $(a^{-1})^{-1}$ can be replaced by any other one (these expressions are to be regarded as being merely different ways of writing the symbol a).

3) The last symbol on the right (left) of one member of a relation can be removed from this position if its inverse is put on the right (left) of the other member of the relation. This will be called *transposition* of the symbol.

For illustrations of the process of deriving equivalent relations from a given relation, consider the relation $ccy^{-1} = 1$, which describes the triangular region of Figure 12. This can be changed into $xxy^{-1} = 1$ by use of (1) of Definition 2.1. Or by use of (3), it can be changed into $cc = 1 \cdot (y^{-1})^{-1}$, and then into $cc = y$ by use of (2). This last relation can be changed into $cc = y \cdot 1$ by use of (2), and then into $y^{-1}cc = 1$ by use of (3). By use of (3) and (2), one can change $cc = y$ into $c = yc^{-1}$, $1 \cdot c = yc^{-1}$, and $1 = yc^{-1}c^{-1}$. Clearly these relations all describe the same triangular region, the differences being essentially differences in the starting point, the direction in which one moves around the boundary of the region, and the letters used to designate sides of the triangle.

A system of relations can be thought of as a *combinatorial presentation of a surface* and two equivalent systems of relations as different presentations of the same surface. The totality of systems equivalent to a given system of relations is an *equivalence class* which serves to define the surface *in the sense of combinatorial topology*. This is stated formally in the following definition. It should be noted that if (a) and (b) of this definition are satisfied by one system of relations, then they are satisfied by any equivalent system.

Definition 2.2. A *surface* is an object associated with an equivalence class of systems of relations for which each system consists of a finite number of relations in a finite number of symbols and has the properties:

a) In the entire system of relations, no symbol occurs more than twice (either the symbol or its inverse is said to be an occurrence of the symbol).

b) If the system of relations is partitioned in any way into two systems (so that each relation belongs to exactly one of the two systems), then there is at least one symbol which occurs once in each of these two systems.

This definition can be given geometric meaning by interpreting each relation as describing a plane polygonal region and the set of relations as describing the surface obtained by identifying edges labeled with the same symbol so that directions correspond.

It should be noted that (a) and (b) of Definition 2.2 are satisfied by the systems of relations for the surfaces of Examples 1.1 and 1.2. The system of relations for Example 1.1 contains some symbols once and some twice, but no symbol occurs more than twice. Since the relation $xyz = 1$ has a symbol in common with each of the other relations, it follows that for any partition of the system of relations into two disjoint systems there is some symbol which occurs in each system. Any one of the systems of relations used to describe the surface of Example 1.2 contains each of its symbols exactly twice. The first system of four relations given for this surface consisted of

$$a_1 b_1 c_1 d_1 = 1, \quad a_1 b_2 c_1 d_2 = 1, \quad a_2 b_2 c_2 d_2 = 1, \quad a_2 b_1 c_2 d_1 = 1.$$

When written in this order, each adjacent pair has a symbol in common. Thus the relations can be thought of as joined in a chain.

When interpreted geometrically, condition (b) of Definition 2.2 is the requirement that the surface consist of one piece. For suppose that the system of relations were partitioned into two nonempty systems. Then each system of relations would define a surface and these surfaces would be entirely disjoint if they have no symbols (edges) in common. It will be shown in Section 4 that any surface is combinatorially equivalent to a surface whose system of relations consists of a single relation. However, in addition to objects consisting of disjoint pieces, there are other objects which are not surfaces in the sense of Definition 2.2. For example, if a torus is squeezed to a point (P in Figure 30), the point is a conical point. Such an object can not be described by a set of relations which are to be interpreted geometrically as has been done, since repeated symbols represent edges to be

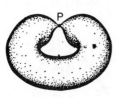

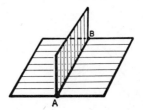

FIG. 30   FIG. 31

identified and there is no means for matching points alone. Also, the restriction (a) that no symbol occur more than twice prevents an edge from belonging to more than two faces (as for AB in Figure 31). The above ideas are consistent with the following definition of surface, as might be given by a point-set topologist. "A *closed surface* is a connected, compact metric space which is homogeneous in the sense that each point has a neighborhood which is a two-cell" (Lefschetz [**14**], page 72). A *surface with boundary curves* would then be defined by changing the neighborhood restriction so that each point on a boundary curve has a neighborhood which is half of a two-cell with the diameter included and lying along the boundary curve. As will be seen in Section 4, Definition 2.2 defines a surface which may have cuffs (and therefore boundary curves). It will be seen (Definition 4.3) that cuffs are present whenever some symbol occurs only once. If a system of relations has the property that each symbol occurs exactly twice, then any equivalent system also has this property (see Definition 2.1). Hence it is reasonable to define a closed surface as follows.

Definition 2.3. A *closed surface* is a surface which has a system of relations in which each symbol occurs exactly twice.

3. Combinatorial equivalence of surfaces.

Let the system of relations for a surface be given geometric meaning by interpreting each relation as describing a plane polygonal region, and the system of relations as describing the object obtained by identifying edges labeled with the same symbol so that directions correspond. This "geometric surface" is changed into a topologically equivalent "surface" by certain "cut-and-paste" operations which have corresponding symbolic operations. There are essentially three separate and distinct cutting operations, each of which has an inverse, a pasting operation. These six operations are enumerated below, with pictures illustrating the changes effected.

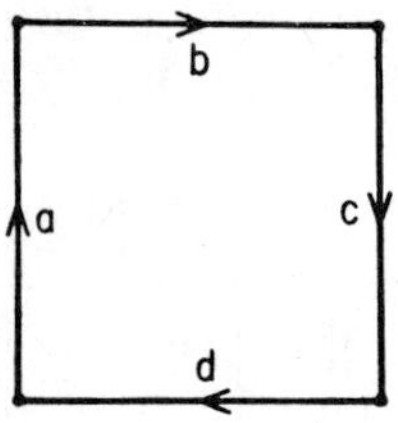

α: "Cut a face in two" (before the cut, or after the paste: $abcd = 1$)

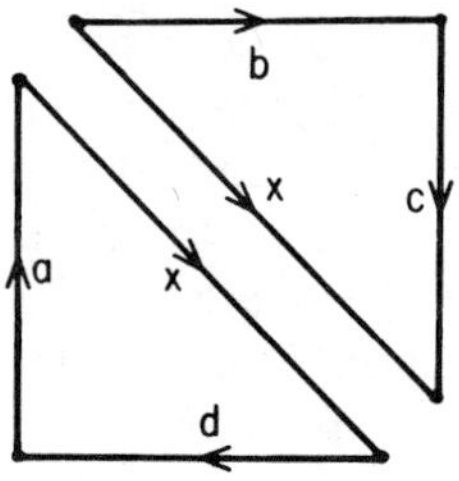

α': "Paste two faces together along a common edge" (after the cut, or before the paste: $axd = 1$; $bc = x$)

FIG. 32

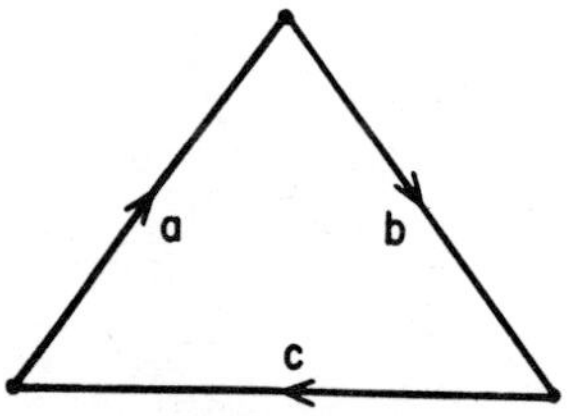

β: "Cut part way into a face" (before the cut, or after the paste: $abc = 1$)

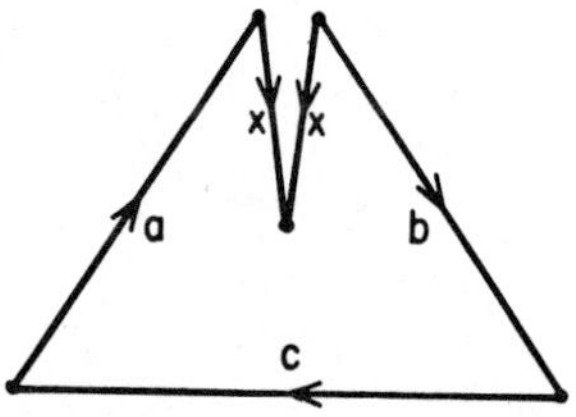

β': "Paste two appropriate edges together" (after the cut, or before the paste: $axx^{-1}bc = 1$)

FIG. 33

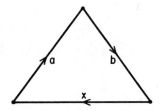

γ: "Break an edge in two (put in a new vertex)" (before the break, or after the weld: $abx = 1$)

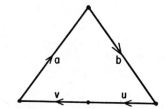

γ': "Weld two appropriate consecutive edges into a single one" (after the break, or before the weld: $abuv = 1$)

FIG. 34

In order to classify the surfaces of Definition 2.2, it is useful to have some concept of equivalence of surfaces. This is given by the following definition. In this definition, capital letters represent blocks consisting of indicated products of symbols and inverses of symbols (or of 1 alone) and lower case letters are single symbols. If A is such a block, then A^{-1} represents the symbols of A in reverse order, with each symbol replaced by its inverse (*e.g.*, if $A = ab^{-1}c$, then $A^{-1} = c^{-1}ba^{-1}$).

Definition 3.1. Two surfaces are *combinatorially equivalent* if a system of relations for one surface can be changed into a system of relations for the other surface by a sequence of operations, each operation being either the replacing of a system of relations by an equivalent system, or the replacing of a system of relations by a new system obtained by use of one of the operations:

1) If x is a new symbol, then any one relation of form $ABC = D$ can be replaced by the two relations $AxC = D$, $B = x$.

1') Any two relations of form $AxC = D$, $B = x$ can be replaced by the single relation $ABC = D$.

2) If x is a new symbol, then 1 can be replaced by xx^{-1}, or by $x^{-1}x$, in any one place.

2') If x and x^{-1} appear side by side in any relation, their product can be replaced by 1.

3) If u and v are new symbols, then any symbol x can be replaced by uv, or x^{-1} by $v^{-1}u^{-1}$, provided this is done wherever x or x^{-1} appears.

3') If x is a new symbol and the symbols u and v occur only in blocks uv or $v^{-1}u^{-1}$, then uv can be replaced by x and $v^{-1}u^{-1}$ by x^{-1}, provided this is done wherever uv or $v^{-1}u^{-1}$ appears.

The operations (1), (2), (3) of this definition correspond to the "cut" operations (α), (β), (γ). The operations (1'), (2'), (3') correspond to the "paste" operations (α'), (β'), (γ'). Figures 32–34 illustrate special cases of these operations.

It should be noted that the indicated multiplication used in a relation is essentially non-commutative and associative, for the operations of Definition 3.1 do not permit interchange of ab and ba in general. But each member of a relation is regarded as a set of symbols written in a definite order with certain operations permitted with individual symbols or adjacent symbols. Thus the order of the symbols is important, but the particular method of grouping into blocks which may be used is merely a notational convenience. It should also be noted that each operation of Definition 3.1 has an inverse operation. In fact, the relation of combinatorial equivalence is an equivalence relation.

4. Canonical forms and classification of surfaces.

Before showing how to reduce a system of relations to canonical form, two transformation rules will be developed. First consider a relation in which a block Q of symbols is adjacent to a symbol x which occurs twice, both times as x rather than x^{-1}. Such a relation can be written as one of

$$AxQBxC = D, \quad AxQB = CxD, \quad AQxBxC = D, \quad AQxB = CxD,$$

where capital letters represent either 1 or blocks consisting of products of symbols and inverses of symbols. Use (1) of Definition 3.1 and a new symbol y to replace these relations by the respective pairs of relations

$$\begin{cases} AyBxC = 1, \\ xQ = y; \end{cases} \quad \begin{cases} AyB = CxD, \\ xQ = y; \end{cases} \quad \begin{cases} AyBxC = 1, \\ Qx = y; \end{cases} \quad \begin{cases} AyB = CxD, \\ Qx = y. \end{cases}$$

By repeated transpositions of symbols (Definition 2.1), $xQ = y$ and $Qx = y$ can be changed into $x = yQ^{-1}$ and $x = Q^{-1}y$, respectively, where Q^{-1} is the block consisting of the symbols of Q multiplied in reverse order with each symbol replaced by its inverse. Now use (1′) of Definition 3.1 to replace the above pairs of relations by the single relations

$$AyByQ^{-1}C = D, \quad AyB = CyQ^{-1}D, \quad AyBQ^{-1}yC = D, \quad AyB = CQ^{-1}yD.$$

Since x did not occur more than twice, it has now been completely eliminated and we can use (1) of Definition 2.1 to replace y by x. This gives

$$AxBxQ^{-1}C = D, \quad AxB = CxQ^{-1}D, \quad AxBQ^{-1}xC = D, \quad AxB = CQ^{-1}xD.$$

Rule I. *If a symbol x occurs twice in the same relation, both times as x, then a block Q which is on one side of x in one position can be removed, provided Q^{-1} is put on the same side of x in the other position.*

The geometric interpretation of this rule is shown in Figure 35. A cut has been made along y. Then the shaded piece has been moved and pasted along x, eliminating x. Then y is replaced by x.

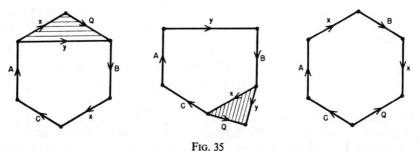

FIG. 35

Rule II. *If a symbol x occurs twice in the same relation, once as x and once as* x^{-1}, *then a block Q which is on one side of x can be moved (without inversion) to the other side of* x^{-1}.

This rule can be derived easily from Rule I. For example, the relation $AxQBx^{-1}C = D$ is equivalent to $AxQ = DC^{-1}xB^{-1}$. By Rule I, this can be replaced by $Ax = DC^{-1}xQ^{-1}B^{-1}$, which is equivalent to $AxBQx^{-1}C = D$.

It should be noted carefully that the symbols of a relation are essentially dummy variables, since (1) of Definition 2.1 enables one to replace a symbol and its inverse by a new symbol and its inverse whenever desired. Rules I and II do not merely move symbols around, but change the meaning of some of them. However, the new system of relations defines a surface which is combinatorially equivalent to the original surface.

These two rules are the principal tools used in the following proof that a system of relations for any surface (in the sense of Definition 2.2) can be reduced to one of four types of canonical forms by use of the transformations permitted by Definition 3.1. The process of reduction to canonical form is outlined in the following series of five steps.

STEP 1. Reduce the system of relations to a single relation. Choose any one of the relations of a system of relations which consists of two or more relations. Then it follows from (b) of Definition 2.2 that there is another relation and a symbol x such that x occurs in each of these relations. By repeated transpositions, these two relations can be written in the forms $Ax = 1$, $B = x$. They can then be replaced by $AB = 1$, using (1′) of Definition 3.1. Any symbol which occurs in $AB = 1$ also occurs in either $Ax = 1$ or $B = x$. Also, any symbol which occurs in one of these two relations and in some third relation also occurs in $AB = 1$. Thus the new set of relations also satisfies (b) of Definition 2.2. Since the system contains a finite number of relations, this process will eventually reduce it to a single relation.

STEP 2. Assemble the crosscaps. Suppose c is any symbol which occurs twice, both times as c rather than c^{-1}. By transposing symbols (if necessary), the relation can be written in the form

$$ABcCcD = 1,$$

where (as before) the capital letters represent blocks of symbols (or simply 1). Geometrically, this relation can be interpreted as representing the polygonal region of Figure 36. If this figure is cut along the curved lines and the edges labeled c pasted together with the directions matching, the shaded region becomes a Moebius band (Figure 13). Now use Rule I to change this relation into $AcCB^{-1}cD$

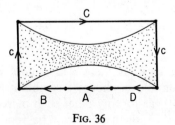

Fig. 36

$=1$ and then into $AccBC^{-1}D=1$. If c_1 is a symbol which occurs twice as c_1, then this process with A taken as 1 produces a relation of form $c_1c_1P=1$. If some other symbol c_2 now occurs twice as c_2, the same process can be used with A representing c_1c_1. This changes the relation into the form $c_1c_1c_2c_2P'=1$. This can be continued until the relation is of the form $c_1c_1\ldots c_qc_qQ=1$, where any symbol x which occurs twice in Q occurs once as x and once as x^{-1}. Now use (1) of Definition 3.1 to replace each c_ic_i by a new symbol y_i and introduce the new relations $y_i=c_ic_i$. We now have a system of relations of the form

$$\begin{cases} y_i=c_ic_i, & i=1,2,\ldots,q; \\ y_1\ldots y_qQ=1. \end{cases}$$

STEP 3. *Assemble the handles.* Let a and b be any two symbols which occur twice in such a way that the relation $y_1\ldots y_qQ=1$ is of the form

$$ABaCbDa^{-1}Eb^{-1}F=1.$$

By repeated use of Rule II, this relation can be successively changed into the relations

$$AaCbDa^{-1}(BE)b^{-1}F=1,$$
$$AaCb(BED)a^{-1}b^{-1}F=1,$$
$$Aa(BEDC)ba^{-1}b^{-1}F=1,$$
$$Aaba^{-1}b^{-1}(BEDCF)=1.$$

If a_1 and b_1 are two symbols which occur in the same way as a and b above and A represents 1, this process changes $y_1\ldots y_qQ=1$ into the form

$$a_1b_1a_1^{-1}b_1^{-1}y_1\ldots y_qQ'=1.$$

If two other symbols a_2 and b_2 occur in the same way as a and b and A represents $a_1b_1a_1^{-1}b_1^{-1}$, the result is of the form

$$a_1b_1a_1^{-1}b_1^{-1}a_2b_2a_2^{-1}b_2^{-1}y_1\ldots y_qQ''=1.$$

This can be continued until the relation is of the form

$$a_1b_1a_1^{-1}b_1^{-1}\ldots a_pb_pa_p^{-1}b_p^{-1}y_1\ldots y_qR=1,$$

where no symbol which occurs twice in R can occur exactly once between two occurrences of another symbol. Now use (1) of Definition 3.1 to replace each $a_ib_ia_i^{-1}b_i^{-1}$ by a new symbol x_i and introduce the new relation $x_i=a_ib_ia_i^{-1}b_i^{-1}$. We now have a system of relations of the form

$$\begin{cases} x_i=a_ib_ia_i^{-1}b_i^{-1}, & i=1,\ldots,p; \\ y_i=c_ic_i, & i=1,\ldots,q; \\ x_1\ldots x_py_1\ldots y_qR=1. \end{cases}$$

STEP 4. *Assemble the cuffs.* If there is a symbol which occurs twice in the block R of the last relation above, choose a symbol d for which there are as few symbols between d and d^{-1} as for any other such paired symbols. If there are no symbols between d and d^{-1}, then dd^{-1} can be removed by operation (2') of

Definition 3.1. Otherwise, the only symbols which can be between d and d^{-1} are symbols which occur only once in the entire relation (for if y and y^{-1} were both between d and d^{-1}, there would be fewer symbols between y and y^{-1} than between d and d^{-1}; while if y were between d and d^{-1} and y^{-1} were not, another handle would have been assembled). Since the symbols between d and d^{-1} occur only once, several applications of (3′) of Definition 3.1 can be used to replace this block of symbols by a single new symbol e. The block R is now of the form $Aded^{-1}B$. By use of Rule II, this can be changed into $ded^{-1}AB$. By continuing this process, R can finally be written as

$$d_1 e_1 d_1^{-1} d_2 e_2 d_2^{-1} \ldots d_r e_r d_r^{-1} S,$$

where S is either 1 or consists entirely of symbols which occur only once. If S is not 1, replace S by a new symbol e. Also use (1) of Definition 3.1 to replace $d_i e_i d_i^{-1}$ by z_i for each i and introduce the new relations $z_i = d_i e_i d_i^{-1}$. The system of relations is now

$$\begin{cases} x_i = a_i b_i a_i^{-1} b_i^{-1}, & i = 1, \ldots, p; \\ y_i = c_i c_i, & i = 1, \ldots, q; \\ z_i = d_i e_i d_i^{-1}, & i = 1, \ldots, r; \\ x_i \ldots x_p y_1 \ldots y_q z_1 \ldots z_r e = 1. \end{cases}$$

If one thinks of the geometric interpretation of these relations, it can be seen that each x_i, y_i, and z_i represents a path whose initial and terminal points are identified. Let W be the path $x_1 \ldots x_p y_1 \ldots y_q z_1 \ldots z_r$. Then, W also has its initial and terminal points identified and We would seem (intuitively) to be equivalent to $Wded^{-1}$, as shown in Figure 37. This change can be interpreted as representing a shift in the point P in a dissection of a surface for which P originally was on the free edge e of a cuff. The introduction of d can be justified formally as follows. Replace the relation $We = 1$ by $Wd^{-1}de = 1$, using (2) of Definition 2.1 and (2) of Definition 3.1.

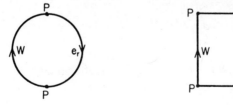

FIG. 37

Then d^{-1} can be moved to the left past each z_i by use of Rule II (replacing z_i by $d_i e_i d_i^{-1}$ before the use of the rule and putting z_i back after using the rule). Similarly, d^{-1} can be moved past each y_i by two applications of Rule I. It can then be moved past each x_i by four applications of Rule II. The relation now is $d^{-1}Wde = 1$, which can be changed into $Wded^{-1} = 1$ by two transpositions. Let r be increased by 1 and z_r be a new symbol. Replace the relation $Wded^{-1} = 1$ by the two relations $Wz_r = 1$ and $z_r = d_r e_r d_r^{-1}$. We now have a system of relations of the form

$$
\begin{cases}
x_i = a_i b_i a_i^{-1} b_i^{-1}, & i = 1, \ldots, p; \\
y_i = c_i c_i, & i = 1, \ldots, q; \\
z_i = d_i e_i d_i^{-1}, & i = 1, \ldots, r; \\
x_1 \ldots x_p y_1 \ldots y_q z_1 \ldots z_r = 1.
\end{cases}
$$

STEP 5. Turn the handles into crosscaps (if there is at least one crosscap). It will be seen that handles can be turned into crosscaps only if there is at least one crosscap with which to work. Thus the original crosscap is a sort of "catalyzer" in the process. Consider a block $aba^{-1}b^{-1}cc$. Apply Rule I, first to the symbol c and the block $a^{-1}b^{-1}$ giving $abcbac$; then to the symbol a and the block bcb, giving $aab^{-1}c^{-1}b^{-1}c$; and finally to the symbol b^{-1} and the block c^{-1}, giving $aab^{-1}b^{-1}cc$. This shows that the handle $x_p = a_p b_p a_p^{-1} b_p^{-1}$ and the crosscap $y_1 = c_1 c_1$ can be substituted into the relation $x_1 \ldots x_p y_1 \ldots y_q z_1 \ldots z_r = 1$, made into three crosscaps, and the relations put back in the above form with q larger by 2 and p smaller by 1. This can be continued until $p = 0$.

By use of the above five steps, the system of relations for a surface can be reduced to a canonical form which is of one of the following four general types.

CANONICAL FORMS

1. *Closed orientable surface.* The surface has no crosscaps and no cuffs ($q = r = 0$) and is said to be of *genus p* ($p \geqslant 0$).

$$
\begin{cases}
x_i = a_i b_i a_i^{-1} b_i^{-1}, & i = 1, \ldots, p; \\
x_1 \ldots x_p = 1.
\end{cases}
$$

2. *Orientable surface with cuffs.* The surface has no crosscaps ($q = 0$), but at least one cuff ($r > 0$).

$$
\begin{cases}
x_i = a_i b_i a_i^{-1} b_i^{-1}, & i = 1, \ldots, p; \\
z_i = d_i e_i d_i^{-1}, & i = 1, \ldots, r; \\
x_1 \ldots x_p z_1 \ldots z_r = 1.
\end{cases}
$$

3. *Closed nonorientable surface.* The surface has no handles and no boundaries ($p = r = 0$), but at least one crosscap ($q > 0$).

$$
\begin{cases}
y_i = c_i c_i, & i = 1, \ldots, q; \\
y_1 \ldots y_q = 1.
\end{cases}
$$

4. *Nonorientable surface with cuffs.* The surface has no handles ($p = 0$), but at least one crosscap ($q > 0$) and at least one cuff ($r > 0$).

$$
\begin{cases}
y_i = c_i c_i, & i = 1, \ldots, q; \\
z_i = d_i e_i d_i^{-1}, & i = 1, \ldots, r; \\
y_1 \ldots y_q z_1 \ldots z_r = 1.
\end{cases}
$$

It should be noted that a surface with $p=q=r=0$ is a closed orientable surface whose canonical form consists of the single relation $1=1$. Such a surface is combinatorially equivalent to a sphere, since a sphere is the geometric interpretation of the relation $xx^{-1}=1$.

It will be shown that the numbers p, q, and r are the same for two combinatorially equivalent surfaces whose systems of relations are in canonical form. The method for doing this will depend on procedures for determining (A) whether a surface is orientable, (B) the number of cuffs, (C) a certain number χ. These procedures will also be used to develop a method for identifying the canonical form for a surface without first reducing the system of relations to canonical form. This complete process will be illustrated by applying it to the example of Figure 38. This surface will be shown to be a nonorientable surface with two crosscaps and two cuffs, that is, a Klein bottle with two cuffs (as shown in Figure 39). The relations for this surface are

$$abdfgm=1, \quad ac^{-1}efhj=1, \quad bcm^{-1}k^{-1}j=1.$$

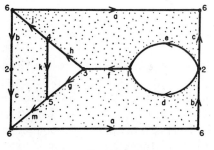

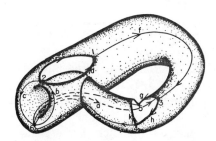

FIG. 38 FIG. 39

(A) In the above testing of canonical forms, a surface whose system of relations is in canonical form was called orientable if $q=0$, that is, if no "piece" of the surface is a Moebius band. This concept of orientability is intuitively equivalent to that discussed in connection with Figure 16. The following definition is an extension of the $q=0$ test and enables one to determine the orientability of a surface from its relations even if they are not in canonical form.

Definition 4.1. A surface is *orientable* if it has a system of relations for which each relation is of the form $A=1$ and any symbol x which occurs twice occurs once as x and once as x^{-1}.

The surface of Figure 38 is nonorientable. Because of the way c and j occur, there is no way of using the operations of Definition 2.1 to obtain relations in the form $A=1$ for which no symbol occurs twice in the same form.

Showing that two combinatorially equivalent surfaces are either both orientable or both nonorientable is equivalent to showing that the property of orientability is preserved by the operations of Definition 3.1, that is, that a system of relations which is equivalent to a system of the type described in Definition 4.1 still has this property after any manipulation by the operations of Definition 3.1. Suppose $ABC=1$ is one of the relations in a system of relations for an orientable surface.

Then this system of relations has the property of being equivalent to a system of relations for which $ABC = 1$ is one of the relations and each symbol x which occurs twice occurs once as x and once as x^{-1} (see Definition 2.1). The system still has this property if operation (1) of Definition 3.1 is used to replace $ABC = 1$ by the two relations $AxC = 1$, $B = x$, and then $B = x$ is changed into $Bx^{-1} = 1$. The inverse operation (1′) can be treated similarly. It is easy to see that the other operations of Definition 3.1 do not affect the orientability of the surface. Thus orientability is combinatorially invariant and a surface is orientable if and only if $q = 0$ for a canonical form of its system of relations, or if and only if no part of the surface is a Moebius band (see Figure 36).

Before discussing the *number of cuffs* for a surface whose system of relations is not in canonical form, it is convenient to introduce the concept of vertices.

Definition 4.2. Each symbol (in a system of relations for a surface) has two vertices. These are not necessarily distinct, but are called the *initial vertex* and the *terminal vertex*. The initial (terminal) vertex of a symbol x is the same as the terminal (initial) vertex of x^{-1}; if two symbols x and y can be made adjacent (in the order xy) by transposing symbols, then the terminal vertex of x is the same as the initial vertex of y; if x occurs in the relation $x = 1$, then the initial vertex of x is also the terminal vertex of x.

To determine the vertices for the surface of Figure 38, let P_1 be the terminal vertex of d. It is then the initial vertex of f and, finally, the terminal vertex of e. This is indicated in the relations below by putting 1 to the right of d; since 1 is to the left of f, it is put to the left of the other occurrence of f; it is then at the right of e; since e is unmatched, the chain is ended. This process can be continued to determine the other vertices. The surface has six vertices, as labeled in Figure 38.

$$a\underline{6}\underline{b}\underline{2}d\underline{1}f\underline{3}g\underline{5}m\underline{6} = 1, \quad a\underline{6}c^{-1}\underline{2}e\underline{1}f\underline{3}h\underline{4}j\underline{6} = 1, \quad b\underline{2}c\underline{6}m^{-1}\underline{5}k^{-1}\underline{4}j\underline{6} = 1.$$

Notice that each of the vertices $1, 2, 3, 4, 5$ of Figure 38 is on two boundary edges and that these boundary edges and boundary vertices fall naturally into the two cyclic sets $\underline{1}d^{-1}\underline{2}e$ and $\underline{3}h\underline{4}k\underline{5}g^{-1}$, which can be obtained by tracing around each closed curve in the boundary. This suggests the following definition and lemma.

Definition 4.3. A vertex is said to be a *boundary vertex* if it is a vertex of a symbol which occurs only once. All other vertices are said to be *interior vertices*. A symbol is said to be a *boundary edge* if it occurs only once and an *interior edge* if it occurs twice.

Lemma. *Each boundary vertex either is a single vertex of exactly two boundary edges or is both the initial and terminal vertex of one boundary edge and is not a vertex of any other boundary edge. The boundary edges can be arranged in cyclic sets so that any two adjacent boundary edges have a common vertex.*

This lemma can be verified as follows. Note that if x is any symbol, then either x occurs only in the relation $x = 1$, or there is a unique symbol which can (by transpositions of symbols) be made adjacent to x on the right, and there is a unique symbol which can be made adjacent to x on the left. Hence if P is the terminal (initial) vertex of an unmatched symbol (boundary edge) x, then P is also an initial (terminal) vertex of some symbol y_1 (possibly y_1 is x). If y_1 occurs twice, then P is a vertex of a symbol y_2 adjacent to y_1 at the other occurrence of y_1. It can be seen that this process terminates as soon as one of the y_i's is an unmatched symbol (possibly x).

(B). If the system of relations for a surface is in canonical form, then the number of cuffs is the number r. To show that the number of cuffs is uniquely defined for any given surface, it will be shown that the number of cuffs (as defined below) is the same for any two combinatorially equivalent surfaces and is equal to r if the set of relations is in canonical form.

Definition 4.4. Given a system of relations for a surface, the *number of cuffs* on the surface is the largest number of sets into which the unmatched symbols (boundary edges) can be divided with the restriction that two symbols must belong to the same set if they have a vertex in common.

It follows from the above lemma that the sets of unmatched symbols of Definition 4.4 must be cyclic sets for which any two adjacent symbols have a common vertex. For the surface of Figure 38, these sets are (d,e), with vertices $(1,2)$, and (g,h,k), with vertices $(3,4,5)$. Thus the surface has two cuffs, and it is expected that r would equal 2 if the relations were put in canonical form.

It will now be shown that the number of cuffs is invariant under the operations of Definition 3.1. For operations (1) or (1′) and the relations $ABC=1$ and $B=x$, the relation $B=x$ can be written as $x^{-1}B=1$, or as $Bx^{-1}=1$. Thus the terminal vertex of A is both the initial vertex of B and the initial vertex of x. The initial vertex of C is both the terminal vertex of B and the terminal vertex of x. Thus each vertex of x is a vertex for some other symbol (Figure 32). Since x occurs twice, there is no change in the set of unmatched symbols and no change in their vertices. This is also true of operations (2) and (2′), although these operations create (or destroy) one vertex, which is either the terminal vertex of x and the initial vertex of x^{-1} (when xx^{-1} occurs), or the initial vertex of x and the terminal vertex of x^{-1} (when $x^{-1}x$ occurs), but is not a vertex for any other symbol (see Figure 33). Operations (3) and (3′) also create (or destroy) a vertex, which is the terminal vertex of u and the initial vertex of v, but is not a vertex of any other symbol. If u and v are unmatched symbols, then the sets of unmatched symbols are changed only by replacing x by both u and v (or u and v by x), since the initial vertex of x is the initial vertex of u, and the terminal vertex of x is the terminal vertex of v.

For a set of relations in canonical form, the unmatched symbols are $e_1 \ldots e_r$. For each e_i, the initial vertex is also the terminal vertex. It is also the terminal vertex of d_i and the initial vertex of d_i^{-1}, but is not a vertex of any other symbol. Thus two different symbols, e_i and e_j, do not have a common vertex, and the number of cuffs is equal to r.

(C). For a surface with a system of relations in canonical form, the number $\chi = 2 - 2p - q - r$ is called the *Euler characteristic* of the surface (the meaning of χ for closed orientable surfaces is discussed by Courant-Robbins [8], pages 236–240,

258–259). The following definition of χ has meaning for any surface in the sense of Definition 2.2. It will be shown that two combinatorially equivalent surfaces have the same value for χ and that $\chi = 2 - 2p - q - r$ for a surface with a system of relations in canonical form.

Definition 4.5. Given a system of relations for a surface, the *Euler characteristic* χ of the surface is the number $\chi = V - E + F$, where V is the number of vertices, E the number of symbols, and F the number of relations.

If the surface is given the geometrical interpretation used previously, then V is the number of vertices, E the number of edges or segments, and F the number of faces or regions. For the surface of Figure 38, $\chi = 6 - 11 + 3 = -2$.

As noted before, operation (1) of Definition 3.1 does not change the number of vertices; however, it increases the number of relations by 1 and increases the number of symbols by 1. It therefore does not change χ. Operation (2) does not change χ, since deleting xx^{-1} (or $x^{-1}x$) deletes the symbol x and the terminal (or initial) vertex of x. Operation (3) increases the number of symbols by 1, but also increases the number of vertices by 1. Operations (1'), (2'), and (3') are the inverses of (1), (2), and (3) and therefore also do not change χ.

For a system of relations in canonical form, there are $r + 1$ vertices: one vertex is both the initial and terminal vertex for each $x_i, a_i, b_i, y_i, c_i, z_i$, and also is the initial vertex for each d_i; the terminal vertex for d_i is the same as the initial and terminal vertices for e_i $(i = 1, \ldots, r)$. There are $3p + 2q + 3r$ symbols and $p + q + r + 1$ relations. Hence

$$\chi = r + 1 - (3p + 2q + 3r) + (p + q + r + 1) = 2 - 2p - q - r.$$

The property of orientability, the number of cuffs (r), and the number χ have now been shown to be the same for any two combinatorially equivalent surfaces. Knowing whether $r = 0$ and whether the surface is orientable or not (whether $q = 0$ or $q > 0$) is enough information to determine uniquely one of the four types of canonical forms. Then the values of χ and r are enough additional information to determine the values of p and q, since $\chi = 2 - 2p, \chi = 2 - 2p - r$, $\chi = 2 - q$, and $\chi = 2 - q - r$, respectively, in each of the four classifications of canonical forms.

For example, the surface of Figure 38 is now known to be nonorientable and to have $r = 2$ and $\chi = -2$. Therefore it is a nonorientable surface with cuffs. Since $\chi = 2 - q - r$, it follows that $q = 2$ and the canonical form is

$$\begin{cases} y_i = c_i c_i, & i = 1, 2; \\ z_i = d_i e_i d_i^{-1}, & i = 1, 2; \\ y_1 y_2 z_1 z_2 = 1. \end{cases}$$

A surface has $\chi = 2$ if and only if $p = q = r = 0$. For such a surface, the system of relations in canonical form is $1 = 1$ and the surface is combinatorially equivalent to a sphere. However, any simple polyhedron is combinatorially equivalent to a sphere. Therefore $V - E + F = 2$ for any simple polyhedron. This is called *Euler's Formula* (Courant-Robbins [8], page 236). The number χ is called *Euler's characteristic*, although Euler only dealt with the case $\chi = 2$ of simple polyhedra (which can be continuously deformed into a sphere).

There are only 4 surfaces for which $\chi=0$, for the only nonnegative integral solutions of $0=2-2p-q-r$ are $(1,0,0)$, $(0,0,2)$, $(0,1,1)$, $(0,2,0)$. These are, respectively, the torus (sphere with one handle), cylinder (sphere with two cuffs), Moebius band, and Klein bottle. These surfaces are shown in Figure 40. They are the only four surfaces which admit a continuous tangent vector field which is nowhere zero. Poincaré [17, page 203] showed that, for a continuous tangent vector field having at most a finite number of simple singularities (zeros), the Euler characteristic is equal to the difference in number between those singularities of positive "index" (nodes, foci) and those of negative "index" (saddle points) (see Lefschetz [14], pages 17–19).

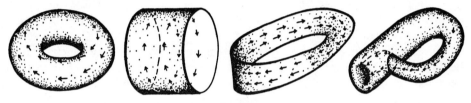

FIG. 40

The Betti number and the connectivity number of a surface are combinatorially invariant numbers which are closely related to χ (see Tucker and Bailey [21]). The *Betti number B* equals $2-\chi$ for a closed surface and to $1-\chi$ for a surface with cuffs*. It is also the maximum number of cuts that can be made on the surface without dividing it into more than one piece, where (1) the cuts are closed paths (or paths joining two points on previous cuts) if the surface is closed (Hilbert and Cohn-Vossen [12], page 294); (2) the cuts are along paths joining two points on free edges (boundaries) of cuffs if the surface is not closed. If a surface has r cuffs, then B is the maximum number of cuts along closed paths which can be made without getting more than r pieces. The idea of connectivity was introduced by Riemann in 1857 (Smith [19], pages 404–410). A surface is said to be *simply connected* if any closed curve in the surface can be continuously deformed to a point without leaving the surface (*e.g.*, the disk and sphere are simply connected); otherwise, the surface is said to be *multiply connected* (Courant and Robbins [8], pages 243–244). The *connectivity number h* is equal to $1+B$, and is 1 for a simply connected surface.

	h	B	χ	p	q	r
disk	1	0	1	0	0	1
plane annulus	2	1	0	0	0	2
Moebius band	2	1	0	0	1	1
projective plane	2	1	1	0	1	0
torus	3	2	0	1	0	0
Klein bottle	3	2	0	0	2	0

---

* Strictly, this is the one-dimensional Betti number mod 2 (Lefschetz [14], pages 68–72, 99–103).

5. Coverings of the sphere

Let R be a system of relations for a surface S, the symbols in R being denoted by letters $a, b, c, \ldots$. Replace these letters by $a_i, b_i, c_i, \ldots$ $(i = 1, \ldots, n)$. This makes n systems of relations $R_1, \ldots, R_n$, which define n copies $S_1, \ldots, S_n$ of the original surface S. Suppose a symbol x occurs twice in R, and that the second occurrence of x_1 in R_1 is replaced by x_2 and the second occurrence of x_2 in R_2 is replaced by x_1. These new systems of relations, R_1' and R_2', together define the surface, which can be thought of as being formed by cutting S_1 and S_2 along x_1 and x_2 and pasting the surfaces together by crossing them so that each side of x_1 is joined to the other side of x_2 (this is the process that was used in Example 1.2). More generally, for each x which occurs twice in R, one can permute the second occurrences (or the first occurrences) of $x_1, \ldots, x_n$ in $R_1, \ldots, R_n$. If this is done so that the new relations $R_1', \ldots, R_n'$ satisfy condition (b) of Definition 2.2 (i.e., so that all n sheets are connected), these relations together define a surface which is said to be an n-sheeted covering surface of S. For example, this can be done by a cyclic permutation of the second occurrences of one letter (the other letters being permuted in any way).

Note that a covering surface of a closed surface is closed, since the above process cannot create unmatched symbols (see Definition 4.4). Since permutations are made among symbols $x_1, \ldots, x_n$ which occur in the same form (either as x_i or as x_i^{-1}), it follows that if the relations for S are written so that no symbol occurs twice in the same form, the relations for the covering surface of S will automatically have this property. It therefore follows that a covering surface of S is orientable if S is orientable (see Definition 4.1). The converse of this is not true (see Theorem 6.1).

Covering surfaces of a sphere are called *Riemann surfaces* (Knopp [13], pages 93–118, 139–142). Riemann, one of the founders of modern topology, studied such surfaces because they served to uniformize multiple-valued functions of a complex variable (see Examples 1.2, 5.2, 5.3). For an excerpt from Riemann's fundamental paper, see Smith [19], pages 404–410. The references made to the theory of complex variables in the following examples are made for the benefit of the readers who have some knowledge of this field. They may be ignored by one who is interested only in the covering surfaces themselves.

Example 5.1. The relation $aa^{-1}bb^{-1}cc^{-1} = 1$ describes a sphere, as shown in Figure 41. Each of the four relations $a_i a_i^{-1} b_i b_i^{-1} c_i c_i^{-1} = 1$ $(i = 1, 2, 3, 4)$ also represents a sphere. The following four relations are obtained by permuting the subscripts on the second occurrences of each letter.

$$\begin{cases} a_1 a_4^{-1} & b_1 b_1^{-1} & c_1 c_4^{-1} = 1, \\ a_2 a_1^{-1} & b_2 b_3^{-1} & c_2 c_1^{-1} = 1, \\ a_3 a_2^{-1} & b_3 b_4^{-1} & c_3 c_2^{-1} = 1, \\ a_4 a_3^{-1} & b_4 b_2^{-1} & c_4 c_3^{-1} = 1. \end{cases}$$

These relations define a 4-sheeted covering surface of the sphere of Figure 41. This

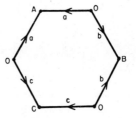

FIG. 41

covering surface is represented by the diagram of Figure 42. In this figure, the convention is used that whenever a symbol without a subscript is written on a face, it is to be given the number of the face as a subscript. Each point of the sphere that is not on one of the edges a,b,c is covered four times—once by each of the similarly located points on the four sheets. The same is true for each point which is on only one of the edges a,b,c, since each edge has four copies.

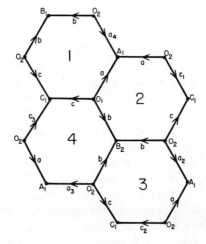

FIG. 42

However, no one of the vertices A, B, C, O is covered by four similarly located points. Thus, the vertex O is the left vertex of each of the symbols a,b,c (Figure 41). Any vertex of the covering surface which is the initial vertex of any of the symbols a_i, b_i, c_i ($i = 1,2,3,4$) is in a similar position and is said to cover O. If the initial vertex of a_1 is O_1, then successive application of the second, fourth, and first of the above relations shows that O_1 is also the intial vertex of b_2, the initial vertex of c_4, and again the initial vertex of a_1. Thus O_1 is a vertex in the covering surface and is the initial point for each of the edges in the *circular system* $a_1b_2c_4$ which surrounds O_1 in Figure 42. The point O_2 has the circular system $(a_4b_1c_1)(a_2b_3c_2)$ $(a_3b_4c_3)$. Figure 43 shows the circular system of edges and faces around O_2. The point O_2 is a *branch point* of order 2, because three sheets of the surface are joined at O_2. This is represented symbolically by the fact that the product abc occurs three

ROBERT C. JAMES

Fig. 43

times in the circular system of edges for O_2. In general, the order of a branch point is one less than the number of sheets joined at the point (see Definition 5.1 below). The remaining vertices and their systems of edges are A_1: $a_2^{-1}a_1^{-1}a_4^{-1}a_3^{-1}$; B_1: b_1^{-1}; B_2: $b_2^{-1}b_3^{-1}b_4^{-1}$; C_1: $c_1^{-1}c_4^{-1}c_3^{-1}c_2^{-1}$. Vertices O_2, A_1, B_2, C_1 are branch points of orders $2, 3, 2, 3$, respectively. The total of these orders is 10 and is the number of vertices "missing" in the sense discussed below.

To determine the Euler characteristic χ of this covering surface, note that the system of relations has four times as many symbols and four times as many relations as the surface itself. If there had been four times as many vertices, χ would also have been four times as large (see Definition 4.5). Since $\chi = 2$ for the sphere and 10 vertices are "missing," $\chi = 4.2 - 10 = -2$ for the covering surface. Since this surface is closed and orientable, $\chi = 2 - 2p$ and $q = r = 0$. Therefore $p = 2$ and the surface is combinatorially equivalent to a sphere with two handles.

If the process of determining vertices for a surface is analyzed, it can be seen that a vertex which has a circular system of edges $(ab...p)$ on the original surface will be "covered" by certain vertices whose circular system is of the type $(ab...p)$ $(ab...p)...(ab...p)$, with subscripts on each symbol. If the surface is n-sheeted, each of the symbols $a, b, ..., p$ has n copies and there are therefore n groupings of type $(ab...p)$ among the circular systems for the covering points. Because of these facts, the number b of the following definition is a non-negative integer. Also, the sum of the orders of the branch points covering a vertex P is equal to the difference between n and the number of covering points of P. Then the number of vertices on an n-sheeted covering surface is equal to n times the number of vertices on the surface itself less the sum of the orders of the branch points. The sum of the orders of the branch points is in this sense the number of vertices "missing" in the covering surface.

Definition 5.1. Let P be a vertex on a surface S and let $\bar{S}$ be an n-sheeted covering surface of S. If P is the initial [terminal] vertex of a symbol x in S, then the initial [terminal] vertices of $x_1, ..., x_n$ in $\bar{S}$ are *covering points* of P. Let $\bar{P}$ be a covering point of P and let $b + 1$ be the number of symbols of which $\bar{P}$ is a vertex divided by the number of symbols of which P is a vertex. If $b \geqslant 1$, then $\bar{P}$ is said to be a *branch point of order b*.

As noted before, an n-sheeted covering surface of a sphere is connected and closed ($q = r = 0$). There are n times as many symbols and n times as many relations as for the sphere. All vertices are covering points of vertices for the sphere and the number of vertices on the covering surface is $nV - \beta$, where V is the number of vertices on the sphere and β is the sum of the orders of the branch points. Thus

$$\chi = 2n - \beta.$$

Since $\chi = 2 - 2p$, the value of χ completely determines the topological nature of a covering surface of a sphere. These methods will be used in analyzing the following examples. Each of Examples 5.2 and 5.3 illustrate the truth of the following theorem (also see Theorem 6.1).

Theorem 5.1. *Any closed orientable surface can be represented as a 2-sheeted Riemann surface.*

Example 5.2. Consider the 2-sheeted Riemann surface which consists of two sheets covering the sphere of Figure 44, with these sheets crossing each other along the paths $b^1, b^2, \ldots, b^m$ (the case $m = 2$ was introduced as Example 1.2). If this surface is cut along the closed path $a^1 b^1 a^2 b^2 \ldots a^m b^m$, each covering sheet of the sphere will be cut into two pieces. The two pieces of the outer [inner] sheet were joined along each a^i, and each hemisphere of the outer sheet was joined to the

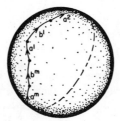

FIG. 44

other hemisphere of the inner sheet along each $b^1, b^2, \ldots, b^m$. The relations for the sphere of Figure 44 are

$$a^1 b^1 a^2 b^2 \ldots a^m b^m = 1; \qquad a^1 b^1 a^2 b^2 \ldots a^m b^m = 1.$$

If two copies of this sphere are made by introducing subscripts 1 for one copy and subscripts 2 for the other, then the relations for the covering surface are obtained by interchanging (for each k) the second occurrence of b_1^k and the second occurrence of b_2^k is the same way b_1 and b_2, and d_1 and d_2, were interchanged in Example 1.2. These four relations for the covering surface are

$$a_1^1 b_1^1 a_1^2 b_1^2 \ldots a_1^m b_1^m = 1, \quad a_1^1 b_2^1 a_1^2 b_2^2 \ldots a_1^m b_2^m = 1,$$
$$a_2^1 b_2^1 a_2^2 b_2^2 \ldots a_2^m b_2^m = 1, \quad a_2^1 b_1^1 a_2^2 b_1^2 \ldots a_2^m b_1^m = 1.$$

For each k, the terminal vertex of a^k has a circular system of edges $(a^k)^{-1} b^k$. For the covering surface, the terminal vertex of a_1^k has the circular system of edges $[(a_1^k)^{-1} b_1^k][(a_2^k)^{-1} b_2^k]$, and the initial vertex of a_1^k has the circular system $[a_1^k (b_2^{k-1})^{-1}][a_2^k (b_1^{k-1})^{-1}]$, where $k - 1$ is replaced by m if $k = 1$. Thus each vertex is covered by a single point and each vertex of the covering surface is a branch point of order 1. The sum of the orders of the branch points is $\beta = 2m$, and therefore $\chi = 4 - 2m = 2 - 2p$. Therefore the covering surface is of genus $p = m - 1$. It is a sphere if $m = 1$ and a torus if $m = 2$.

The Riemann surface of Example 5.2 uniformizes the function

$$w^2 = \left[1 - (k_1 z)^2\right]\left[1 - (k_2 z)^2\right]\dots\left[1 - (k_m z)^2\right],$$

where $k_1,\dots,k_m$ are distinct complex numbers. This surface can be formed as a 2-sheeted covering of the complex plane which crosses itself along cuts $b^1, b^2,\dots,b^m$, where the path $a^1 b^1 a^2 b^2 \dots a^m b^m$ is a closed polygonal path which joins the points $\pm 1/k_1,\dots, \pm 1/k_m$ in some order (but does not cross itself). By stereographic projection, this surface becomes the 2-sheeted covering surface of the sphere shown in Figure 44.

Example 5.3. Consider the 2-sheeted Riemann surface which consists of two sheets covering the sphere of Figure 45, with the sheets crossing each other along the paths $a^1, a^2,\dots,a^n$. The sphere has the relation

$$a^1 (a^1)^{-1} a^2 (a^2)^{-1} \dots a^n (a^n)^{-1} = 1.$$

The covering surface has the relations

$$\begin{cases} a_1^1 (a_2^1)^{-1} a_1^2 (a_2^2)^{-1} \dots a_1^n (a_2^n)^{-1} = 1, \\ a_2^1 (a_1^1)^{-1} a_2^2 (a_1^2)^{-1} \dots a_2^n (a_1^n)^{-1} = 1. \end{cases}$$

If n is odd, the vertex O, which is the initial vertex of each a^i, has a covering point O_1 which has the circular system of edges $[a_1^1 a_2^2 a_1^3 a_2^4 \dots a_1^n][a_2^1 a_1^2 \dots a_2^n]$ and is a branch point of order 1. If n is even, O is covered by the two points, O_1 and O_2, where O_1 has the circular system of edges $[a_1^1 a_2^2 \dots a_2^n]$, and O_2 has the circular system of edges $[a_2^1 a_1^2 a_2^3 \dots a_1^n]$ (in this case O_1 and O_2 are not branch points). Each of the other vertices of the covering surface is a branch point of order 1, for Z_k is covered by only one point, with the circular system of edges $[(a_1^k)^{-1}][(a_2^k)^{-1}]$. Therefore the sum of the orders of the branch points is $\beta = n + 1$ if n is odd and $\beta = n$ if n is even. But $\chi = 4 - \beta = 2 - 2p$, so the surface is of genus $p = (n-1)/2$ if n is odd and $p = (n-2)/2 = m - 1$ if $n = 2m$. If $n = 1$ or $n = 2$, the Riemann surface is a sphere. For $n = 3$ or $n = 4$, the surface is a torus.

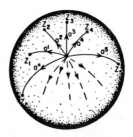

FIG. 45

The Riemann surface of Example 5.3 uniformizes the function

$$w^2 = (z - z_1)(z - z_2)\ldots(z - z_n),$$

where $z_1,\ldots,z_n$ are distinct complex numbers (Knopp [13], pages 112–118). This surface can be formed as a 2-sheeted covering of the complex plane, which crosses itself along cuts a^i from ∞ to z_i. By stereographic projection with $z = \infty$ at O, this surface becomes the 2-sheeted covering of the sphere shown in Figure 45.

Example 5.4. In 1809 Poinsot described two polyhedra, called the *great dodecahedron* and the *great icosahedron*, which have properties of regularity like the five classical regular polyhedra (see Ball and Coxeter [3], pages 143–145, and Figures 34, 30 opposite page 134).

The great dodecahedron may be obtained from a regular icosahedron (Figure 46) by deleting its 20 triangular faces and inserting 12 pentagonal faces to fit the 12 regular pentagonal perimeters (such as $\overline{ADECF}$). It can be thought of as a Riemann surface [the sphere is defined by 20 relations (one for each face of the icosahedron) and the Riemann surface is based on these relations by radial projection onto the icosahedron; e.g., the pentagonal face $\overline{ADECF}$ is represented by 5 relations corresponding to 5 triangles having a common vertex below (but covering) $\overline{B}$]. The surface is 3-sheeted and has 12 branch points of order 1, since two sheets of the great dodecahedron are joined at each vertex of the icosahedron. Thus the sum β of the orders of the branch points is 12 and $\chi = 3 \cdot 2 - 12 = -6 = 2 - 2p$. Therefore $p = 4$ (using the formula $\chi = V - E + F$ gives $\chi = 12 - 30 + 12 = -6$). The great dodecahedron is therefore a closed orientable surface of genus 4 (a sphere with 4 handles).

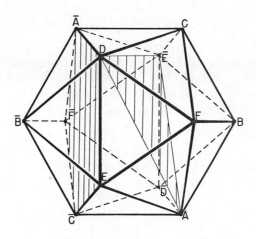

FIG. 46

The great icosahedron may be obtained from a regular icosahedron by joining each vertex by a line segment (edge) to each of the other five vertices which are neither adjacent nor antipodal to it and then inserting 20 triangular faces to fit the

20 equilateral triangular perimeters (such as $D\overline{EA}$) to be found in the network formed by the 30 new edges and the 12 original vertices (the original faces and edges are deleted). This is also a Riemann surface. It is 7-sheeted with 12 branch points of order 1, so $\chi = 7 \cdot 2 - 12 = 2$ and $p = 0$ (using $\chi = V - E + F$ gives $\chi = 12 - 30 + 20 = 2$). The great icosahedron is therefore combinatorially equivalent to a sphere.

As shown before, the Euler characteristic of an n-sheeted covering surface of a sphere is $\chi = 2 \cdot n - \beta$, where β is the sum of the orders of the branch points. Since $\chi \leqslant 2$ for any surface, it follows that $\beta = 0$ only if $n = 1$. Thus any non-trivial covering of the sphere has branch points. Such a covering surface is closed and orientable, so χ is even and the value of β is at least $2n - 2$.

6. Unbranched coverings.

As described above, a covering can be formed for any surface for which some symbol occurs twice. This is done by making copies and permuting in one occurrence of one or more symbols in such a way that the sheets are connected. It has been seen that a covering surface is orientable (or closed) if the covered surface is orientable (or closed) and that a covering surface of the sphere has branch points if it has more than one sheet. The two classes of coverings which have been of most interest are Riemann surfaces and unbranched coverings. Because of the possibility of reducing a surface to canonical form, it will be possible to establish some general results about unbranched coverings by means of examples. The whole process of forming and analyzing a covering surface can be carried out in a completely abstract symbolic fashion, but if the permutation is done at random the covering is likely to have branch points. Unbranched coverings are the result of a special form of weaving and are usually gotten with a specific geometric scheme in mind, as illustrated by Examples 6.4 and 6.5 (also see Seifert and Threlfall [18], Chapter 8).

Example 6.1. The projective plane can be thought of as a disk with diametrically opposite points identified. It can be described by the relation $cc = 1$. The two copies $c_1 c_1 = 1$, $c_2 c_2 = 1$ can be woven together in the following way (and only this way): $c_1 c_2 = 1$, $c_2 c_1 = 1$. Thus the sphere is an orientable "double" of the projective plane (a 2-sheeted covering without branch points).

Example 6.2. The Klein bottle is combinatorially equivalent to a sphere with two crosscaps. It can be described by the relation $aba^{-1}b = 1$. Figure 48 shows the 2-sheeted unbranched covering defined by the relations given below. This "double" of the Klein bottle is a torus.

$$\begin{cases} a_1 b_1 a_1^{-1} b_2 = 1, \\ a_2 b_2 a_2^{-1} b_1 = 1. \end{cases}$$

Examples 6.1 and 6.2 show that any closed nonorientable surface has a 2-sheeted orientable unbranched covering (this is also one of the conclusions which follows from Example 6.3). For in the case of a sphere with an odd number of crosscaps, all but one of the crosscaps can be changed into handles (the inverse of Step 5 of the reduction to canonical form). Two copies of the surface can then be woven

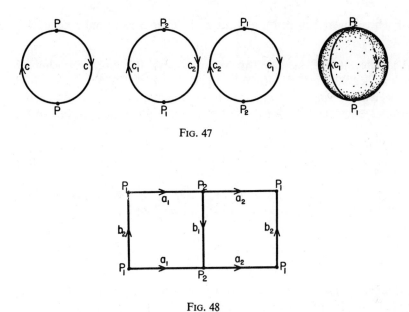

FIG. 47

FIG. 48

together, as in Example 6.1. If the sphere has an even number of crosscaps, all but two crosscaps can be changed into handles and the remaining two form a Klein bottle. Two copies of the surface can then be woven together, as in Example 6.2.

Example 6.3. Four closed surfaces are defined by the following four sets of relations.

$$P: \quad ccxx^{-1} = 1. \qquad S: \quad c_1c_2x_1x_1^{-1} = 1. \qquad c_2c_1x_2x_2^{-1} = 1.$$

$$C_n : \begin{cases} c^1c^1x^1(x^2)^{-1} = 1, \\ c^2c^2x^2(x^3)^{-1} = 1, \\ c^3c^3x^3(x^4)^{-1} = 1, \\ \quad\vdots \\ c^nc^nx^n(x^1)^{-1} = 1. \end{cases} H_{n-1} : \begin{cases} c_1^1c_2^1x_1^1(x_1^2)^{-1} = 1, \quad c_2^1c_1^1x_2^1(x_2^2)^{-1} = 1, \\ c_1^2c_2^2x_1^2(x_1^3)^{-1} = 1, \quad c_2^2c_1^2x_2^2(x_2^3)^{-1} = 1, \\ c_1^3c_2^3x_1^3(x_1^4)^{-1} = 1, \quad c_2^3c_1^3x_2^3(x_2^4)^{-1} = 1, \\ \quad\vdots \qquad\qquad\qquad\qquad \vdots \\ c_1^nc_2^nx_1^n(x_1^1)^{-1} = 1. \quad c_2^nc_1^nx_2^n(x_2^1)^{-1} = 1. \end{cases}$$

The surface P is the projective plane and S is a sphere. Since the block c^1c^1 occurs in the relations for C_n, the surface C_n is nonorientable. The surface H_{n-1} is orientable, since by inverting the relations in the right column each symbol can be made to occur once inverted and once not inverted (see Definition 4.1). As indicated by the notation, S is a 2-sheeted covering of P; C_n and H_{n-1} are n and $2n$-sheeted coverings of P; H_{n-1} is an n-sheeted covering of S and a 2-sheeted covering of C_n. The two vertices of P have the circular systems of edges $c^{-1}xc$ and

x^{-1}. Each is covered twice by vertices of S. Thus S is an unbranched covering of P. Also:

(1′). The surface C_n has two vertices, with the circular systems of edges $[(c^1)^{-1}x^1c^n][(c^n)^{-1}x^nc^{n-1}]\ldots[(c^2)^{-1}x^2c^1]$ and $[(x^1)^{-1}][(x^2)^{-1}]\ldots[(x^n)^{-1}]$. Thus C_n is an n-sheeted covering of P with two branch points of order $n-1$. Then $\chi = n\cdot 1 - 2(n-1) = 2-n$, since $\chi = 1$ for P. Hence C_n is a closed nonorientable surface with n crosscaps.

(1″). The surface H_{n-1} has four vertices, with the circular systems of edges $[(c_2^1)^{-1}x_1^1c_1^n][(c_2^n)^{-1}x_1^nc_1^{n-1}]\ldots[(c_2^2)^{-1}x_1^2c_1^1]$, $[(c_1^1)^{-1}x_2^1c_2^n][(c_1^n)^{-1}x_2^nc_2^{n-1}]\ldots[(c_1^2)^{-1}x_2^2c_2^1]$, $[(x_1^1)^{-1}][(x_1^2)^{-1}]\ldots[(x_1^n)^{-1}]$, and $[(x_2^1)^{-1}][(x_2^2)^{-1}]\ldots[(x_2^n)^{-1}]$. Thus H_{n-1} is a $2n$-sheeted covering of P with four branch points, each of order $n-1$. Hence $\chi = (2n)\cdot 1 - 4(n-1) = 2-2(n-1)$ and H_{n-1} is a closed orientable surface with $n-1$ handles.

(2) Since each vertex of C_n is covered twice by vertices of H_{n-1}, H_{n-1} is a 2-sheeted covering of C_n with no branch points.

(3) Since each vertex of S is covered only once by vertices of H_{n-1}, H_{n-1} is an n-sheeted covering of S with four branch points of order $n-1$.

These facts can be summarized as the following theorem.

Theorem 6.1. (1) *Any closed surface (orientable or not) can be represented as a covering surface of the projective plane.*

(2) *Any closed nonorientable surface has a 2-sheeted unbranched covering which is closed and orientable.*

(3) *Any closed orientable surface can be represented as a Riemann surface with four branch points of equal order.*

Example 6.4. The relation $aba^{-1}b^{-1}cc = 1$ defines a sphere with one handle and one crosscap, which is also a torus with one crosscap (Figure 1 without the cuff). Several examples of covering surfaces will be developed simultaneously by use of the following sets of relations. The notation indicates that S_2, S_3, and S_6 are 2, 3, and 6-sheeted coverings of S_1, and that S_6 is a 3-sheeted covering of S_2 and a 2-sheeted covering of S_3.

$$S_1: \quad aba^{-1}b^{-1}cc = 1. \quad S_2: \quad a^1b^1(a^1)^{-1}(b^1)^{-1}c^1c^2 = 1. \quad a^2b^2(a^2)^{-1}(b^2)^{-1}c^2c^1 = 1.$$

$$S_3: \begin{cases} a_1b_1a_1^{-1}b_2^{-1}c_1c_1 = 1, \\ a_2b_2a_2^{-1}b_3^{-1}c_2c_2 = 1, \\ a_3b_3a_3^{-1}b_1^{-1}c_3c_3 = 1. \end{cases} \quad S_6: \begin{cases} a_1^1b_1^1(a_1^1)^{-1}(b_2^1)^{-1}c_1^1c_1^2 = 1, \quad a_1^2b_1^2(a_1^2)^{-1}(b_2^2)^{-1}c_1^2c_1^1 = 1, \\ a_2^1b_2^1(a_2^1)^{-1}(b_3^1)^{-1}c_2^1c_2^2 = 1, \quad a_2^2b_2^2(a_2^2)^{-1}(b_3^2)^{-1}c_2^2c_2^1 = 1, \\ a_3^1b_3^1(a_3^1)^{-1}(b_1^1)^{-1}c_3^1c_3^2 = 1. \quad a_3^2b_3^2(a_3^2)^{-1}(b_1^2)^{-1}c_3^2c_3^1 = 1. \end{cases}$$

It can be easily verified that the surfaces S_1, S_2, S_3, and S_6 have 1, 2, 3, and 6 vertices, respectively. Therefore no one of the covering surfaces has branch points. The Euler characteristic of an n-sheeted unbranched covering surface is equal to $n\chi$, where χ is the Euler characteristic of the base surface. Since $\chi = -1$ for S_1, the values of χ for S_2, S_3, and S_6 are -2, -3, and -6. Of course, the number of symbols (edges or segments) and the number of relations (faces of regions) in these four surfaces are also proportional to 1, 2, 3, 6. All four surfaces are closed; S_1 and S_3 are nonorientable, since the blocks cc and c_1c_1 occur in their relations; S_2 and S_6 are orientable, since by inverting the right-hand relations each symbol can be made to appear once inverted and once not inverted (see Definition 4.1).

For surface S_3, $\chi = -3 = 2 - 2p - q$. Thus S_3 can be interpreted as a torus ($p = 1$) with 3 crosscaps (or as a sphere with 5 crosscaps). With this interpretation, it can be represented as in Figure 49. The cuts shown divide the surface into the three pieces described by the above relations. It can be seen to be a 3-sheeted unbranched covering of a torus with one crosscap by imagining that it is cut around a circle between two crosscaps and rejoined along the cut after being pulled through itself for two revolutions so that the three crosscaps are superimposed.

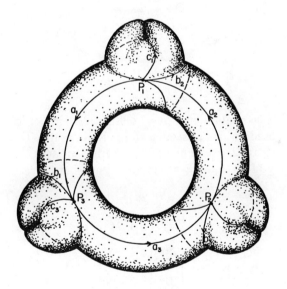

FIG. 49

The surface S_6 has $\chi = -6 = 2 - 2p$ and is a sphere with four handles. It is represented in Figure 50 with the cuts which produce the pieces described by the above relations. It can be seen to be a 3-sheeted unbranched covering of a torus with one handle (S_2) by the same method of pulling it through itself as used for Figure 49.

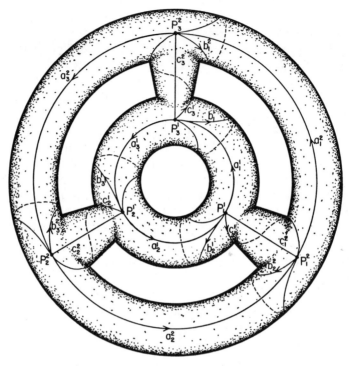

FIG. 50

Example 6.5. The relation $aba^{-1}b = 1$ defines a Klein bottle. The following sets of relations are based on this dissection of the Klein bottle. The notation indicates that R_2, R_3, and R_6 are 2, 3, and 6-sheeted covering surfaces of R_1, and that R_6 is a 3-sheeted covering of R_2 and a 2-sheeted covering of R_3.

$$R_1: \quad aba^{-1}b = 1. \quad R_2: \quad a^1b^1(a^1)^{-1}b^2 = 1. \quad a^2b^2(a^2)^{-1}b^1 = 1.$$

$$R_3: \begin{cases} a_1b_1a_2^{-1}b_3 = 1, \\ a_2b_2a_3^{-1}b_2 = 1, \\ a_3b_3a_1^{-1}b_1 = 1. \end{cases} \quad R_6: \begin{cases} a_1^1b_1^1(a_2^1)^{-1}b_3^2 = 1, & a_1^2b_1^2(a_2^2)^{-1}b_3^1 = 1, \\ a_2^1b_2^1(a_3^1)^{-1}b_2^2 = 1, & a_2^2b_2^2(a_3^2)^{-1}b_2^1 = 1, \\ a_3^1b_3^1(a_1^1)^{-1}b_1^2 = 1. & a_3^2b_3^2(a_1^2)^{-1}b_1^1 = 1. \end{cases}$$

It can be easily verified that the surfaces R_1, R_2, R_3, and R_6 have 1, 2, 3, and 6 vertices, respectively. Therefore no one of the covering surfaces has branch points. Since $\chi = 0$ for R_1, it follows that $\chi = 0$ for all four surfaces. If the right-hand relations for R_2 and R_6 are inverted, then no symbol occurs twice in the same form. Hence R_2 and R_6 are closed orientable surfaces with $\chi = 2 - 2p = 0$. Thus $p = 1$ and R_2 and R_6 are each combinatorially equivalent to a torus. In the relations for R_3, each a_i occurs once as a_i and once as a_i^{-1}, but no b_i can be made to have this property without destroying it for one of the a_i's. Hence R_3 is nonorientable (Definition 4.1) and has $\chi = 2 - q = 0$. Thus $q = 2$, and R_1 and R_3 are each combinatorially equivalent to a Klein bottle. Thus we have a Klein bottle (R_3) which is a "triple" of a Klein bottle (R_1), a torus $(R_2$ or $R_6)$ which is a "double" of a Klein bottle $(R_1$ or $R_3)$, and a torus (R_6) which is a "triple" of a torus. These coverings can be represented as in Figure 51.

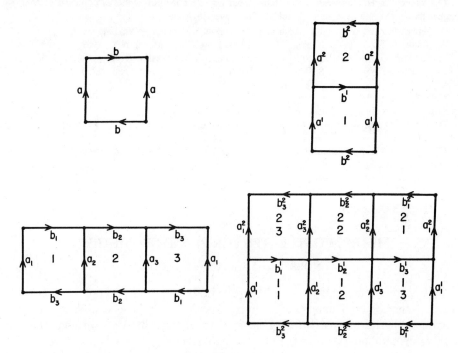

FIG. 51

References

1. Aleksandrov, P. S., Combinatorial Topology, vol. 1, Graylock Press, Rochester, N.Y., 1956.

2. Arnold, B. H., Intuitive Concepts in Elementary Topology, Prentice-Hall, Englewood Cliffs, N.J., 1962.

3. Ball, W. W. R., Mathematical Recreations and Essays, revised by H. S. M. Coxeter, Macmillan, N.Y., 1947.

4. Barr, S., Experiments in Topology, Thomas Y. Crowell Co., N.Y., 1964.

5. Blackett, D. W., Elementary Topology: A combinatorial and algebraic approach, Academic Press, N.Y., 1967.

6. Brahana, H. R., "Systems of circuits on two-dimensional manifolds," Annals of Mathematics, vol. 23 (1921–22), pages 144–168.

7. Cairns, S. S., Introductory Topology, Ronald Press, N.Y., 1968.

8. Courant, R., and H. Robbins, What is Mathematics?, Oxford University Press, N.Y., 1941.

9. Curtiss, D. R., Analytic Functions of a Complex Variable, Carus Mathematical Monographs No. 2, 1948.

10. Fréchet, M., and Ky Fan, Initiation to Combinatorial Topology, Prindle, Weber, and Schmidt, Boston, 1967.

11. Gramain, A., Topologie des Surfaces, Presses Universitaires de France, 1971.

12. Hilbert, D., and S. Cohn-Vossen, Geometry and the Imagination, Chelsea, N.Y., 1952.

13. Knopp, K., Theory of Functions (part two), Dover Publications, N.Y., 1947.

14. Lefschetz, S., Introduction to Topology, Princeton University Press, 1949.

15. Massey, W. S., Algebraic Topology: An Introduction, Harcourt, Brace, and World, 1967.

16. Newman, M. H. A., Elements of the Topology of Plane Sets of Points, Cambridge University Press, 1954.

17. Poincaré, J. H., "Sur les courbes définies par les équations différentielles", Journal de Mathématiques Pures et Appliqueés, series 4, vol. 1, 1885, pages 167–244.

18. Seifert, H., and W. Threlfall, Lehrbuch der Topologie, Chelsea, NY., 1945.

19. Smith, D. E., A Source Book in Mathematics, Dover Publications, N.Y., 1959.

20. Springer, G., Introduction to Riemann Surfaces, Addison-Wesley, Reading, Mass., 1957.

21. Tucker, A. W., and H. S. Bailey, Jr., "Topology," Scientific American, vol. 182 (1950), pages 18–24.

22. Tuckerman, B., "A non-singular polyhedral Möbius band whose boundary is a triangle," AMERICAN MATHEMATICAL MONTHLY, vol. 55 (1948), pages 309–311.

23. Veblen, O., Analysis Situs, American Mathematical Society Colloquim Publications, vol. 5, 1931.

24. Whittlesey, E. F., "Finite Surfaces: A study of finite 2-complexes," MATHEMATICS MAGAZINE, vol. 34 (1960), pages 11–22, 67–80.

METRIC POSTULATES FOR PLANE GEOMETRY*

SAUNDERS Mac LANE, University of Chicago

1. Introduction. The current interest in better high school mathematics directs attention to Euclidean geometry, as that part of the high school curriculum with the most subtle logical structure. For this very reason, there is little agreement on the arrangement of geometry for the schools. Even in its best form, the Euclidean arrangement suffered from various imperfections centering in the neglect of betweenness and separation properties; in recent years this arrangement has undergone steady attrition from all hands—those who would transfer training and those who wouldn't; those who would use geometry as a vehicle for reasoning in everyday life and those who prepare lists of "essential theorems." It is small wonder that the idea of a logical structure for elementary geometry is sometimes lost from sight.

A break with these difficulties is provided by the *Basic Geometry* of Birkhoff and Beatley [2]. These authors observe that the real number system is at hand in the schools; assuming it, they take the measures of distance and of angle as primitives for geometry; the result is that they build up the traditional system of geometry much more quickly than did Euclid [7], Hilbert [8], or Veblen [12]. The basic reason for the speed is simple: the subtleties of the betweenness relation are mastered by the use of the order relation for numbers. The particular axioms to be used for the measures of angles and distance can be varied and perhaps improved from Birkhoff's original form [1]. The problem

---

* From AMERICAN MATHEMATICAL MONTHLY, vol. 66 (1959), pp. 543–555.

is easily stated: to find a simple and intuitive set of facts on distance and angle which suffice to characterize plane geometry. An efficient metric foundation of geometry in this manner would be of such manifest utility that I present herewith a statement of one such variant, in the hope that others may be encouraged to try their hands at improving this type of axiomatics.

That axioms on distance and angle measure will suffice is clear, for distance can be used to define betweenness and this latter concept alone is a sufficient primitive for geometry in Veblen's system [11]. Blumenthal [4] and Gillam [6] have also considered axiom systems based just on distance. Another such system for plane geometry appears as support for a philosophical thesis in Reidemeister [10]; however, he takes the Pythagorean theorem as an axiom, while we would rather deduce this from more evident geometric facts.

2. Primitive notions. This axiomatics applies to the plane. Primitive terms are point, distance, line, and angle measure. The distance $d(AB)$ is intended to be the usual (nonnegative) distance from the point A to the point B. Practically, it is the quantity measured by a ruler. One sees at once that the points P on the segment from A to B are exactly those points for which $d(AB)$ is the sum $d(AP)+d(PB)$. Furthermore, the points on the ray AB consist of these points P on the segment, plus the points Q for which B is on the segment AQ. In other words, segments, rays, and, for the matter of that, lines may be defined in terms of distance.

Finally, if r and s are any two rays from the same point, the angle $\angle rs$ is to be the angle measured in degrees counterclockwise from r to s. Practically, it is the quantity measured (in degrees) by a protractor. Note especially that, with Birkhoff [1], we take angles to be directed. This fact, which is not emphasized in the book [2], seems useful for this type of axiomatics, since it gives immediate access to separation properties. For example, to show that a line l separates the plane, we take any ray r on l; the side to the left of r is then the union of all these rays s such that the $\angle rs$ is less than 180°. More detail is given below, but the basic idea is clear. Directed angles are fundamental to geometry and to elementary trigonometry, which we regard as part of geometry; they are the basis of the important facts about orientation; they should appear in beginning geometry. Even the most fuzzy-minded teenager can understand that a left is a left, not a right, glove, while those to whom this example may not appeal can surely distinguish a left turn from a right one.

3. The axioms on distance. There are four axioms on points and distances, as follows.

D1. *There are at least two points.*

D2. *If A and B are points, $d(AB)$ is a nonnegative number.*

D3. *For points A and B, $d(AB)=0$ if and only if $A=B$.*

D4. *If A and B are points, $d(AB)=d(BA)$.*

Axioms D2, D3, and D4 are those for a semimetric space. If the usual triangle axiom is added, we would have the axioms for a metric space; however we do not need the triangle axiom.

A point B is said to lie *between* the points A and C if all three points are different and $d(AC)=d(AB)+d(BC)$. If $A \neq C$, the interval, Int AC, is defined to be the set of all points B which lie between A and C. Thus Int AC is intended to be the usual open interval from A to C. If $A \neq O$ are points we now define

$$\text{Ray } OA = \{ P \mid P \neq O \text{ and } d(AP) = \mid d(OA) - d(OP) \mid \}.$$

Any such set $r = \text{Ray } OA$ is called a *ray* from O; thus each ray from O is a set of points not containing O. Ray OA could also be described as the union of the interval OP with the set of all points P such that either $P = A$ or A lies between O and P. For that matter, Int OA can now be described as the set of all points P on Ray OA for which $d(OP) < d(OA)$. Also, Int OA is the intersection of Ray OA with Ray AO.

4. The axioms on lines. There are four axioms on lines, as follows:

L1. *A line is a set of points containing more than one point.*

L2. *Through two distinct points there is one and only one line.*

L3. *Three distinct points lie on a line if and only if one of them is between the other two.*

L4. *On each ray from a point O and to each positive real number b there is a point B with $d(OB) = b$.*

For distinct points $A \neq B$, we shall denote by Line (AB) the unique line l containing A and B. In view of Axiom L3 this line is the union of the following four disjoint sets: Int (AB); the set of all P with A between P and B; the set of all Q with B between A and Q, and the set consisting of A and B. In particular, Ray $AB \subset$ Line AB.

LEMMA 1. *If A, B, and C are collinear, $d(AC) \leq d(AB)+d(BC)$.*

This asserts that the triangle axiom holds at least for the points of any one line. The proof is direct from Axiom L3. Either B is between A and C, in which case the equality holds, or one of A or C lies between, in which case the inequality applies.

LEMMA 2. *If B is between O and A, then Int $OB \subset$ Int OA.*

Proof. Take any $P \in$ Int OB; then $d(OB)=d(OP)+d(PB)$. By hypothesis, B is between O and A, so $O \neq A$ and $d(OA)=d(OB)+d(BA)$. Combining these equations and applying the triangle axiom (which holds because all points concerned lie on Line (OA)) we have

$$d(OA) = d(OP) + d(PB) + d(BA) \geq d(OP) + d(PA).$$

But the triangle axiom also gives the reverse inequality $d(OA) \leq d(OP) + d(PA)$. Hence equality holds, and P is between O and A.

LEMMA 3. *On each ray from a point O and to each positive real number b there is exactly one point B with $d(OB) = b$.*

Proof. Suppose, instead, that there were two such points B and B' on some Ray OA. Then $O, B, B',$ and A all lie on a line, to wit, the line OA. For the purposes of this proof let us write $[x, y, z]$ when the three real numbers $x, y,$ and z are such that one of them is the sum of the other two. Then O, B, B' collinear implies by Axiom L3 that $[b, b, d(BB')]$; since $B \neq B'$ and $d(BB') \neq 0$, this gives $d(BB') = 2b$. Set $a = d(OA)$.

Case 1: $b > a$. Then A is between O and B and also between O and B'; hence $d(AB) = b - a = d(AB')$. Then A, B, B' collinear gives $[b-a, b-a, d(BB')]$; hence $d(BB') = 2b - 2a$. Since $a \neq 0$, this contradicts the previous value $d(BB') = 2b$.

Case 2: $b = a$. Then $B \in$ Ray OA gives either $B \in$ Int OA or $A \in$ Int OB. If the former, then $a = d(OA) = d(OB) + d(BA)$, whence $d(BA) = 0$ and $B = A$ by Axiom D3. The same conclusion holds in the latter case. All told, $B = A = B'$, in contradiction to the assumption $B \neq B'$.

Case 3: $b < a$. Here B and B' are both between O and A, so that $d(BA) = a - b = d(B'A)$. Then A, B, B' collinear gives $[a-b, a-b, d(BB')]$; hence $d(BB') = 2a - 2b$. The previous value was $d(BB') = 2b$. Hence $a = 2b$ and $d(OB) = d(BA) = d(AB') = d(B'O) = b$, $d(OA) = d(BB') = 2b$. In other words, O, B, A, B' are, in order, the vertices of a "square" with sides b with both "diagonals" of length $2b$.

This "square" has the property that any *three* of its points can be embedded in the ordinary real line, preserving distance, while all four cannot be so embedded. Such a configuration is known in distance geometry (*cf.* [4]) as a pseudo-Euclidean quadruple.

To show this impossible, we use Axiom L4 to choose a point C on Ray OA with $d(OC) = 3b$. Then A is between O and C, and $B, B' \in$ Int $OA \subset$ Int OC, by Lemma 2. Hence $d(BC) = 2b = d(B'C)$. Then B, B' and C collinear yields $[2b, 2b, 2b]$, a contradiction.

This completes the proof in all cases.

THEOREM 1. *If $O \neq A$ and $B \in$ Ray OA, then Ray $OB =$ Ray OA.*

In proving this we may assume $B \neq A$. Then $B \in$ Ray OA asserts that either $B \in$ Int OA or $A \in$ Int OB. Since our conclusion is symmetric in A and B, it will suffice to prove, say, that $B \in$ Int OA implies Ray $OB =$ Ray OA. Also, Ray OB contains a point at any given distance from O, while by Lemma 3, Ray OA contains only one such point. Hence it will suffice to prove that

$$B \in \text{Int } OA \quad \text{implies} \quad \text{Ray } OB \subset \text{Ray } OA.$$

To show this, take $P \in$ Ray OB and consider the three possible cases for P. If $P \in$ Int OB, then by Lemma 2, Int $OB \subset$ Int $OA \subset$ Ray OA, and so $P \in$ Ray OA. If $P = B$, then $P \in$ Ray OA. There remains the case when $B \in$ Int OP, so that $d(OP) = d(OB) + d(BP)$. Since $B \in$ Int OA, we also have $d(OA) = d(OB) + d(BA)$. Suppose now that $P \in$ Ray OA is false; since P is on the line OA we then have O between P and A and hence

$$
\begin{aligned}
d(PA) &= d(PO) + d(OA) \\
&= d(OB) + d(BP) + d(OB) + d(BA) \\
&\geq 2d(OB) + d(PA),
\end{aligned}
\qquad \text{(Lemma 1),}
$$

a manifest contradiction to $d(OB) \neq 0$.

COROLLARY. *Any two rays from a fixed point O are equal or disjoint.*

This is immediate.

For elementary instruction some of these arguments, notably that for Lemma 3, will seem too artificial. This can be handled by making Lemma 3 a part of Axiom L4. Going further, we can add Theorem 1 as an axiom. From this Lemmas 2 and 3 can be proved easily. For example, to prove Lemma 3 observe that Theorem 1 gives Ray $OA =$ Ray OB, so that the proof of Lemma 3 need only be given in Case 2.

5. Coordinates on a line.

THEOREM 2. *If O is any point of the line l, then l contains exactly two disjoint rays from O, and every point $B \neq O$ on l is on one of these rays.*

Proof. By Axiom L1 there is on l a point $A \neq O$. By Axiom L4, there is on Ray AO a point A' with $d(AA') = 2d(OA)$. Since A' cannot be between O and A, we must then have O between A and A'. If $a = d(OA)$, then $d(AA') = 2a$. Let $r =$ Ray OA, $r' =$ Ray OA'. Take $B \neq O$ on l. If B is not in r, then by Axiom L3 and the definition of a ray, O is between A and B, and $d(AB) = a + d(OB)$. If also B is not in r', then O is between A' and B, and $d(A'B) = a + d(OB)$. But A, A', and B are collinear and distinct, which implies that one of the three numbers $a + d(OB)$, $a + d(OB)$, and $2a$ is the sum of the other two, which is impossible. Therefore B is either on r or on r'. Since $A' \in r'$ and $A' \notin r$, Theorem 1 shows that r and r' can have no points in common. Any ray from O contained in l must have a point on one of r or r'; hence by Theorem 1 it must be either r or r'. We may call r' the ray *opposite* r.

We can now define coordinates on a line l. If $O \in l$, choose one of the two rays from O in l, and call it the "positive" ray r; call the other ray from O in l the "negative" ray r'. Define a function x on the set l to the real numbers by setting $x(O) = 0$, $x(P) = d(OP)$, if $P \in r$, and $x(P) = -d(OP)$ if $P \in r'$. We call x the *coordinate function* on l with origin O and positive direction r. One readily proves

THEOREM 3. *Each coordinate function x on the line l is a one-one mapping of l onto the set of real numbers such that $d(PQ) = |x(P) - x(Q)|$ for all P, Q on l.*

Birkhoff assumed as an axiom the existence of a coordinate function with the property stated in Theorem 3. Since such a coordinate function is by no means unique, we have regarded it as appropriate to analyze his axiom into more primitive assertions about distance as formulated in our axioms above.

From this theorem it follows that Q lies between P and R if and only if either $x(P) < x(Q) < x(R)$, or $x(P) > x(Q) > x(R)$. Hence, also, the two rays from any point A on l are the sets of all P with $x(P) > x(A)$ and $x(P) < x(A)$, respectively. It follows that if x and y are any two coordinate functions on l, there is a real number a and a number $\epsilon = \pm 1$ such that $x(P) = \epsilon y(P) + a$ for all $P \in l$. The choice $\epsilon = -1$ amounts to a change in the "direction" of the line.

The notion of directions can be treated formally if one wishes. A *direction* D on a line l can be described as a function which assigns to each point $P \in l$ a ray $D(P)$ from P and contained in l, in such fashion that for any two points P and Q in l, the intersection $D(P) \cap D(Q)$ is one of $D(P)$ or $D(Q)$. We may then say that P comes *before* Q in the direction D if $D(P) \cap D(Q) = D(Q)$. This relation "before" yields a transitive order of the points on the line. Using the results above, one may easily prove

THEOREM 4. *A line has exactly two directions D and D'.*

Given one of these directions D, the second is described by letting $D'(P)$ be the ray opposite $D(P)$ at every point P. A direction is determined by any one of its positive rays; that is, given any ray r from a point O contained in the line l, there is exactly one direction D of l with $D(O) = r$.

6. One-dimensional geometry. The discussion hitherto has been essentially one-dimensional. It yields a set of metric axioms sufficient to characterize the one-dimensional Euclidean space: the Axioms D1, D2, D3, D4, the definition of "betweenness" and of "ray," the Axiom L4, and a rephrased Axiom L3:

L3′. *Given three distinct points, one of them lies between the other two.*

There are many other ways of setting up such axiom systems for a line; for example, one is given by Reidemeister in [**10**].

7. The axioms on angles. We regard an angle as the figure (or ordered triple) consisting of a point O and two rays r and s from O. Since we measure angles in degrees, it is convenient for each real number c to write $c°$ for the number c taken modulo 360.

There are four axioms on angle measure.

A1. *If r and s are rays from the same point, $\angle rs$ is a real number modulo* 360.

A2. *If r, s, and t are three rays from the same point, then $\angle rs + \angle st = \angle rt$.*

A3. *If r is a ray from O and c is a real number, then there is a ray s from O such that $\angle rs = c°$.*

If $A \neq O \neq B$ are points, we write $\angle AOB$ as usual for $\angle$ (Ray OA)(Ray OB). Then

A4. *If* $A \neq O \neq B$, *then* $\angle AOB = \angle BOA \neq 0°$ *if and only if* $d(AB) = d(AO)$ $+d(OB)$.

There are only two more axioms to come later, a *similarity* axiom and a *continuity* axiom. For the moment we observe that there are three existence axioms: the existence of points (D1); of points on a ray (L4); and of rays on a point (A3).

Axiom A2 giving the additivity of angles is possible only because we deal with directed angles. It is a most useful axiom; for example, it implies at once that $\angle rr = 0$ and that $\angle rs = -\angle sr$. In Axiom A4, observe that always $\angle AOB$ $+\angle BOA = 360° = 0°$; hence $\angle AOB = \angle BOA \neq 0°$ implies that $\angle AOB = 180°$, so the axiom is equivalent to the following requirement;

A4'. *If* $A \neq O \neq B$, *then* $\angle AOB = 180°$ *if and only if* O *is between* A *and* B.

THEOREM 5. *Let* r *and* s *be two rays from the same point* O. *If* $\angle rs = 180°$, *then* r *and* s, *together with* O, *form a straight line. If* $\angle rs = 0°$, *then* $r = s$.

Proof. Suppose first that $\angle rs = 180°$; then $r \neq s$ and hence $r \cap s = 0$. Pick points A on r and B on s; then $A \neq O \neq B$, and $\angle AOB = \angle rs = 180°$; hence, by Axiom A4, O lies between A and B and therefore on the line AB. This line contains r and s; by Theorem 2 it must consist of r, s, and O, as asserted.

Suppose next that $\angle rs = 0°$. By Axiom A3 there is a ray t from O with $\angle st = 180°$; hence s, t, and O form a line. But then $\angle rt = \angle rs + \angle st = 180°$, so that again r, t, and O form a line. It is the same line, hence r is one of s and t, by Theorem 2. Since $r \neq t$, we get $r = s$, as asserted.

COROLLARY. *If* $A \neq O$, *Ray* OA *consists of all points* $B \neq O$ *with* $\angle AOB = 0°$.

We can now introduce polar coordinates with O as origin and prove the usual facts about the one-one correspondence between points and their polar coordinates.

8. Similar triangles. With Birkhoff we deduce the theory of parallel lines from that of similar triangles. One reason for this order of events is that the classical order, with parallels first, involves a theorem that an exterior angle of a triangle exceeds either remote interior angle; the usual proof makes tacit assumptions about sides of a line.

By an *ordered triangle* we mean the figure $\triangle ABC$ formed by three distinct points A, B, and C; if you wish, the three intervals which they determine can be regarded as part of the figure. But notice that we name the vertices in order. This saves trouble, for the same reason that the use of ordered simplices in the modern version of singular homology theory saves trouble (see, *e.g.*, S. Eilenberg, *Singular Homology Theory*, Ann. of Math. vol. 45, 1944, pp. 407–447). A less sophisticated reason is better: ordered triangles are what actually occur in the theorems of elementary geometry.

Two ordered triangles $\triangle ABC$ and $\triangle A'B'C'$ are similar if there is a positive real number k and number $\epsilon = \pm 1$ such that $\angle ABC = \epsilon \angle A'B'C'$, $\angle BCA$

$= \epsilon \angle B'C'A'$, $\angle CAB = \epsilon \angle C'A'B'$, $kd(AB) = d(A'B')$, $kd(BC) = d(B'C')$, $kd(CA)$ $= d(C'A')$. The triangles are congruent if this holds with $k = 1$. We assume the side-angle-side (SAS) case of similarity as an axiom, in the following form.

SIMILARITY AXIOM. *If two ordered triangles $\triangle ABC$ and $\triangle A'B'C'$ have $\angle ABC = \epsilon \angle A'B'C'$, $d(AB) = kd(A'B')$, and $d(BC) = kd(B'C')$ (for $\epsilon = \pm 1$, k positive) they are similar.*

Recall that Hilbert [8] assumed a weaker version of the congruence case $(k = 1)$ of this axiom. We assume the case of similarity as well, so as to avoid the sophistication of an "incommensurable case" argument (*i.e.*, a continuity argument). This stronger assumption is in keeping with standard practice; for example, in a real vector space the distributive law $k(V + W) = kV + kW$ for multiples by a real scalar could be proved by continuity from the laws of addition, but we usually simply assume the general law.

Reinhold Baer has pointed out to me that the similarity axiom may be interpreted in terms of "motions." Take $\epsilon = \pm 1$, $k > 0$, and let two rays $r = $ Ray BC, $r' = $ Ray $B'C'$ be given. There is then a "motion" M which takes r into r' with scale factor k; explicitly, M is defined as that transformation of the plane which takes each point P into the point $P' = M(P)$ with $\angle PBC = \epsilon \angle P'B'C'$ and $d(P'B') = kd(PB)$; the Axioms A3 and L3 and Theorems 1 and 5 insure that there is a unique such point P'. The similarity axiom is then the assertion that M multiplies distances by k and angles by ϵ: $d(M(P)M(Q)) = kd(PQ)$, $\angle M(P)M(O)M(Q) = \epsilon \angle POQ$. By representing this motion M as the composite of several simpler motions, one can deduce the general similarity axiom from the following three special cases: *congruence* $(k = 1$, $\epsilon = \pm 1$, $r \neq r')$, *reflection* $(k = 1$, $\epsilon = -1$, $r = r')$, and *stretch* $(k \neq 1$, $\epsilon = 1$, $r = r')$.

A triangle $\triangle ABC$ is said to be *degenerate* if its vertices A, B, and C are collinear. It suffices to postulate the similarity axioms for a nondegenerate triangle, since in the case of a degenerate triangle the axioms can be proved from the previous axioms.

From the similarity axiom one now derives the other congruence and similarity theorems in the following order. First ASA (angle-side-angle) for a nondegenerate triangle, then base angles in an isosceles triangle are equal, then SSS (three sides given), and finally the theorem that the sum of the angles in a triangle is 180°. Note that ASA may be false for a degenerate triangle. The details are as in Birkhoff [1]; note especially the elegant proof of the angle-sum proposition made by bisecting the sides of the given triangle. Each similarity theorem (SAS, ASA, *etc.*) includes the corresponding congruence theorems, though for elementary purposes it may be better to first prove the congruence theorems and then the corresponding ones on similarity.

9. The continuity axiom. Let X move along the line AB, and let O be on one side of AB. Our continuity axiom is to assert that $\angle AOX$ is a monotonic and continuous function of $x = d(AX)$. This can be put in more elementary form by speaking of "proper" angles—an angle $c°$ is *proper* if $0 < c < 180°$, and *im-*

proper if $180° < c < 360°$. Clearly the supplement of a proper angle is proper, and the (additive) inverse of a proper angle is improper.

CONTINUITY AXIOM. *Let $\angle AOB$ be proper (Fig. 1). If D is between A and B, then $0 < \angle AOD < \angle AOB$. Conversely, if $0 < \angle AOC < \angle AOB$, then the ray OC meets the interval AB.*

The statement that $\angle AOX$ is continuous (as above) follows easily; it is indeed a very simple and illuminating exercise in elementary epsilontics. Birkhoff takes this continuity as an axiom (part of his Postulate III), though it is not quite clear whether he means to assume just continuity or continuity and monotonicity. In the book [2], "continuous" is defined to mean continuous and monotonic, and it is then assumed that $\angle AXB$ varies "continuously" when X varies along a line *or* a curve. This is troublesome, for if A and B are the endpoints of the diameter of a circle and X moves along a curve crossing the circumference of that circle several times, $\angle AXB$ does not vary monotonically.

We now sketch briefly how the continuity axiom gives the separation of the plane by lines. First observe the cyclic order of angles in a triangle, as in:

THEOREM 6. *If an ordered triangle $\triangle ABC$ has one angle $\angle ABC$ proper, then the remaining angles $\angle BCA$ and $\angle CAB$ are proper.*

For school use this might well be an axiom; the following proof, though simpler than that presented by Birkhoff, is still somewhat subtle. Take X on $\{B\} \cup$ Ray BC (Fig. 2) and set $x = d(BX)$. Now set $f(x) = 180° - \angle ABC - \angle XAB$. For all $x \geq 0$, $f(x)$ is a continuous function. For $x > 0$, the angle-sum theorem gives $f(x) = \angle BXA$, which is neither $0°$ nor $180°$. For $x = 0$, $f(0) = 180° - \angle ABC$; hence is proper. Thus $f(x)$ is a continuous function on the segment $0 \leq x \leq d(BC)$; it is never $0°$ nor $180°$; therefore $f(x)$ is always proper; therefore $f(d(BC)) = \angle BCA$ is proper.

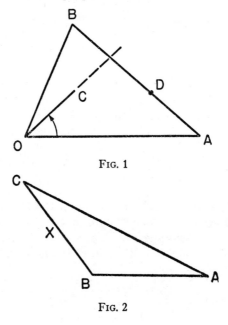

FIG. 1

FIG. 2

We can now define the "sides" of a line l. Take a direction D on the line l, as described in Theorem 4. If A is before B in the direction D, we say that P is to the *left* of l, as directed by D, if $\angle BAP$ is proper, and to the *right* of l if $\angle BAP$ is improper. A simple application of Theorem 6 shows that this definition is independent of the choice of the points A and B; that is, if also A' is before B' in the direction D, then $\angle B'A'P$ is proper precisely when $\angle BAP$ is proper. We now can show that each line separates the plane into its left and right side.

THEOREM 7. *If l is a directed line, then points P and Q not on l lie on the same side of l if and only if l does not meet the interval PQ.*

Proof. Suppose first that $l \cap \text{Int } PQ \neq 0$, and that P is, say, to the left of l in the direction D. We must then show Q to the right of l. Let A be the point in which l meets Int PQ, and take some $B \in l$ with A before B. Then $\angle BAP$ is proper. But by Axiom A4, $\angle QAB + \angle BAP = 180°$; hence $\angle QAB$ is proper and therefore $\angle BAQ = -\angle QAB$ is improper. Hence Q is to the right of l in the direction D.

Suppose next that $l \cap \text{Int } PQ = 0$, and that P is again to the left of l. We must show that also Q is to the left of l. Again, take A before B on l in the direction D; then $\angle BAP$ is proper. For $X \in \text{Int } PQ$ the angle $\angle BAX$ is a continuous function of $d(PX)$; since $l \cap \text{Int } PQ = 0$, this function is never 0° nor 180°. Hence $\angle BAX$ is always proper; in particular, $\angle BAQ$ is proper and Q is also to the left of l.

COROLLARY (Axiom of Pasch). *Given $\triangle ABC$ and a line l with $A \notin l$, if l meets Int BC, then l also meets one of Int AB or Int AC.*

Proof. Since l meets Int BC, B and C lie on opposite sides of l. Since $A \notin l$, A lies on one of these sides, say with C. Thus A and B are on opposite sides, so that by the theorem, l meets Int AB.

10. Parallels and perpendiculars. It is convenient to start with the existence of a triangle for given ASA.

THEOREM 8. *Given an interval AB and two numbers c and d between 0 and 180 such that $c + d < 180$, there exists a point C such that the ordered triangle ABC has $\angle CAB = c°$, $\angle ABC = d°$.*

Theorem 5 shows that the ray making a given angle with a given ray is unique; hence the conclusion of this theorem could be rephrased as follows.

On the interval AB (Fig. 3), erect at B a ray BE' with $\angle ABE' = d°$ and at A, a ray AE with $\angle EAB = c°$. The rays BE' and AE must then meet.

Proof. We first show that there exists some triangle with angles c and d. Set $e = 180 - c - d$ and construct any triangle $\triangle LMN$ with $\angle LMN = c°$. If $\angle MNL = d°$, we are done. If $\angle MNL > d°$, the continuity axiom provides a point $X \in \text{Int } LM$ with $\angle MNX = d°$, and $\triangle MNX$ is the desired triangle with angles c and d. Otherwise $\angle MNL < d°$; the angle-sum proposition then shows that $\angle NLM > e°$. In this case the continuity axiom provides a point Y on Int MN

with $\angle YLM = e°$, hence with $\angle MYL = 180° - e° - c° = d°$. In any case, we then have a triangle RST with $\angle TRS = c°$, $\angle RST = d°$.

Now return to the given interval AB. By polar coordinates, there is a point C with $\angle ABC = d°$ and $d(BC)/d(BA) = d(ST)/d(SR)$. Then $\triangle ABC$ is similar to $\triangle RST$ by the axiom of similarity; hence $\triangle ABC$ has the desired angles.

Perpendicular lines are defined as usual. Axiom A3 includes the fact that one can erect a perpendicular to a given line at any point of that line. We also prove

THEOREM 9. *If $C \notin l$, there is through C a unique line $\perp l$.*

Proof. That there is at most one such perpendicular follows from the angle-sum proposition for a triangle. To construct one, direct l from A to B so that C lies to the left of l. Then $\angle BAC$ is proper. If $\angle BAC = 90°$, we are done. Otherwise take $B' \in l$ so that ray AB is opposite ray AB'; then one of the angles $\angle BAC$, $\angle CAB'$ is acute. If $\angle BAC$ is acute, Theorem 10 gives a triangle EAC with $\angle EAC = \angle BAC$, $\angle ACE = 90° - \angle BAC$. Then $\angle CEA = 90°$, and CE is the desired perpendicular. If $\angle CAB'$ is acute, there is a similar construction.

Two lines are said to be *parallel* if they do not meet.

THEOREM 10. *If $C \notin l$, there is through C a unique line parallel to l.*

Proof. By Theorem 9, take m through C with $m \perp l$; then take a line n through C with $n \perp m$. By the angle-sum proposition n and l cannot meet; by Theorem 8 any other line through C will meet l.

The Theorem of Pythagoras now follows, say, by dropping a perpendicular from the vertex of the right angle to the hypotenuse, and using similar triangles. With this, we can introduce rectangular coordinates, prove the distance formula, and the usual description of straight lines by linear equations. This shows that the plane described by our axioms is the classical one. The trigonometric functions may also be defined. The fact that our angle measure agrees with the classical one rests on the usual proportionality between inscribed angles and arc; the proof, as in Birkhoff [1], employs the continuity axiom.

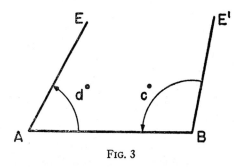

FIG. 3

11. Problems and alternatives. Our axioms suffer from at least one blemish; in pure Euclidean geometry, distance should be a magnitude with no fixed unit of measure. Such a notion of magnitude can be easily described axiomatically, but is probably not suitable for high schools; at any rate, we can easily observe that all our constructions and theorems are unaltered under change of scale (*i.e.*, when all distances involved are multiplied by some fixed positive real number). A similar remark applies to a change in the unit of angle measure.

More important is the observation that our axioms apply to an *oriented* plane, hence cannot at once be extended to three-space. This is because there is no continuous way of choosing directions for all angles $\angle rs$ in three-space so that the addition axiom A2 will hold. It would be interesting to have an axiom system of this general type for the unoriented plane and the extension of this axiom system to space. This would allow easy comparison with the ordinary axiom systems for geometry, which are normally formulated for three dimensions.

We have not studied the independence of all of our axioms, but we can present examples to show the independence of the similarity axioms and of the continuity axiom.

To show the independence of the similarity axiom use the example of [8], page 47. Take the points, lines, and angle measures to be the usual points, lines, and angle measures of a plane II, situated in three-space, but take $d(AB)$ to be the Euclidean distance $d(A'B')$, where A' and B' are the respective projections of A and B on some other plane II' cutting II at an acute angle. It is then easily seen that all the axioms except the similarity axiom hold. In view of this example it is convenient to speak of a system of points, lines, distances, and angle measures which satisfy the axioms D1—D4, L1—L4, and A1—A4 as a *pre-Euclidean plane*.

To show the independence of the continuity axiom use a Hamel basis of the real numbers to construct an automorphism ϕ of the additive group of real numbers such that $\phi(a) = a$ for every rational number but $\phi(b) = 3b$ for some irrational number b. Now take the points, lines, and distances to be the points, lines, and distances of an ordinary plane, but define a new angle measure by $\alpha(rs) = \phi(\angle(rs))$ for any two rays r and s from the same point. Since ϕ is the identity on rational numbers, the new angle measure α has $\alpha(rs) = 180 \pmod{360}$ if and only if $\angle rs = 180°$. Hence the elementary axioms, and especially Axioms A1–A4 on angle measure, are still valid. That the similarity axiom continues to hold follows from the observation that two angles are congruent in the old measure if and only if they are congruent in the new measure. However, the continuity axiom is obviously violated because sometimes $\phi(b) = 3b$.

Many variants of the axioms are possible. Birkhoff's system consisted essentially of the following axioms: D1–D4, L1, L2, A1, the similarity and continuity axioms, and the following two:

AXIOM L. *On each line l there is a function x which is a one-one correspondence between the points P on l and the real numbers, such that $d(PQ) = |x(P) - x(Q)|$ for all P and Q on l.*

AXIOM A. *For each point O there is a function θ which is a one-one correspondence between the rays r from O and the real numbers modulo 360 such that $L(rs) = \theta(s) - \theta(r)$ for all rays r and s from O.*

These axioms are exactly Birkhoff's four postulates (he used D1–D4, L1, and A1 without numbering them) with the unessential changes that his angles are measured mod 2π, and that his primitive for angles is $\angle AOB$ and not $\angle rs$ (while his rays are defined as half-lines). Since both axiom systems give the (cartesian) plane, his system is equivalent to ours; however it is not until the Theorem of Pythagoras is proved that it follows that his line AB contains all points metrically between A and B.

In our system the line need not be taken as a primitive concept, since one can define Line AB as the set of all points C such that $C=A$, $C=B$, or one of A, B, and C is between the other two. This definition does not seem to allow us to reduce the number of axioms.

Another variant of our system without lines as primitive notion is the following. Take, point, distance, and angle measure $\angle rs$ as primitive, with rays defined as above. As axioms take D1–D4, A1–A4, and D5 (the present Theorem 1) and D6 (the present L4). Then define a line to be the union of a point O with two rays r, r', from O such that $\angle rr' = 180°$. It can be shown that these axioms also yield a pre-Euclidean plane.

Still other variants can be set up. For example, one may replace the angle between two rays r and s by the angle $\angle AOB$ given for any three points $A \neq O \neq B$. The ray OA can then be defined as the set of all points $B \neq O$ with $\angle AOB = 0°$. As axioms for a pre-Euclidean plane, one may then take D1-D4, L4 (with uniqueness required for the point B) and the following:

A1′. *If* $A \neq O \neq B$, *then* $\angle AOB$ *is a real number modulo* 360.

A2′. *If* A, B, *and* C *are all different from* O, *then* $\angle AOB + \angle BOC = \angle AOC$.

A3′. *If* $A \neq O$, *and* c *is a real number, there is a point* $B \neq O$ *with* $\angle AOB = c°$.

A4′. *If* $A \neq O \neq B$, *then* $\angle AOB = 180°$ *if and only if* $d(AB) = d(AO) + d(OB)$.

A5. *If* $A \neq O \neq B$, *then* $\angle AOB = 0°$ *if and only if* $d(AB) = \left| d(AO) - d(BO) \right|$.

References

1. G. D. Birkhoff, A set of postulates for plane geometry based on scale and protractor, Ann. of Math., vol. 33, 1932, pp. 239–345; reprinted in George David Birkhoff, Collected Mathematical Papers, vol. III, Amer. Math. Soc., New York, 1950, pp. 365–381.

2. G. D. Birkhoff and Ralph Beatley, Basic Geometry, New York, 1941.

3. ———, Manual to Basic Geometry.

4. L. M. Blumenthal, Theory and Application of Distance Geometry, New York, 1953.

5. H. G. Forder, The Foundations of Euclidean Geometry, London, 1927.

6. B. E. Gillam, A new set of postulates for Euclidean geometry, Revista de Ciencia, vol. 42, 1940, pp. 869–899.

7. T. L. Heath, The Thirteen Books of Euclid's Elements, London, 1925.

8. David Hilbert, Grundlagen der Geometrie, Leipzig, 1930.

9. R. L. Moore, Sets of metrical hypotheses for geometry, Trans. Amer. Math. Soc., vol. 9, 1908, pp. 487–512.

10. Kurt Reidemeister, Ueber den Unterschied der Gegenden im Raum, Z. Philosophische Forschung, vol. 2, 1947, pp. 131–150.

11. Oswald Veblen, A system of axioms for geometry, Trans. Amer. Math. Soc., vol. 5, 1904, pp. 343–384.

12. ———, The Foundations of Geometry; Ch. I in Monographs on Topics of Modern Mathematics, J. W. A. Young, ed., New York, 1955.

Note*: Since the time of this paper, considerable changes have been made in the teaching of high school geometry, especially by the School Mathematics Study Group. Their approach [17] is a combination of the Hilbert geometry axioms and the Birkhoff-Beatley approach and hence distantly related to that described in the above article. This has appeared in a commercial text by Moise and Down, and is described in a college level text by Moise [15].

Another important development has been the revival of the approach of Felix Klein, who emphasized a treatment of geometry in terms of groups of transformations (e.g., similarity transformations). This is described in the references by Bachmann and Guggenheimer.

13. F. Bachmann, Aufbau der Geometrie aus dem Spiegelungsbegriff, Berlin, Springer-Verlag, 1959.

14. Henrich Guggenheimer, Plane Geometry and Its Groups, San Francisco, Holden-Day, 1967.

15. E. E. Moise, Elementary Geometry from an Advanced Standpoint, Reading, Mass., Addison-Wesley, 1963.

16. E. E. Moise and Floyd Downs, Geometry, Reading, Mass., Addison-Wesley, 1964.

17. School Mathematics Study Group, Mathematics for High School, Geometry, New Haven, Conn., Yale University Press, 1961.

THE STEINER-LEHMUS THEOREM†

G. GILBERT, B.I.C.C. Research Organization, London and
D. MacDONNELL, Fenlow Electronics, Weybridge, England

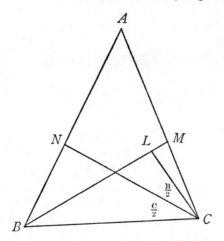

THEOREM. *Any triangle having two equal internal angle bisectors (each measured from a vertex to the opposite side) is isosceles.*

---

* Prof. Mac Lane provided these remarks and additional references for the present volume. – ED.

† From AMERICAN MATHEMATICAL MONTHLY, vol. 70 (1963), pp. 79–80.

Proof. Let ABC be the triangle with equal angle bisectors BM and CN, as in the figure. If the angles B and C are not equal, one must be less, say $B < C$. Take L on BM so that $\angle LCN = \frac{1}{2}B$. Since this is equal to $\angle LBN$, the four points L, N, B, C are concyclic (lie on a circle). Since

$$B < \tfrac{1}{2}(B + C) < \tfrac{1}{2}(A + B + C),$$

$\angle CBN < \angle LCB < 90°$. Since smaller chords of a circle subtend smaller acute angles, and $BL < CN$,

$$\angle LCB < \angle CBN.$$

We thus have a contradiction.

Editorial Note: Martin Gardner, in his review of Coxeter's *Introduction to geometry* (Scientific American, 204 (1961) 166–168) described this famous theorem in such an interesting manner that hundreds of readers sent him their own proofs. He took the trouble to refine this massive lump of material until only the above gem remained. This theorem was proposed in 1840 by C. L. Lehmus, and proved by Jacob Steiner. For its history until 1940 see J. A. McBride, Edinburgh Math. Notes, 33 (1943) 1–13. For its more recent history, see Mordechai Levin, On the Steiner-Lehmus theorem, in MATHEMATICS MAGAZINE, 47 (1974), 87–9.

A POLYGON PROBLEM*

E. R. BERLEKAMP, E. N. GILBERT, and F. W. SINDEN,
Bell Telephone Labs., Murray Hill, N. J.

Let P_0 be a closed polygon. Let P_1 be the polygon obtained by joining the midpoints of P_0's sides in order (see Figure 1). We write

$$P_1 = TP_0.$$

Let $P_2 = TP_1$, $P_3 = TP_2$, etc. We will call P_1, P_2, P_3, $\cdots$ the *descendants* of P_0.

We consider the question: for what polygons P_0 does there exist a *convex* descendant P_n?

It is easy to show that all descendants of convex polygons are convex. Hence, the existence of one convex descendant implies the existence of infinitely many.

For polygons with fewer than five sides the results are immediate:

2-gons: all descendants have coincident vertices.

3-gons: all descendants are similar to the parent P_0, which is necessarily convex.

4-gons: all descendants are parallelograms, hence convex, though the parent P_0 may be nonconvex.

(To see the last result observe that the sides of P_1 are parallel to the diagonals of P_0.)

---

* From AMERICAN MATHEMATICAL MONTHLY, vol. 72 (1965), pp. 233–241.

Polygons with five or more sides are more complicated, but in many examples we found that repeated application of T led quickly to convex polygons. This did not seem surprising, for intuitively T is a smoothing operator. We began with the conjecture that every polygon ultimately has convex descendants.

Figure 2 shows this conjecture to be false. The regular pentagram begets only its own kind.

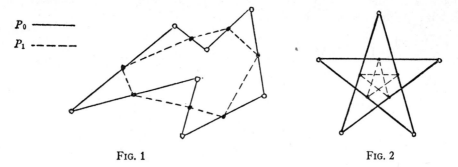

P_0 ———

P_1 - - - - -

FIG. 1   FIG. 2

This example, though, may seem unfair since the pentagram intersects itself. Figure 3 shows a non-selfintersecting polygon without convex descendants.

This polygon, in fact, is less innocent than it looks. Not only are its descendants nonconvex but infinitely many of them are self-intersecting. (To see this, note that the relation

$$\frac{X_2 - X_1}{X_3 - X_2} = \frac{1 + \sqrt{5}}{2}$$

between vertex projections X_1, X_2, X_3 holds for all descendants as well as for the polygon itself. P_i is self-intersecting whenever P_{i-1} is not.)

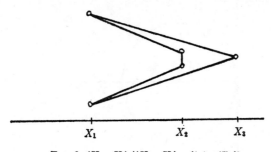

X_1 X_2 X_3

FIG. 3. $(X_2 - X_1)/(X_3 - X_2) = (1 + \sqrt{5})/2$

In spite of these examples, the conjecture that all polygons have convex descendants is not entirely empty. It turns out that *almost* all polygons have convex descendants.

This problem originated with G. R. MacLane, to whom we are indebted for some helpful suggestions. We have recently heard of an independent solution by A. M. Gleason.

Planar polygons. To analyze planar polygons it is convenient to use complex numbers. Let the N-tuple $\mathbf{z} = (z_1, \cdots, z_N)$ represent the vertices of a closed polygon in order. We will refer to the polygon simply as $\mathbf{z}$. The first descendant of $\mathbf{z}$ is a polygon $\mathbf{w}$ with vertices

$$w_1 = \tfrac{1}{2}(z_1 + z_2), \cdots, w_N = \tfrac{1}{2}(z_N + z_1).$$

This may be written $\mathbf{w} = T\mathbf{z}$, where

$$T = \frac{1}{2}\begin{bmatrix} 1 & 1 & 0 & \cdots & 0 \\ 0 & 1 & 1 & \cdots & 0 \\ & & \cdots & & \\ 0 & & \cdots & 1 & 1 \\ 1 & & \cdots & 0 & 1 \end{bmatrix}$$

The eigenvalues and eigenvectors of T, i.e., the solutions of $T\mathbf{u} = \lambda\mathbf{u}$, are of particular interest, for if $\mathbf{u}$ is an eigenvector then the polygon represented by $\mathbf{u}$ is merely rotated and uniformly diminished by T. Specifically it is rotated about the origin through the angle arg λ and diminished by the factor $|\lambda|$.

The eigenvalues and eigenvectors can be written down explicitly in terms of Nth roots of unity $\alpha^k = e^{i2\pi k/N}$. The eigenvalues are

$$\lambda_k = \tfrac{1}{2}(1 + \alpha^k), \, k = 0, 1, \ldots, N-1;$$

(see Figure 4) and the corresponding eigenvectors are

$$u_k = (\alpha^k, \alpha^{2k}, \ldots, \alpha^{Nk}).$$

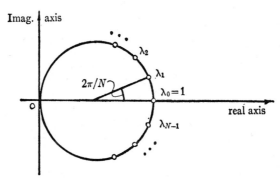

FIG. 4. Eigenvalues

The normalization $\mathbf{u}_k \cdot \mathbf{u}_k{}^* = N$ seems a little more convenient in the present context than the usual $\mathbf{u}_k \cdot \mathbf{u}_k{}^* = 1$. Note that the components of the eigenvectors are all roots of unity. Thus the polygons represented by the eigenvectors (eigen-polygons) are all inscribed in the unit circle. The kth one is obtained by joining in order the points on the unit circle with arguments

$$k1\phi,\ k2\phi,\ \cdots,\ kN\phi, \qquad\qquad \phi = 2\pi/N.$$

The eigenpolygons for $N = 5,\ 6$ are shown in Figure 5. It is evident geometrically that these polygons are only rotated and diminished by the transformation T.

We will call any polygon that is similar to the kth eigen-N-gon a *regular N-gon of order k*. Thus, we include among regular polygons self-intersecting, multiply-traversed and degenerate polygons.

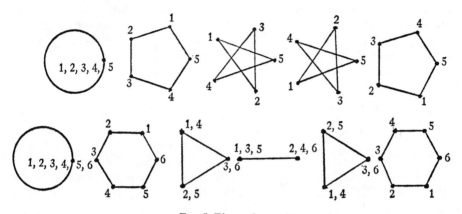

FIG. 5. Eigenpolygons

We observe that regular N-gons have the following properties: (1) The regular N-gons of order 0 are totally degenerate, i.e., all vertices coincide. (2) The regular N-gons of orders 1 and $N-1$ are the regular convex N-gons, counterclockwise and clockwise, respectively. We will regard all self-intersecting, multiply-traversed and degenerate polygons as nonconvex. (3) The regular N-gons of orders k and $N-k$ are the same except for the sense in which they are traversed. (4) The regular N-gon of order k is multiply-traversed if and only if k and N are not relatively prime.

The crucial fact is that an arbitrary N-gon can be written as a sum of regular N-gons. This is the meaning of the following eigenvector expansion, whose validity follows readily (see [1]) from the fact that the eigenvalues are distinct.

(1) $$\mathbf{z} = \sum_{k=0}^{N-1} a_k \mathbf{u}_k \quad (a_k \text{ complex})$$

or explicitly:

$$z_j = \sum_{k=0}^{N-1} a_k \alpha^{jk}.$$

The coefficients are given by the formula

$$a_k = \frac{1}{N} \sum_{j=1}^{N} z_j \alpha^{-jk}.$$

This formula can be given a geometric interpretation as follows. Consider the transformation $\mathbf{w} = W\mathbf{z}$, where

$$w_j = z_j \alpha^{-j}.$$

The effect of W is to rotate the jth vertex about the origin through the angle $-j(2\pi/N)$. The vertex z_N makes a complete revolution, returning to its starting position, while the other vertices turn through a fraction j/N of a revolution proportional to their indices. This diminishes the angle between successive vertices by the constant amount $2\pi/N$. We will say that W *winds* polygon $\mathbf{z}$ about the origin. Winding carries the kth eigenpolygon into the $(k-1)$-st mod N. In particular it carries the convex eigenpolygon $\mathbf{u}_1$ into the degenerate eigenpolygon $\mathbf{u}_0 = (1, 1, \cdots, 1)$ and it carries $\mathbf{u}_0$ into the convex eigenpolygon $\mathbf{u}_{N-1}$. Applied to an arbitrary polygon, winding simply permutes cyclically the coefficients of the component eigenpolygons:

$$a_1 \rightarrow a_0, \; a_2 \rightarrow a_1, \; \cdots, \; a_0 \rightarrow a_{N-1}.$$

The centroid of a polygon $\mathbf{z}$ is $\sum z_j/N$. All of the eigenpolygons except $\mathbf{u}_0$ have centroid O. Hence, the centroid of an arbitrary polygon is the centroid of its zeroth component, namely a_0. It follows that the centroid of the polygon obtained by winding $\mathbf{z}$ is a_1.

The coefficients a_k, then, can be given this geometric interpretation:

The coefficient a_k is the centroid of the polygon obtained by winding $\mathbf{z}$ k times.

We return now to the midpoint-joining transformation T. Consider a polygon $\mathbf{z}$ in decomposed form:

$$\mathbf{z} = \sum_{k=0}^{N-1} a_k \mathbf{u}_k.$$

Since $T\mathbf{u}_k = \lambda_k \mathbf{u}_k$, the first descendant of $\mathbf{z}$ is

$$\mathbf{z}_1 = T\mathbf{z} = \sum_{k=0}^{N-1} a_k \lambda_k \mathbf{u}_k$$

and the nth descendant is

$$\mathbf{z}_n = T^n \mathbf{z} = \sum_{k=0}^{N-1} a_k \lambda_k^n \mathbf{u}_k.$$

As n increases, the components with the largest eigenvalues in absolute value become relatively dominant. As can be seen from Figure 4, the eigenvalues are ordered as follows:

$$1 = \lambda_0 > |\lambda_1| = |\lambda_{N-1}| > |\lambda_2| = |\lambda_{N-2}| > \cdots$$

Since $\lambda_0 = 1$, the centroids (zeroth coefficients) of $\mathbf{z}_n$ remain fixed. The other components die out. Thus all vertices of $\mathbf{z}_n$ converge to the centroid. Relatively, though, the convex components ($k = 1$, $N-1$) die out least rapidly.

The normalized polygons

$$\mathbf{z}_n' = \frac{1}{|\lambda_1|^n} (\mathbf{z}_n - a_0 \mathbf{u}_0)$$

have centroids O and convex components of constant magnitude. As n increases all nonconvex components go to zero.

The sum of the two convex components of $\mathbf{z}$ is the polygon with vertices

$$w_j = a_1 \alpha^j + a_{N-1} \alpha^{(N-1)j}.$$

This polygon is the affine image of a regular convex N-gon. Its vertices all lie on the ellipse

$$w(t) = a_1 e^{it} + a_{N-1} e^{-it}$$

at the points $w_j = w(t_j)$, $t_j = j(2\pi/N)$. The ellipse $w(t)$ circumscribes not only the sum $\mathbf{w}$ of the two convex components of $\mathbf{z}$, but also the sums $\mathbf{w}_n'$ of the two convex components of $\mathbf{z}$'s normalized descendants $\mathbf{z}_n'$. As n increases, $\mathbf{z}_n'$ approaches $\mathbf{w}_n'$. If the circumscribing ellipse $w(t)$ is nondegenerate, then for sufficiently large n $\mathbf{z}_n'$ is convex. If, on the other hand, $w(t)$ is degenerate, then all $\mathbf{z}_n'$ are self-intersecting or degenerate, hence by our definition nonconvex. A necessary and sufficient condition for $\mathbf{z}$ to have convex descendants, therefore, is that $w(t)$ be nondegenerate.

The semimajor axis of $w(t)$ has length $|a_1| + |a_{N-1}|$ and the semiminor axis has length $||a_1| - |a_{N-1}||$. The ellipse degenerates if and only if $|a_1| = |a_{N-1}|$, i.e., if and only if the two convex components of $\mathbf{z}$ are of equal magnitude.

We can now state the answer to the question posed at the beginning of the paper as follows:

Represent the vertices of polygon P_0 by the complex numbers $z_1, \cdots, z_N$. Let $a_1, \cdots, a_N$ be the complex numbers given by:

$$a_k = \frac{1}{N} \sum_{j=1}^{N} z_j \alpha^{-jk}.$$

P_0 has a convex descendant P_n if and only if

$$(2) \qquad\qquad |a_1| \neq |a_{N-1}|.$$

Stated geometrically the condition is this: P_0 has a convex descendant if and only if the centroids of the polygons P_+ and P_-, obtained by winding P_0 positively and negatively about the origin, lie at different distances from the origin.

Comments on planar polygons.

1. In many cases the descendants of a polygon display a curious alternation. The descendants of a rectangle, for example, are alternately rhombuses and

rectangles. In the general case, the even descendants P_2, P_4, P_6, $\cdots$, if normalized in size, approach a fixed polygon P and the odd descendants approach another fixed polygon P'. P and P' are affine images of regular polygons and are inscribed in the same ellipse.

2. Without essentially complicating the problem the transformation T, which had the formula $w_j = \frac{1}{2}(z_j + z_{j+1})$, can be replaced by the more general transformation T' with the formula

$$w_j' = A_0 z_j + A_1 Z_{j+1} + \cdots + A_{N-1} z_{j+N-1}$$

where $A_0, A_1, \cdots$ are any constants and where the N subscripts on the z_{j+k} are to be computed modulo N. The eigenvectors are the same $\mathbf{u}_k$ as before but the eigenvalues are now

$$\lambda_k = \sum_r A_r \alpha^{rk}.$$

If $|\lambda_1|$ or $|\lambda_{N-1}|$ exceeds all of $|\lambda_2|, \cdots, |\lambda_{N-2}|$ then, for almost all polygons P, all but a finite number of descendants of P are convex. Two examples of this kind are

(i) $$w_j' = \alpha z_j + \beta z_{j+1} \qquad (\alpha > 0, \beta > 0, \alpha + \beta = 1)$$

and

(ii) $$w_j' = \frac{1}{3}(z_j + z_{j+1} + z_{j+2}).$$

The midpoints of the sides are replaced by points which subdivide the sides in ratio α/β in (i) and by the centroids of successive vertex triples in (ii).

3. There exist orphan polygons which have no ancestors. Any quadrilateral that is not a parallelogram is such a one. Orphan polygons are those even-sided polygons containing a nonzero component of order $N/2$. The vanishing eigenvalue $\lambda_{N/2} = 0$ wipes out this component in the first generation. Thus no polygon containing the $N/2$ component can be the descendant of another polygon.

Similar remarks apply to the more general transformation T' of the preceding paragraph. If T' has no zero eigenvalue (as in example (i) if $\alpha \neq \beta \neq \frac{1}{2}$) then all polygons have unique ancestors. In this case the family tree may be traced backwards as well as forwards. It is interesting to note that almost all convex polygons have remote ancestors which are not convex.

Nonplanar polygons. The nonplanar case, surprisingly, is hardly more complicated than the planar case. To make the generalization it is convenient to write the eigenpolygon decomposition in a new form. Combining the kth and $(N-k)$th terms in (1) we obtain a sum of $[N/2] + 1$ terms each representing an affine image of a regular polygon; here $[N/2]$ means the integer part of $N/2$. For d-dimensional space this sum generalizes as follows. If X is a closed N-gon in d-space then X can be written

$$X = \sum_{k=0}^{[N/2]} W_k,$$

where again each component polygon W_k is an affine image of a regular polygon. *This differs from the planar case only in that the W_k need not all lie in the same plane.*

This may be derived explicitly as follows. Let

$$\mathbf{x}(t) = \sum_{k=0}^{[N/2]} \mathbf{u}_k \cos kt + \mathbf{v}_k \sin kt,$$

where the symbols $\mathbf{x}$, $\mathbf{u}$, $\mathbf{v}$ stand for vectors with d real components. It is always possible [2] to choose $\mathbf{u}_k$ and $\mathbf{v}_k$ so that $\mathbf{x}(t)$ assumes given values at N given points $t_1, \cdots, t_N$, in particular so that

$$\mathbf{x}\left(j\,\frac{2\pi}{N}\right) = \mathbf{x}_j,$$

where $\mathbf{x}_j$ are the vertices of the given polygon X. Then X can be written as

$$X = \sum_{k=0}^{[N/2]} W_k,$$

where W_k is the polygon with vertices

$$\mathbf{w}_{kj} = \mathbf{u}_k \cos kj\,\frac{2\pi}{N} + \mathbf{v}_k \sin kj\,\frac{2\pi}{N}\,.$$

Polygon W_k lies in the plane determined by $\mathbf{u}_k$ and $\mathbf{v}_k$ and is inscribed in the ellipse

$$\mathbf{w}(t) = \mathbf{u}_k \cos t + \mathbf{v}_k \sin t.$$

W_k may be regarded as the sum of a clockwise and a counterclockwise regular N-gon of order k as was done in the first section. Alternatively W_k may be regarded as the projection onto a plane of a regular N-gon of order k. In general, a d-dimensional polygon may be regarded as the projection onto d-space of a $(d+1)$-dimensional polygon all of whose components are regular.

The midpoint-joining transformation T may be analyzed in terms of the decomposition in the same way as before. The only new element is the dimension of the descendants. (A polygon has the dimension of the smallest space it can be imbedded in.) The following example shows that T can reduce the dimension: Let X be a 3-dimensional polygon with vertices alternately in the planes $z=1$ and $z=-1$. The first descendant of X lies in the x, y-plane. This, however, is as far as the dimension reduction can go. Further application of T leaves the dimension unchanged. In general the situation is this:

All descendants $X_1, X_2, \cdots$ of a d-dimensional polygon X_0 have the same dimension. X_1 may be of dimension one less than X_0, but this can occur only if X_0 is an orphan, i.e., only if X_0 contains a nonzero component of order $N/2$.

Proof. Suppose X_n has dimension d and X_{n+1} has dimension $d' < d$. Let $\overline{X}_n$ and $\overline{X}_{n+1}$ be the projections of X_n and X_{n+1} onto a line orthogonal to the d'-space containing X_{n+1}. All vertices of $\overline{X}_{n+1}$ lie in a single point p. Since p must be the midpoint of all sides of $\overline{X}_n$, the vertices of $\overline{X}_n$ must lie alternately in two points p_1, p_2 equidistant from p. $\overline{X}_n$, then, is a regular N-gon of order $N/2$. If X_n has the decomposition

$$X_n = \sum_{k=0}^{[N/2]} W_k$$

then $\overline{X}_n$ has the decomposition

$$\overline{X}_n = \sum_{k=0}^{[N/2]} \overline{W}_k.$$

But in the latter decomposition only $\overline{W}_{N/2}$ can be nonzero. If $\overline{W}_{N/2}$ is nonzero then so is $W_{N/2}$. In this case X_n is an orphan, hence $n = 0$. This completes the proof.

In conclusion, we may say that almost all nonplanar polygons lack planar descendants. If the first descendant is nonplanar then so are all the rest.

On the other hand, all nonplanar polygons have descendants which differ arbitrarily little (relative to their size) from planar polygons. And *almost* all nonplanar polygons have descendants which differ arbitrarily little (relative to their size) from planar *convex* polygons.

The first author is now at the University of California, Berkeley.

References

1. F. R. Gantmacher, Theory of Matrices, vol. 1, Chelsea, New York, 1959, p. 72.

2. Ch. de la Vallée Poussin, Leçons sur l'approximation des fonctions d'une variable réelle, Gauthier-Villars, Paris, 1919, p. 93.

GENERALIZATIONS OF THEOREMS ABOUT TRIANGLES*

CARL B. ALLENDOERFER, University of Washington

1. Introduction. Since one of the most powerful methods in mathematical research is the process of generalization, it is certainly desirable that young students be introduced to this process as early as possible. The purpose of this article is to call attention to the usually untapped possibilities for generalizing theorems on the triangle to theorems about the tetrahedron. Some of these, of course, do appear in our textbooks on solid geometry; but here I shall describe two situations where the appropriate generalizations seem to be generally un-

* From MATHEMATICS MAGAZINE, vol. 38 (1965), pp. 253–259.

known. The questions to be answered are: (1) What is the generalization to a tetrahedron of the angle-sum theorem for a triangle? (2) What is the corresponding generalization of the laws of sines and cosines for a triangle? Expressed in this form, the questions are certainly vague; for surely there are many generalizations. From these we are to select the ones which are most satisfying and which have a clear right to be called *the generalizations*. In attacking these problems we will need to reexamine the theorems as they are stated for a triangle, and perhaps to reformulate them so that the generalizations appear to be natural. Thus we have a bonus in that we learn additional ways of thinking about triangles.

2. The angle-sum theorem. Since this theorem is one of the most familiar in Euclidean geometry, it is strange that its three-dimensional generalization is not part of the classical literature on geometry. I ran across this generalization some years ago and have been putting the question to mathematicians wherever I find them. Only one of them, Professor Pólya, knew of it. He attributes it to Descartes [1].

The first question to be settled is that of the type of angles in a tetrahedron to be considered. It would be most natural to consider the inner solid angles and their sum. I remind you that the measure of a solid angle is the area of the region on the unit sphere which is the intersection of the sphere with the interior of the solid angle whose vertex is at the center. Thus the measure of the solid angle at a corner of a room is $4\pi/8 = \pi/2$, and the measure of a "straight" solid angle is $4\pi/2 = 2\pi$. By considering a few cases, we conclude that the sum of the measures of the inner solid angles of a tetrahedron is not a constant. For example consider the situation in Figure 1, where all the points lie in a plane. If D is raised slightly, we have a tetrahedron the sum of whose interior solid angles is very near to 2π. On the other hand let us raise segment AB in the plane Figure 2 a small amount. Then we have a tetrahedron the sum of whose inner solid angles is very near to zero. Hence the obvious generalization is incorrect. As a matter of fact it has been proved [2] that the sum of the solid angles of a tetrahedron can take any value between 0 and 2π.

In order to make a fresh start, let us reformulate the triangle theorem in the statement: The sum of the outer angles of a triangle equals 2π. There are two possible definitions of an outer angle. The usual one is that it is the angle between a pair of successive directed sides (Fig. 3). This clearly does not generalize to three dimensions. Less familiar is the definition that an outer angle at a vertex is the angle between the two outward drawn normals to the two edges which meet at this vertex (Fig. 4).

Using this second definition, we can construct an elegant proof of the theorem. Choose any point P in the interior of the triangle and draw the perpendiculars from P to the three sides (Fig. 5). Then the outer angles α, β, and γ are equal to the three angles formed at P. Hence $\alpha + \beta + \gamma = 2\pi$.

Now we can generalize at once. To find the corresponding theorem on the tetrahedron, first define the outer angle at a vertex as the trihedral angle formed by the three outer normals to the three faces meeting at this vertex. Choose an

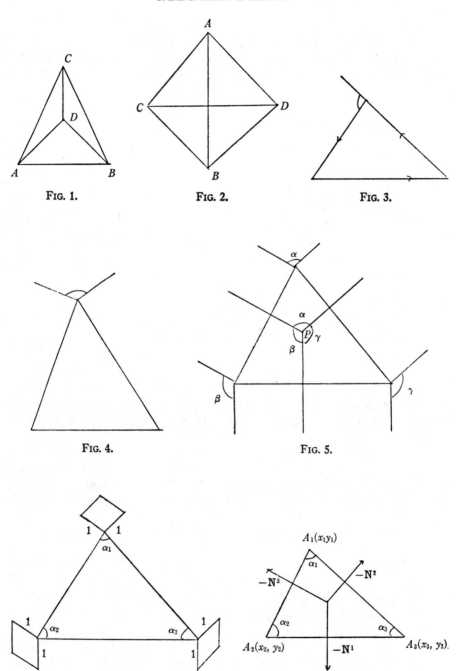

FIG. 1.   FIG. 2.   FIG. 3.

FIG. 4.   FIG. 5.

FIG. 6.   FIG. 7.

interior point P and draw the perpendiculars from P to the four faces. By the same argument that we used for the triangle, we find that   .

THEOREM 1. *The sum of the outer angles of a tetrahedron is* 4π.

By a straightforward generalization of the notion of an outer angle, we can similarly prove that

THEOREM 2. *The sum of the outer angles of any convex polyhedron is equal to* 4π.

There is also an immediate generalization to higher dimensions.

3. The Laws of Sines and Cosines. Before considering the generalization of these laws to a tetrahedron, let me give unfamiliar proofs of them which will suggest the proper generalization.

First, consider the Law of Sines. At each vertex (Fig. 6) draw the unit outer normals to the sides meeting at that vertex and complete the parallelograms determined by these pairs. By a familiar theorem of trigonometry the areas of these parallelograms are respectively $\sin(\pi - \alpha_1) = \sin \alpha_1$, $\sin(\pi - \alpha_2) = \sin \alpha_2$, and $\sin(\pi - \alpha_3) = \sin \alpha_3$. We shall proceed to compute these areas in terms of the coordinates of the vertices of the triangle (Fig. 7), choosing the notation appropriately so that $A_1 A_2 A_3$ are labeled in a counterclockwise fashion.

The equation of side $A_2 A_3$ is

$$x N_x^1 + y N_y^1 + (x_2 y_3 - x_3 y_2) = 0$$

where N_x^1 and N_y^1 are respectively the cofactors of x_1 and y_1 in the determinant

$$\Delta = \begin{vmatrix} x_1 & y_1 & 1 \\ x_2 & y_2 & 1 \\ x_3 & y_3 & 1 \end{vmatrix}.$$

Thus the vector $\mathbf{N}^1$ with components (N_x^1, N_y^1) is normal to $A_2 A_3$; $-\mathbf{N}^1$ is an outer normal; and $\mathbf{U}^1 = -\mathbf{N}^1/a_1$ is the unit outer normal (where a_1 is the length of $A_2 A_3$). More generally $\mathbf{U}^i = -\mathbf{N}^i/a_i$ ($i = 1, 2, 3$) are the three outer normals, where N_x^i and N_y^i are the cofactors of x_i and y_i respectively and a_i is the length of the side to which $\mathbf{U}^i$ is normal.

The area of the outer parallelogram at A_1 of which two sides are $\mathbf{U}^2$ and $\mathbf{U}^3$ is

$$\sin \alpha_1 = \begin{vmatrix} U_x^2 & U_y^2 \\ U_x^3 & U_y^3 \end{vmatrix} = \frac{1}{a_2 a_3} \begin{vmatrix} N_x^2 & N_y^2 \\ N_x^3 & N_y^3 \end{vmatrix}.$$

By a classical theorem on determinants (Bôcher, Introduction to Higher Algebra,

p. 31) it follows that

$$\begin{vmatrix} N_x^2 & N_y^2 \\ N_x^3 & N_y^3 \end{vmatrix} = \Delta \cdot 1.$$

Hence

$$\sin \alpha_1 = \frac{\Delta}{a_2 a_3} \quad \text{and} \quad \frac{\sin \alpha_1}{a_1} = \frac{\Delta}{a_1 a_2 a_3} \cdot$$

In a similar fashion we prove that

$$\frac{\sin \alpha_1}{a_1} = \frac{\sin \alpha_2}{a_2} = \frac{\sin \alpha_3}{a_3} = \frac{\Delta}{a_1 a_2 a_3}$$

which is the familiar Law of Sines.

To arrive at the Law of Cosines, we begin with a theorem of Möbius.

THEOREM 3. $\mathbf{N}^1 + \mathbf{N}^2 + \mathbf{N}^3 = 0.$

This theorem follows from the facts that $N_x^1 + N_x^2 + N_x^3 = 0$ and $N_y^1 + N_y^2 + N_y^3 = 0$. These may be computed directly, or they may be proved by expanding the determinants

$$\begin{vmatrix} 1 & y_1 & 1 \\ 1 & y_2 & 1 \\ 1 & y_3 & 1 \end{vmatrix} = 0 \quad \text{and} \quad \begin{vmatrix} x_1 & 1 & 1 \\ x_2 & 1 & 1 \\ x_3 & 1 & 1 \end{vmatrix} = 0.$$

This theorem can be rewritten in the form:

$$\mathbf{N}^1 = -\mathbf{N}^2 - \mathbf{N}^3.$$

Now take the scalar product of each side of this equation with itself. The result is

$$\mathbf{N}^1 \cdot \mathbf{N}^1 = \mathbf{N}^2 \cdot \mathbf{N}^2 + \mathbf{N}^3 \cdot \mathbf{N}^3 + 2\mathbf{N}^2 \cdot \mathbf{N}^3.$$

Since $\mathbf{N}^i \cdot \mathbf{N}^i = a_i^2$, and $\mathbf{N}^2 \cdot \mathbf{N}^3 = -a_2 a_3 \cos \alpha_1$, this becomes $a_1^2 = a_2^2 + a_3^2 - 2a_2 a_3 \cos \alpha_1$.

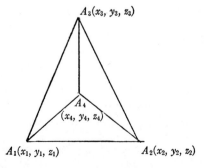

FIG. 8.

4. The Generalized Laws of Sines and Cosines. These generalizations are due to Grassmann, but are relatively unfamiliar. Their proofs follow the lines just given in Section 3.

Consider a tetrahedron (Fig. 8) whose vertices are ordered so that

$$\Delta = \begin{vmatrix} x_1 & y_1 & z_1 & 1 \\ x_2 & y_2 & z_2 & 1 \\ x_3 & y_3 & z_3 & 1 \\ x_4 & y_4 & z_4 & 1 \end{vmatrix} > 0.$$

Then the vector $\mathbf{N}^1$ whose components (N_x^1, N_y^1, N_z^1), are the cofactors of x_1, y_1, z_1 respectively in Δ, is normal to the face $A_2A_3A_4$. The length of $\mathbf{N}^1$, namely a_1, is equal to twice the area of this face. The vector $\mathbf{U}^1 = -\mathbf{N}^1/a_1$ is the unit outer normal to this face. Other normals $\mathbf{N}^i$ and $\mathbf{U}^i$ are defined in a similar fashion.

We now define the generalized sine ("G-sin") of the inner trihedral angle at A_1 to be the volume of the parallelopiped whose edges are $\mathbf{U}^2$, $\mathbf{U}^3$, and $\mathbf{U}^4$. Thus

$$G\text{-}\sin \alpha_1 = \begin{vmatrix} U_x^2 & U_y^2 & U_z^2 \\ U_x^3 & U_y^3 & U_z^3 \\ U_x^4 & U_y^4 & U_z^4 \end{vmatrix} = \frac{-1}{a_2 a_3 a_4} \begin{vmatrix} N_x^2 & N_y^2 & N_z^2 \\ N_x^3 & N_y^3 & N_z^3 \\ N_x^4 & N_y^4 & N_z^4 \end{vmatrix} = \frac{(-1)\Delta^2(-1)}{a_2 a_3 a_4} = \frac{\Delta^2}{a_2 a_3 a_4}.$$

By a continuation of this argument, we obtain the Generalized Law of Sines:

THEOREM 4.

$$\frac{G\text{-}\sin \alpha_1}{a_1} = \frac{G\text{-}\sin \alpha_2}{a_2} = \frac{G\text{-}\sin \alpha_3}{a_3} = \frac{G\text{-}\sin \alpha_4}{a_4} = \frac{\Delta^2}{a_1 a_2 a_3 a_4}.$$

To establish the Generalized Law of Cosines, we observe that we can prove the following generalization of the Theorem of Möbius.

THEOREM 5. $\mathbf{N}^1 + \mathbf{N}^2 + \mathbf{N}^3 + \mathbf{N}^4 = 0$.

Then writing

$$\mathbf{N}^1 = -\mathbf{N}^2 - \mathbf{N}^3 - \mathbf{N}^4$$

and $f_i = a_i/2 =$ area of the ith face, we prove as above the result:

THEOREM 6.

$$f_1^2 = f_2^2 + f_3^2 + f_4^2 - 2[f_2 f_3 \cos (f_2, f_3) + f_2 f_4 \cos (f_2, f_4) + f_3 f_4 \cos (f_3, f_4)],$$

where (f_i, f_j) is the inner dihedral angle of the tetrahedron between the faces whose areas are f_i and f_j respectively.

We also have another, rather novel, generalization if we start from $\mathbf{N}^1+\mathbf{N}^2$ $=-\mathbf{N}^3-\mathbf{N}^4$. The result is

THEOREM 7.

$$f_1^2 + f_2^2 - 2f_1f_2 \cos (f_1, f_2) = f_3^2 + f_4^2 - 2f_3f_4 \cos (f_3, f_4).$$

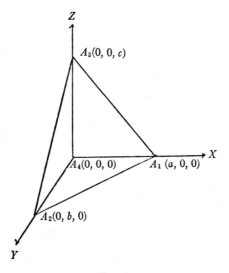

FIG. 9.

5. Supplementary matters. Another approach to the Generalized Law of Sines is to begin with a right tetrahedron (Fig. 9). Then it would be reasonable to define

$$G\text{-}\sin \alpha_1 = \frac{\text{Area } A_2A_3A_4}{\text{Area } A_1A_2A_3} = \frac{bc}{\{b^2c^2 + a^2c^2 + a^2b^2\}^{1/2}}.$$

Let us show that this agrees with our previous definition of G-sin α_1. We have:

$$\Delta = \begin{vmatrix} a & 0 & 0 & 1 \\ 0 & b & 0 & 1 \\ 0 & 0 & c & 1 \\ 0 & 0 & 0 & 1 \end{vmatrix}.$$

Then

$$G\text{-}\sin \alpha_1 = \frac{\Delta^2}{a_2a_3a_4} = \frac{a^2b^2c^2}{(ac)(ab)\{b^2c^2 + a^2c^2 + a^2b^2\}^{1/2}} = \frac{bc}{\{b^2c^2 + a^2c^2 + a^2b^2\}^{1/2}}.$$

Also we have the reassuring result that for our right tetrahedron:

$$(G\text{-}\sin \alpha_1)^2 + (G\text{-}\sin \alpha_2)^2 + (G\text{-}\sin \alpha_3)^2 = 1.$$

It is natural to ask whether G-$\sin \alpha_1$ is actually the sine of the measure of the inner or the outer solid angle at A_1; the answer is "no". To give an elementary counter-example we consider the right tetrahedron with $a=b=c=1$. Then G-$\sin \alpha_1 = 1/\sqrt{3}$; sine (measure of inner solid angle at A_1) $= 1/3$; and sine (measure of outer solid angle at A_1) $= \sin 7\pi/6 = -1/2$.

As a matter of fact, G-$\sin \alpha$ is not even a functon of either the inner or the outer solid angles at the given vertex. Rather it depends directly on the face angles of the outer trihedral angle. If these angles are λ, μ, ν and $s = (\lambda+\mu+\nu)/2$, then

$$G\text{-}\sin\alpha = 2\{\sin s \sin(s-\lambda)\sin(s-\mu)\sin(s-\nu)\}^{1/2}.$$

References

1. R. Descartes, Oeuvres, vol. 10, pp. 257–276.

2. J. W. Gaddum, The sums of the dihedral and trihedral angles of a tetrahedron, Amer. Math. Monthly, 59 (1952) 370–375.

HIGHER CURVATURES OF CURVES IN EUCLIDEAN SPACE*

HERMAN GLUCK, Harvard University†

1. Introduction. It is the object of this note to point out a simple algorithm, based on the Gram-Schmidt orthonormalization process, for computing the curvatures of a curve in Euclidean n-space. This algorithm has a number of agreeable features. There is just one formula involved for all the curvatures, and it is very simple. There is no duplication of effort, in the sense that if one calculates the Frenet frame associated with a curve at a given point, then the various curvatures can be obtained immediately from the by-products of this calculation. Finally, the entire procedure is no more involved if the parametrization is not by arc length.

2. The Frenet frame and the curvatures. Let I be an interval in R^1 and $F: I \rightarrow R^n$ a C^k-parametrization by arc length. This means that the arc length along the curve from $F(s_1)$ to $F(s_2)$ is $|s_1 - s_2|$, or equivalently that $|F'(s)| = 1$ for all $s \in I$. Suppose that for each $s \in I$, the vectors

$$F'(s), F''(s), \cdots, F^{[r]}(s) \qquad\qquad r < k$$

are linearly independent. Applying the Gram-Schmidt orthonormalization process to these vectors, one obtains an orthonormal r-tuple of vectors,

$$(V_1(s), V_2(s), \cdots, V_r(s)),$$

* From AMERICAN MATHEMATICAL MONTHLY, vol. 73 (1966), pp. 699–704.

† Prof. Gluck is now at the University of Pennsylvania.

called the *Frenet r-frame* associated with the curve at the point $F(s)$. $V_i(s)$ is easily seen to be of class C^{k-i}.

Formulas for the derivatives $V_i'(s)$ can be obtained by first differentiating the orthonormality relations $V_i(s) \cdot V_j(s) = \delta_{ij}$, to get

$$V_i'(s) \cdot V_j(s) = - V_j'(s) \cdot V_i(s).$$

Combining this with the fact that, for $1 \leq i \leq r-1$, $V_i'(s)$ is a linear combination of $V_1(s)$, $V_2(s)$, $\cdots$, $V_{i+1}(s)$, one concludes that

$$V_i'(s) \cdot V_j(s) = 0,$$

except possibly for $j=i-1$ and $j=i+1$. The derivative formulas can therefore be written

$$V_1'(s) = k_1(s) V_2(s)$$
$$V_i'(s) = - k_{i-1}(s) V_{i-1}(s) + k_i(s) V_{i+1}(s) \qquad 2 \leq i \leq r - 1.$$

There is a bit of a problem with $V_r'(s)$, since there may be no $V_{r+1}(s)$. Given $s_0 \in I$, if $F^{[r+1]}(s_0)$ is linearly independent of $F'(s_0)$, $F''(s_0)$, $\cdots$, $F^{[r]}(s_0)$, then this will also be true in some neighborhood of s_0 in I. For s in such a neighborhood, $V_{r+1}(s)$ can be defined as above and one will have

$$V_r'(s) = - k_{r-1}(s) V_{r-1}(s) + k_r(s) V_{r+1}(s).$$

If $F^{[r+1]}(s_0)$ happens to be linearly dependent upon $F'(s_0)$, $F''(s_0)$, $\cdots$, $F^{[r]}(s_0)$, then

$$V_r'(s_0) = - k_{r-1}(s_0) V_{r-1}(s_0).$$

The coefficients appearing above, $k_1(s)$, $k_2(s)$, $\cdots$, $k_{r-1}(s)$, are the *curvatures* associated with the given curve at the point $F(s)$. The rth curvature, $k_r(s)$, may be defined similarly when $F^{[r+1]}(s)$ is independent of $F'(s)$, $F''(s)$, $\cdots$, $F^{[r]}(s)$, and to be zero in the dependent case. It turns out that $k_i(s) > 0$ for $1 \leq i \leq r-1$, and $k_r(s) \geq 0$. For $1 \leq i \leq r-1$, $k_i(s)$ will be of class C^{k-i-1}. $k_r(s)$ will be of class C^{k-r-1} wherever it does not vanish, but overall can only be guaranteed to be continuous. There is a standard way out of this dilemma in the special case $r=n-1$, for then there is a natural choice for $V_{r+1}(s)$. We do not go into details, but simply remark that the corresponding $k_r(s)$ will then be of class C^{k-r-1} on all of I, but will not necessarily be a positive function.

3. The algorithm for parametrizations by arc length. The Gram-Schmidt process is actually carried out as follows. Let

$$E_1(s) = F'(s) \qquad \text{and} \qquad V_1(s) = \frac{E_1(s)}{|E_1(s)|}.$$

If $V_1(s)$, $V_2(s)$, $\cdots$, $V_{i-1}(s)$ have already been determined, let

$$E_i(s) = F^{[i]}(s) - \sum_{j<i} [F^{[i]}(s) \cdot V_j(s)] V_j(s)$$

and

$$V_i(s) = \frac{E_i(s)}{|E_i(s)|} \cdot$$

This works for $i = 2, 3, \cdots, r$. The process can actually be carried a half step further by computing

$$E_{r+1}(s) = F^{[r+1]}(s) - \sum_{j<r+1}[F^{[r+1]}(s) \cdot V_j(s)]V_j(s),$$

since $r < k$. On the other hand, there is no guarantee that $E_{r+1}(s) \neq 0$, so in general we can not form $V_{r+1}(s)$. The vectors $E_1(s)$, $E_2(s)$, $\cdots$, $E_{r+1}(s)$ are conveniently referred to as the *excess vectors*, since $E_i(s)$ is nothing but the component of $F^{[i]}(s)$ orthogonal to the subspace spanned by $F'(s)$, $F''(s)$, $\cdots$, $F^{[i-1]}(s)$.

The algorithm for computing the curvatures derives from the following

THEOREM 3.1. $k_i(s) = |E_{i+1}(s)| / |E_i(s)|$ *for* $1 \leq i \leq r$.

First assume $i < r$. Then

$$k_i(s) = V_i'(s) \cdot V_{i+1}(s) = \left(\frac{E_i(s)}{|E_i(s)|}\right)' \cdot V_{i+1}(s)$$

$$= \frac{E_i'(s) \cdot V_{i+1}(s)}{|E_i(s)|} + \left(\frac{1}{|E_i(s)|}\right)' E_i(s) \cdot V_{i+1}(s).$$

Now $E_i(s)$ and $V_{i+1}(s)$ are orthogonal, so the second term on the right above is zero. Hence

$$k_i(s) = \frac{E_i'(s) \cdot V_{i+1}(s)}{|E_i(s)|} \cdot$$

To verify the theorem, we must show that $E_i'(s) \cdot V_{i+1}(s) = |E_{i+1}(s)|$.

Differentiating the equation $E_i(s) = F^{[i]}(s) - \sum_{j<i}[F^{[i]}(s) \cdot V_j(s)]V_j(s)$ yields

$$E_i'(s) = F^{[i+1]}(s) - \sum_{j<i}[F^{[i]}(s) \cdot V_j(s)]'V_j(s) - \sum_{j<i}[F^{[i]}(s) \cdot V_j(s)]V_j'(s).$$

Every vector on the right hand side of this last equation, except for $F^{[i+1]}(s)$, is a linear combination of $V_1(s)$, $V_2(s)$, $\cdots$, $V_i(s)$, and these are all orthogonal to $V_{i+1}(s)$. Therefore

$$E_i'(s) \cdot V_{i+1}(s) = F^{[i+1]}(s) \cdot V_{i+1}(s).$$

But $E_{i+1}(s) = F^{[i+1]}(s) - \sum_{j<i+1}[F^{[i+1]}(s) \cdot V_j(s)]V_j(s)$, so

$$F^{[i+1]}(s) \cdot V_{i+1}(s) = E_{i+1}(s) \cdot V_{i+1}(s) = |E_{i+1}(s)|,$$

completing the proof in the case $i < r$.

If $i = r$ and $E_{i+1}(s) \neq 0$, the same proof works. If $E_{i+1}(s) = 0$, so is $k_i(s) = 0$, and the theorem is then certainly true.

4. The algorithm for arbitrary parametrizations. If $F^*: I^* \to R^n$ is a C^k-immersion (i.e., $F^{*\prime}(t) \neq 0$ for all $t \in I^*$) which is not necessarily a parametrization by arc length, the situation is only a trifle more involved. As before, we suppose that for each $t \in I^*$, the vectors

$$F^{*\prime}(t), F^{*\prime\prime}(t), \cdots, F^{*[r]}(t) \qquad\qquad r < k$$

are linearly independent. As in the preceding section, the Gram-Schmidt process yields the orthogonal excess vectors

$$E_1^*(t), E_2^*(t), \cdots, E_r^*(t), E_{r+1}^*(t)$$

and the orthonormal frame vectors

$$V_1^*(t), V_2^*(t), \cdots, V_r^*(t).$$

Let I be an interval in R^1 and $t: I \to I^*$ an orientation-preserving C^k-diffeomorphism such that $F = F^* t$ is a C^k-parametrization by arc length. Such a change of parameter always exists. Evaluating the various derivatives of F in terms of those of F^* via the chain rule, one easily sees that for $1 \leqslant i \leqslant r$,

$$F'(s), F''(s), \ldots, F^{[i]}(s) \text{ and } F^{*\prime}(t(s)), F^{*\prime\prime}(t(s)), \ldots, F^{*[i]}(t(s))$$

generate the same subspace of R^n. In particular, $F'(s), F''(s), \cdots, F^{[r]}(s)$ are linearly independent for all $s \in I$, so we are in a position to calculate the orthogonal excess vectors

$$E_1(s), E_2(s), \cdots, E_r(s), E_{r+1}(s)$$

and the orthonormal frame vectors

$$V_1(s), V_2(s), \cdots, V_r(s)$$

for the parametrization F. Looking carefully at the chain rule formulas for the derivatives of F in terms of those of F^* leads one to the following conclusions.

THEOREM 4.1. *For* $1 \leq i \leq r+1$, $E_i(s) = E_i^*(t(s)) (dt/ds)^i$ *and therefore* $|E_i(s)| = |E_i^*(t(s))| / |E_1^*(t(s))|^i$. *For* $1 \leq i \leq r$, $V_i(s) = V_i^*(t(s))$.

Let $k_i^*(t)$ denote the ith curvature of the curve defined by F^* at the point $F^*(t)$. Then from Theorems 3.1 and 4.1 we immediately get

THEOREM 4.2.

$$k_i^*(t) = \frac{|E_{i+1}^*(t)|}{|E_1^*(t)| \, |E_i^*(t)|} \qquad for \quad 1 \leq i \leq r.$$

This gives the algorithm for computing the curvatures in the case of a parametrization not necessarily by arc length.

5. An example. Let p and q be real numbers, $0 < p < q$. Consider the curve given by $F: R^1 \to R^4$, where $F(t) = (\cos pt, \sin pt, \cos qt, \sin qt)$. This curve lies on the torus

$$T = \left\{ (x_1,\, x_2,\, x_3,\, x_4) : x_1^2 + x_2^2 = 1,\, x_3^2 + x_4^2 = 1 \right\}$$

in four-space. The curve will be closed if and only if p/q is rational. Given any two points on the curve, there is an isometry of R^4 onto itself which takes the curve onto itself and takes the one point onto the other. Thus the various curvatures are the same at all points of the curves. They may be calculated according to the algorithm given by Theorem 4.2, as follows.

$$F'(0) = (0,\, p,\, 0,\, q) \qquad\qquad F'''(0) = (0,\, -p^3,\, 0,\, -q^3)$$

$$F''(0) = (-p^2,\, 0,\, -q^2,\, 0) \qquad F^{\mathrm{iv}}(0) = (p^4,\, 0,\, q^4,\, 0).$$

Beginning the Gram-Schmidt process, we get

$$E_1(0) = (0,\, p,\, 0,\, q)$$

$$\left| E_1(0) \right| = \sqrt{(p^2 + q^2)}$$

$$V_1(0) = \frac{1}{\sqrt{(p_2 + q^2)}} (0,\, p,\, 0,\, q).$$

Next stage:

$$E_2(0) = (-p^2,\, 0,\, -q^2,\, 0)$$

$$\left| E_2(0) \right| = \sqrt{(p^4 + q^4)}$$

$$V_2(0) = \frac{1}{\sqrt{(p^4 + q^4)}} (-p^2,\, 0,\, -q^2,\, 0).$$

Third Stage:

$$E_3(0) = \frac{pq(q^2 - p^2)}{p^2 + q^2} (0,\, q,\, 0,\, -p)$$

$$\left| E_3(0) \right| = \frac{pq(q^2 - p^2)}{\sqrt{(p^2 + q^2)}}$$

$$V_3(0) = \frac{1}{\sqrt{(p^2 + q^2)}} (0,\, q,\, 0,\, -p).$$

Final stage:

$$E_4(0) = \frac{p^2 q^2 (q^2 - p^2)}{p^4 + q^4} (-q^2,\, 0,\, p^2,\, 0)$$

$$\left| E_4(0) \right| = \frac{p^2 q^2 (q^2 - p^2)}{\sqrt{(p^4 + q^4)}}$$

$$V_4(0) = \frac{1}{\sqrt{(p^4 + q^4)}} (-q^2,\, 0,\, p^2,\, 0).$$

The curvatures are then given by:

$$k_1(0) = \frac{|E_2(0)|}{|E_1(0)| \, |E_1(0)|} = \frac{\sqrt{(p^4 + q^4)}}{p^2 + q^2} = \frac{\sqrt{(1 + r^4)}}{1 + r^2},$$

where $r = p/q$, and hence $0 < r < 1$.

$$k_2(0) = \frac{|E_3(0)|}{|E_1(0)| \, |E_2(0)|} = \frac{pq(q^2 - p^2)}{(p^2 + q^2)\sqrt{(p^4 + q^4)}} = \frac{r(1 - r^2)}{(1 + r^2)\sqrt{(1 + r^4)}},$$

$$k_3(0) = \frac{|E_4(0)|}{|E_1(0)| \, |E_3(0)|} = \frac{pq}{\sqrt{(p^4 + q^4)}} = \frac{r}{\sqrt{(1 + r^4)}}.$$

6. Miscellany. Several nice formulas follow easily from Theorems 3.1 and 4.2. We give them here only for parametrizations by arc length and leave the derivation of the corresponding formulas in the general case—actually no more difficult—to the reader.

Denote by $v_i(s)$ the hypervolume of the parallelepiped with "leading edges" $F'(s), F''(s), \cdots, F^{[i]}(s)$, so that

$$v_i(s) = |E_1(s)| \, |E_2(s)| \, \cdots \, |E_i(s)|.$$

Since $k_i(s) = |E_{i+1}(s)|/|E_i(s)|$, we have $|E_i(s)| = k_1(s)k_2(s) \cdots k_{i-1}(s)$. Hence

$$v_i(s) = [k_1(s)]^{i-1}[k_2(s)]^{i-2} \cdots [k_{i-2}(s)]^2[k_{i-1}(s)].$$

Also, since $|E_i(s)| = v_i(s)/v_{i-1}(s)$, we get

$$k_i(s) = \frac{v_{i+1}(s)v_{i-1}(s)}{[v_i(s)]^2}.$$

This last equation is essentially the same as formula (6.2-17) in J. C. H. Gerretsen, *Lectures on Tensor Calculus and Differential Geometry*, Noordhoff, Groningen, 1962.

Editorial Note: For a sequel to this paper, see Prof. Gluck's article in this MONTHLY, 74 (1967), 1049–56.

LATTICE POINTS AND POLYGONAL AREA*

IVAN NIVEN, University of Oregon, and H. S. ZUCKERMAN, University of Washington

1. By a simple polygon is meant a polygon that is topologically equivalent to a circle. By lattice points in the plane we mean those points with integral coordinates. H. Steinhaus [**1**, p. 76] states the result that the area of a simple polygon all of whose vertices are lattice points is equal to the number of interior lattice points, plus half the number of lattice points on the boundary, minus one. This result is often called Pick's Theorem because George Pick was apparently first in discovering this formula, in 1899. Steinhaus offers so brief a sketch of the proof of Pick's Theorem that it is not easy to formulate a complete argument. Because of this, and because this elegant theorem deserves to be known more widely, we offer here a full proof that we feel may also be of interest for its own sake.

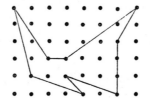

2. Triangles.

THEOREM 1. *Let (a, b) and (c, d) be lattice points such that (a, b), (c, d) and the origin are not collinear. Let T denote the triangle with vertices $(0, 0)$, (a, b), (c, d). Then T has area $\frac{1}{2}$ if and only if*

    (I) *T contains no lattice points other than its three vertices.*

Proof. The proof we give is not the shortest one possible but it is interesting because of the various ideas that it combines. Let P be the parallelogram having vertices $(0, 0)$, (a, b), $(a+c, b+d)$, (c, d). First we prove that (I) is equivalent to:

    (II) *P contains no lattice points other than its four vertices.*

It is obvious that (II) implies (I). To prove the converse suppose that P contains a lattice point (e, f) different from its vertices. If (e, f) is not in the triangle (a, b), $(a+c, b+d)$, (c, d) then it is in T and (I) is false. Therefore we can suppose that (e, f) is in that triangle. Translating P by the amount $-a-c$ in the x direction, $-b-d$ in the y direction, we see that $(e-a-c, f-b-d)$ is a lattice point in the triangle $(-c, -d)$, $(0, 0)$, $(-a, -b)$. Reflecting through the origin we find that $(-e+a+c, -f+b+d)$ is a lattice point in T, and hence (I) is false. Therefore (I) and (II) are equivalent.

Now the entire plane can be covered by replicas of P, each replica being P translated by the amount $ma+nc$ in the x direction, $mb+nd$ in the y direction,

---

* From AMERICAN MATHEMATICAL MONTHLY, vol. 74 (1967), pp. 1195–1200.

$m = 0, \pm 1, \pm 2, \cdots, n = 0, \pm 1, \pm 2, \cdots$. The edges of the parallelograms join to create a network N of lines, one set of lines parallel to the side $(0, 0)$ (a, b) of T, the other set parallel to $(0, 0)$ (c, d). Call the points at which these lines intersect, the vertices of N. The vertices of N are just the vertices of the replicas of P. They are all lattice points. If some lattice point is not a vertex of N it will lie in some replica of P and be distinct from the vertices of the replica of P. Translating, we would then have a lattice point in P, not a vertex of P. Thus we see that (II) is equivalent to:

(III) *All lattice points are vertices of N.*

We now get an algebraic statement equivalent to (III). It is clear that the vertices of N are just the points $(ma + nc, mb + nd)$. Therefore (III) is equivalent to

(IV) *For each pair of integers u, v there is a pair of integers m, n such that $am + cn = u$, $bm + dn = v$.*

But the solution of the equations in (IV) is

$$m = (du - cv)/\Delta, \qquad n = (av - bu)/\Delta, \qquad \Delta = ad - bc.$$

The fact that $(0, 0)$, (a, b), (c, d) are not collinear ensures that $\Delta \neq 0$. Obviously $\Delta = \pm 1$ implies (IV). Conversely (IV) implies that Δ divides both $du - cv$ and $av - bu$ for all pairs of integers u, v. In particular, taking first $u = 1$, $v = 0$, and then $u = 0$, $v = 1$, we have that (IV) implies that Δ divides each of d, b, c, a, hence Δ^2 divides $ad - bc = \Delta$. Since if Δ^2 divides Δ then $\Delta = \pm 1$, we see that (IV) is equivalent to $\Delta = \pm 1$.

Finally, it is a well-known, and easily proved, fact that the area of T is $\frac{1}{2}|\Delta|$. Therefore the theorem is proved.

3. Polygons. In order to extend the result of Section 2 we now define a certain set S. We let S consist of all simple polygons having all vertices at lattice points. Thus if U belongs to S then U is a polygon that consists of a single piece, has no holes, has all its vertices at lattice points, and is such that its boundary is a simple closed polygonal curve. If U and V belong to S and are such that the sets have in common a single connected part of the boundary of each, not just a single point, and no other points, then the union of the sets U and V, minus the interior of their common boundary, forms a polygon of S. We denote this polygon by $U + V$. Notice that $U + V$ is not defined for all pairs U, V of S, but just for certain pairs.

Let $f(U)$ be a real valued function defined for all U of S. If $f(U) + f(V) = f(U + V)$ for all U, V for which $U + V$ is defined, then we say that $f(U)$ is an additive function.

THEOREM 2. *Let $I(U)$ denote the number of lattice points that are in the interior of U, not on the boundary. Let $B(U)$ denote the number of lattice points on the boundary of U. Let $f(U) = \alpha I(U) + \beta B(U) + \gamma$. Then $f(U)$ is additive if and only if $\beta = \alpha/2$, $\gamma = -\alpha$.*

Proof. Suppose that $f(U)$ is additive. Let U be the square $(0, 0)$, $(1, 0)$, $(1, 1)$, $(0, 1)$, V the square $(1, 0)$, $(2, 0)$, $(2, 1)$, $(1, 1)$, and W the rectangle $(0, -1)$, $(2, -1)$, $(2, 0)$, $(0, 0)$. Then both $U+V$ and $(U+V)+W$ are defined. Also $I(U)=0$, $B(U)=4$, $f(U)=4\beta+\gamma$. Similarly $f(V)=4\beta+\gamma$, $f(U+V)=6\beta+\gamma$, $f(W)=6\beta+\gamma$, $f((U+V)+W)=\alpha+8\beta+\gamma$. Therefore $8\beta+2\gamma=6\beta+\gamma$ and $12\beta+2\gamma=\alpha+8\beta+\gamma$, from which we find $\beta=\alpha/2$, $\gamma=-\alpha$.

Conversely, suppose that $\beta=\alpha/2$, $\gamma=-\alpha$ and that U and V are members of S such that $U+V$ is defined. Now U and V have in common just a part of the boundary of each. Let C denote the common boundary, and k the number of lattice points on C. Since C is more than just a point it contains at least two lattice points, the ends of C. Any lattice point on C other than the two end-points is an interior point of $U+V$. Interior points of U and of V are interior points of $U+V$. We find $I(U+V)=I(U)+I(V)+k-2$. Similarly

$$B(U + V) = B(U) - k + B(V) - k + 2 = B(U) + B(V) - 2k + 2,$$

hence

$$f(U + V) = \alpha I(U) + \beta B(U) + \alpha I(V) + \beta B(V) + \alpha(k - 2) - 2\beta(k - 1) + \gamma$$
$$= f(U) + f(V) + \alpha(k - 2) - 2\beta(k - 1) - \gamma$$
$$= f(U) + f(V)$$

since $2\beta=\alpha$, $\gamma=-\alpha$. This completes the proof of the theorem.

The function $A(U)$, defined as the area of U, is also an additive function. Suppose that V is a triangle having no lattice points except its vertices. A translation by integral amounts will move V into a position with one vertex at the origin. The functions $A(U)$, $I(U)$ and $B(U)$ are invariant under such a transformation. Therefore the translated V is a T of section 2 and we have $A(V)=\frac{1}{2}$. Also if V is any triangle in S having area $\frac{1}{2}$, then V contains no lattice points other than its three vertices.

THEOREM 3. *If W is any triangle in S then $A(W)=I(W)+\frac{1}{2}B(W)-1$.*

Proof. If $A(W)=\frac{1}{2}$ then we have just seen that $I(W)=0$, $B(W)=3$, and we see that the theorem holds in this case. For a general triangle W in S, let the vertices be (x_1, y_1), (x_2, y_2), (x_3, y_3), then

$$A(W) = \frac{1}{2}\left| (x_2 - x_1)(y_3 - y_1) - (y_2 - y_1)(x_3 - x_1) \right|$$

which is one half an integer. Therefore the theorem will follow by induction on the size of $A(W)$ if we show that any W can be expressed in the form $W=W_1+W_2$ or $W=W_1+W_2+W_3$ where the W_i are triangles of S having area less than $A(W)$. But this is easy. If there is a lattice point, other than a vertex, on a side of W, we break W into two triangles by means of a line from that point to the opposite vertex. If W contains an interior lattice point, we connect that point to the three vertices and break W into three triangles. The additivity of the functions A and f completes the induction.

COROLLARY. *If C is a convex polygon in S then $A(C)=I(C)+\frac{1}{2}B(C)-1$.*

Proof. The polygon C can be broken into pieces, each a triangle of S. Then we use additivity and the preceding theorem.

THEOREM 4. *If P is any polygon in S then $A(P) = I(P) + \frac{1}{2}B(P) - 1$.*

REMARK. The proof of the preceding corollary depends on the obvious fact that a convex polygon can be dissected into suitable triangles. Theorem 4 can be proved in the same way using the fact [2, p. 238] that any polygon can be dissected into triangles of the proper sort. However this result is not really needed. For the sake of completeness we give the following independent proof of Theorem 4.

Proof. We use induction on the number of vertices of P. If $n = 3$ then P is a triangle and the theorem is true for P. Suppose that $n > 3$ and that the theorem is true for all polygons with $n - 1$ or fewer vertices. Let P have the vertices $A_1, A_2, \cdots, A_n$ and sides $A_1A_2, A_2A_3, \cdots, A_{n-1}A_n, A_nA_1$. Let P_0 be the smallest convex polygon containing P. Then P_0 is a convex polygon in S. The vertices of P_0 form a subset of $A_1, A_2, \cdots, A_n$. Some of the A_i may be interior points of P_0 and some may be boundary points but not vertices of P_0. If all the vertices of P are vertices of P_0, then P is convex and the theorem follows from the preceding corollary.

Now we can assume that some vertex of P is not a vertex of P_0, in fact that some vertex is actually an interior point of P_0. By possibly relabelling the subscripts of $A_1, A_2, \cdots, A_n$ in a cyclical fashion, we may presume that A_1 is an interior point of P_0. Now let $s \geq 2$ be the smallest subscript such that A_s is on the boundary of P_0 and let $r \leq n$ be the largest subscript such that A_r is on the boundary of P_0. Since P_0 has at least three vertices, we observe that at least three of the points $A_1, A_2, \cdots, A_n$ lie on the boundary of P_0, and hence $r - s \geq 2$. Let P_1 be the polygon formed by connecting $A_s, A_{s+1}, \cdots, A_r, A_s$ in turn, and let P_2 be the polygon formed by connecting

$$A_r, A_{r+1}, \cdots, A_n, A_1, A_2, \cdots, A_s, A_r$$

in turn. Then P_1 and P_2 are in S and each has fewer than n vertices. Also $A(P) = A(P_1) - A(P_2)$ and hence by the induction hypothesis

$$
\begin{aligned}
(1) \qquad A(P) &= I(P_1) + \tfrac{1}{2}B(P_1) - 1 - I(P_2) - \tfrac{1}{2}B(P_2) + 1 \\
&= I(P_1) - I(P_2) + \tfrac{1}{2}B(P_1) - \tfrac{1}{2}B(P_2).
\end{aligned}
$$

Let j denote the number of lattice points on the segment A_rA_s. These j points are on the boundary of both P_1 and P_2. The remaining $B(P_2) - j$ boundary points of P_2 are interior points of P_1. All interior points of P and P_2 are also interior points of P_1. Using this we find

$$(2) \qquad I(P_1) = I(P) + I(P_2) + B(P_2) - j.$$

Of the boundary points of P_1, all are boundary points of P except for the $j - 2$ points on A_rA_s different from A_r and A_s. All the rest of the boundary points of P are precisely those boundary points of P_2 that do not lie on A_rA_s. These number $B(P_2)-j$. We find

(3) $$B(P) = B(P_1) - j + 2 + B(P_2) - j.$$

Combining (1), (2) and (3) we find $A(P) = I(P) + \frac{1}{2}B(P) - 1$, and our proof is complete.

4. Generalizations. It may be noted that, although there is no restriction that the polygon be convex, Theorem 4 does not generalize to polygons that are not simple. In fact the formula for the area is not correct for a generalized polygon constructed from two triangles meeting at a common vertex.

For a generalization of Theorem 4 to 3-dimensional space in a form containing the Euler-Poincaré characteristic from algebraic topology, see Reeve [3]. However, the formula must be more complicated than the type

(4) $$V(P) = c_1 I(P) + c_2 F(P) + c_3 E(P) + c_4 W(P) + c_5,$$

where P is a polygon whose vertices are lattice points and where $V(P)$ denotes its volume, $I(P)$ the number of interior lattice points, $F(P)$ the number of lattice points in the interior of the faces, $E(P)$ the number of lattice points other than the vertices lying on the edges, and $W(P)$ the number of vertices. We now give a simple example to show that no such formula can hold, even for tetrahedra.

The tetrahedron T_1 with vertices $(0, 0, 0)$, $(1, 0, 0)$, $(0, 1, 0)$, $(0, 0, 1)$ has $V(T_1) = 1/6$, $I(T_1) = F(T_1) = E(T_1) = 0$, $W(T_1) = 4$. The tetrahedron T_2 with vertices $(0, 0, 0)$, $(1, 0, 0)$, $(0, 1, 0)$, $(0, 0, 2)$ has $V(T_2) = 1/3$, $I(T_2) = F(T_2) = 0$, $E(T_2) = 1$, $W(T_2) = 4$. The tetrahedron T_3 with vertices $(0, 0, 0)$, $(2, 0, 0)$, $(0, 2, 0)$, $(0, 0, 2)$ has $V(T_3) = 4/3$, $I(T_3) = F(T_3) = 0$, $E(T_3) = 6$, $W(T_3) = 4$. When these values are substituted in (4) we find we have three equations involving c_3, c_4, c_5 and these equations are inconsistent. Thus no constants c_i exist that make (4) correct, even just for tetrahedra.

References*

1. H. Steinhaus, Mathematical Snapshots, Oxford University Press, New York, 1950.
2. Howard Eves, A Survey of Geometry, vol. 1, Allyn and Bacon, Boston, 1963.
3. J. E. Reeve, On the volume of lattice polyhedra, Proc. London Math. Soc. (3), 7 (1957) 378–395.
4. H. S. M. Coxeter, Introduction to Geometry, Wiley, New York, 1961, pp. 208, 209.
5. R. Honsberger, Ingenuity in Mathematics, vol. 23 of the New Mathematical Library of the M.A.A., 1970, pp. 27–31, 35–37.
6. W. W. Funkenbusch, From Euler's formula to Pick's formula using an edge theorem, this MONTHLY, 81 (1974), 647–648.
7. R. W. Gaskell, M. S. Klamkin, and P. Watson, Triangulations and Pick's theorem, MATHEMATICS MAGAZINE 49 (1976), 35–37.

* Prof. Niven has provided References 5, 6, and 7 to update this article for the present volume.— ED.

A PROOF OF MINKOWSKI'S INEQUALITY
FOR CONVEX CURVES*

HARLEY FLANDERS, Purdue University†

1. Introduction. There is a beautiful proof of the isoperimetric inequality in the plane by L. A. Santaló [7, pp. 38–39]. This is based on the following ideas. Given an oval in the plane, move a circle of suitable radius, counting for each position how many times it intersects the oval. The average number of intersections can be computed in two ways, and the result is the isoperimetric inequality with an explicit error term.

This proof of Santaló has been reproduced several times and is fairly well known. What is not so widely known is that the proof may be modified to yield a proof of the Minkowski inequality on mixed areas, and actually yields an improvement of this inequality which is due to Bonnesen. See Blaschke [2, pp. 33–36].

The purpose of this paper is to give an exposition of the theory of closed convex plane curves, mixed area, and the Minkowski inequality. The prerequisites are a slight acquaintance with convex bodies in the plane and the beginnings of the differential geometry of plane curve theory now included in most vector oriented calculus books. In the next section we review the elementary differential geometry of closed convex curves.

In the last section we state the form taken by the Minkowski inequality in three-space. Alas, no analogue of the Santaló proof is known and each of the known proofs is much harder than the one we give shortly for the plane case. Those wishing to explore the subject further will find plenty in the list of the references at the end which includes more than the works cited here.

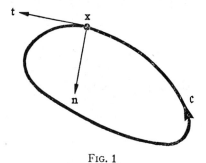

FIG. 1

2. Preliminaries. Let c be a smooth closed convex curve in E^2 with positive curvature. Let s be the arc length, $x = x(s)$ the moving point in c, $t = t(s)$ the moving unit tangent, and $n = n(s)$ the moving unit normal (Fig. 1). The Frenet formulas are

$$(2.1) \qquad \frac{dx}{ds} = t, \qquad \frac{dt}{ds} = \kappa n, \qquad \frac{dn}{ds} = -\kappa t.$$

---

* From AMERICAN MATHEMATICAL MONTHLY, vol. 75 (1968), pp. 581–593.

† Prof. Flanders is now a Visiting Professor at Florida Atlantic University.

Here $\kappa = \kappa(s)$ is the curvature which we are assuming satisfies $\kappa > 0$ everywhere. If the total length of the curve is L, then x is a periodic vector function of s with fundamental period L and the other functions t, n, κ also have period L, although not necessarily as a fundamental period. Note that n is the inward drawn normal so that t, n is a right-handed frame.

Of course we have

$$(2.2) \qquad\qquad L = \oint_c ds.$$

There is also a line integral for the area A enclosed by c. If we write $dx = (dx, dy) = t\,ds$, then a rotation by angle $\pi/2$ leads to

$$(-dy, dx) = n\,ds,$$
$$(x, y) \cdot (-dy, dx) = x \cdot n\,ds,$$
$$y\,dx - x\,dy = x \cdot n\,ds.$$

Because of the well-known relation

$$A = \frac{1}{2}\oint_c (x\,dy - y\,dx),$$

an immediate consequence of Green's Theorem, we have

$$(2.3) \qquad\qquad A = -\frac{1}{2}\oint_c x \cdot n\,ds.$$

This formula is valid for any simply closed curve and has nothing whatever to do with the convexity. We exploit the fact that c is convex by introducing the parameter θ, the angle the outward drawn normal $-n$ makes with the fixed x-axis. It is convenient to take the origin inside c. Since the curve c is turning continuously $(dt/ds = \kappa n, \kappa > 0)$, each point x of the curve has a unique θ (modulo 2π) associated with it and θ makes a complete circuit, $0 \leq \theta \leq 2\pi$ as $0 \leq s \leq L$. (Fig. 2). The *support function* p is the distance of the tangent line at x from 0.

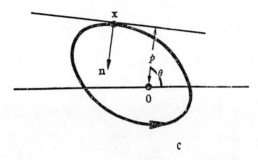

FIG. 2

We write $p = p(\theta)$ and

(2.4)
$$p = -x \cdot n.$$

The function $p = p(\theta)$ has period 2π.

Analytically (this means as usual that, picturesque drawings to the contrary, all vectors start at 0) we have

(2.5)
$$t = (-\sin\theta, \cos\theta),$$
$$n = (-\cos\theta, -\sin\theta).$$

The second formula is a direct consequence of the definition of θ and the first follows from this by rotating $\pi/2$.

After this we shall always denote $d/d\theta$ by $(')$. Clearly

(2.6)
$$t' = n, \qquad n' = -t.$$

Thus $dt = nd\theta$. But $dt = \kappa n ds$ by (2.1). Hence

(2.7)
$$d\theta = \kappa ds.$$

Of course this is the standard interpretation of the curvature at the rate of turning of the tangent. Next we differentiate (2.4): $p' = -x' \cdot n + x \cdot t$. Since x' is parallel to t $(x' = s' dx/ds = s't)$, $x' \cdot n = 0$ and so

(2.8)
$$x \cdot t = p'.$$

Differentiating again, $x' \cdot t + x \cdot n = p''$. But $x \cdot n = -p$ and $x' \cdot t = (s't) \cdot t = s'$, hence

(2.9)
$$s' = p + p''.$$

But (2.7), $s' = 1/\kappa = \rho$, the radius of curvature so this formula may be rewritten

(2.9′)
$$\rho = p + p''.$$

We now transform the integrals (2.2), (2.3):

$$L = \oint ds = \int_0^{2\pi} \rho d\theta = \int_0^{2\pi} (p + p'') d\theta.$$

Since $p'' d\theta = d(p')$ is exact and p' has period 2π, $\oint p'' d\theta = 0$ and we have $L = \int_0^{2\pi} p d\theta$. For the area we have

$$A = -\frac{1}{2} \oint x \cdot n ds = \frac{1}{2} \int_0^{2\pi} p(p + p'') d\theta.$$

Since $d(pp') = pp'' d\theta + p'^2 d\theta$ we have $\int_0^{2\pi} pp'' d\theta = -\int_0^{2\pi} p'^2 d\theta$. We have obtained

(2.10)
$$L = \int_0^{2\pi} p d\theta, \qquad A = \frac{1}{2} \int_0^{2\pi} (p^2 - (p')^2) d\theta.$$

It is worth noting that p determines c completely. In fact, by (2.4), (2.8), and (2.5),

$$x = p't - pn = (-p' \sin \theta + p \cos \theta, \ p' \cos \theta + p \sin \theta).$$

Geometrically, the tangents to c envelop c. To say it another way, the convex set K whose boundary is c is the intersection of the half-planes including K determined by the lines of support (tangents). This means

(2.11) $$K = \{v \in E^2 \mid\ -v \cdot n(\theta) \leq p(\theta) \text{ for all } \theta \}.$$

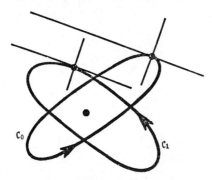

FIG. 3

MIXED AREAS. For this concept we work with two closed convex curves c_0 and c_1. Thus we have two vector functions $x_0 = x_0(\theta)$, $x_1 = x_1(\theta)$ of θ. By writing them this way we automatically set up a one-one correspondence between the curves whereby corresponding points have the same normal $n = n(\theta)$. (See Fig. 3.) The support functions are

$$p_0(\theta) = -x_0 \cdot n, \qquad p_1(\theta) = -x_1 \cdot n.$$

We now study a new convex curve obtained by translating c_1 to all possible positions such that the origin of c_1 is on c_0. To understand this (Fig. 4) we think of c_1 as the boundary of a rigid lamina which we are free to slide by translations only over the plane in which c_0 is fixed. We let the origin of c_1 in this lamina slide along c_0. Then c_1 envelops a curve (two curves actually; we take the outside one as illustrated). It is clear that the moving point of contact of this new curve c with a particular translate of c_1 has the same direction as its moving origin has on c_0. We immediately conclude that our new curve c has support function

$$p = p_0 + p_1.$$

Another way to look at this is through the Minkowski sum of convex sets. Let $c_0 = \partial K_0$, $c_1 = \partial K_1$ so that K_0, K_1 are convex regions bounded by c_0 and c_1 respectively. The Minkowski sum is the set K defined by

(2.12) $$K = K_0 + K_1 = \{v_0 + v_1 \mid v_0 \in K_0, v_1 \in K_1 \}.$$

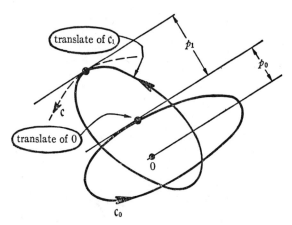

Fig. 4

By (2.11) applied to both K_0 and K_1, if $v_0 \in K_0$ and $v_1 \in K_1$, then

$$-(v_0 + v_1) \cdot n(\theta) = -v_0 \cdot n(\theta) - v_1 \cdot n(\theta) \leqq p_0(\theta) + p_1(\theta).$$

But $x_0(\theta) \in K_0$, $x_1(\theta) \in K_1$, hence $x_0(\theta) + x_1(\theta) \in K$ and

$$-[x_0(\theta) + x_1(\theta)] \cdot n(\theta) = p_0(\theta) + p_1(\theta).$$

This shows that K is the intersection of all the half-planes

$$\{v \mid -v \cdot n(\theta) \leqq p_0(\theta) + p_1(\theta)\}$$

so that K is convex and its boundary $c = \partial K$ has support function $p = p_0 + p_1$.

   (2.13) LEMMA. *The length and area of c are given by*

$$L = L_0 + L_1, \qquad A = A_0 + 2A_{01} + A_1,$$

where L_i, A_i are the length and area of c_i ($i = 0, 1$) and $A_{01} = \frac{1}{2}\int_0^{2\pi} [p_0 p_1 - p_0' p_1'] d\theta$ is the mixed area *of c_0 and c_1.*

   Proof.

$$L = \int_0^{2\pi} p \, d\theta = \int_0^{2\pi} (p_0 + p_1) d\theta = \int_0^{2\pi} p_0 d\theta + \int_0^{2\pi} p_1 d\theta = L_0 + L_1.$$

$$A = \tfrac{1}{2} \int_0^{2\pi} (p^2 - p'^2) d\theta$$

$$= \tfrac{1}{2} \int_0^{2\pi} [(p_0^2 + 2p_0 p_1 + p_1^2) - (p_0'^2 + 2p_0' p_1' + p_1'^2)] d\theta$$

$$= \tfrac{1}{2} \left\{ \int_0^{2\pi} (p_0^2 - p_0'^2) d\theta + 2 \int_0^{2\pi} (p_0 p_1 - p_0' p_1') d\theta + \int_0^{2\pi} (p_1^2 - p_1'^2) d\theta \right\}$$

$$= A_0 + 2A_{01} + A_1.$$

It is clear that everything in sight is symmetric. If we slide the origin of c_0 along c_1 we get the same curve c simply because $p_1 + p_0 = p_0 + p_1$. Also $A_{10} = A_{01}$. We may obtain an unsymmetric formula for A_{01} by integrating the exact differential

$$d(p_0 p_1') = (p_0' p_1' + p_0 p_1'')d\theta.$$

This yields

$$\oint p_0' p_1' \, d\theta = -\oint p_0 p_1'' \, d\theta.$$

Thus

$$(2.14) \qquad A_{01} = \tfrac{1}{2}\int_0^{2\pi} p_0(p_1 + p_1'')d\theta.$$

By (2.9) and (2.9') we have

$$(2.15) \qquad A_{01} = \tfrac{1}{2}\int_0^{2\pi} p_0\rho_1 d\theta = \tfrac{1}{2}\oint p_0 ds_1,$$

and similarly

$$(2.16) \qquad A_{01} = \tfrac{1}{2}\int_0^{2\pi} p_1\rho_0 d\theta = \tfrac{1}{2}\oint p_1 ds_0.$$

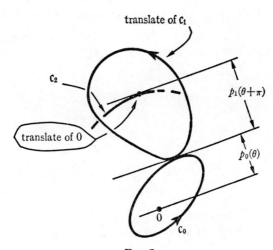

FIG. 5

Instead of translating c_1 so that the origin moves along c_0, let us try the following. We translate c_1 so that it is in contact with c_0 externally. Then the locus of the translated origin traces a new curve c_2 (Fig. 5). From the figure, this is a convex curve with support function p_2 given by

$$p_2(\theta) = p_0(\theta) + p_1(\theta + \pi).$$

Now $p_1(\theta+\pi)$ is the support function of the curve c_1^* obtained by reflecting c_1 in the origin or, equivalently, by rotating c_1 through angle π. Thus the curve c_2 is obtained simply by applying our previous construction to c_0 and c_1^* and (2.13) applies.

(2.17) LEMMA. *The length and area of c_2 are given by*

$$L_2 = L_0 + L_1, \qquad A_2 = A_0 + 2A_{01}^* + A_1$$

where

$$A_{01}^* = \tfrac{1}{2} \int_0^{2\pi} [p_0(\theta)p_1(\theta + \pi) - p_0'(\theta)p_1'(\theta + \pi)]d\theta$$

$$= \tfrac{1}{2} \oint p_1(\theta + \pi)ds_0(\theta).$$

Proof. It is clear both geometrically and analytically that c_1^* and c_1 have the same length and same area so the formula is a consequence of (2.13). The last expression comes from (2.16) applied to $p_1^*(\theta) = p_1(\theta+\pi)$.

3. The Minkowski inequality. The main result we are after is the Minkowski inequality which is the two-dimensional version of the Brunn-Minkowski inequality. We refer to Bonnesen-Fenchel [3, Sect. 49, 51] and Hadwiger [6, Chapt. 4] for other treatments of this and more general results.

(3.1) MINKOWSKI INEQUALITY. *If c_0 and c_1 are closed convex curves with areas A_0 and A_1 respectively and mixed area A_{01}, then $A_{01}^2 \geqq A_0A_1$.*

Note that in case c_1 is the unit circle we have $A_1 = \pi$, $p_1 = 1$, and by (2.16)

$$A_{01} = \tfrac{1}{2} \oint ds_0 = \tfrac{1}{2}L_0,$$

so the Minkowski inequality specializes to

(3.2) $$\tfrac{1}{4}L_0^2 \geqq \pi A_0$$

which is the classical isoperimetric inequality.

We also can state precisely when there is equality in (3.1).

(3.3) SUPPLEMENT. *If $A_{01}^2 = A_0A_1$ then c_0 and c_1 are homothetic, i.e., they differ by a dilatation and translation.*

Our proof will yield the following stronger form of (3.1) which makes (3.3) obvious. To state it we need the ideas of relative inradius and relative circumradius. The curves c_0 and c_1 are given and we write as before $c_0 = \partial K_0$, $c_1 = \partial K_1$.

The *inradius of c_0 relative to c_1* is the largest real number r_0 such that a translate of r_0K_1 is in K_0 (Fig. 6). The *circumradius of c_0 relative to c_1 is the* smallest real number R_0 such that a translate of R_0K_1 contains K_0. Obviously $R_0 \geqq r_0$ with equality if and only if K_0 is a translate of r_0K_1 or, to say it another way, K_0 and K_1 are similar and similarly placed. Note that if c_1 is the unit circle then r_0 and R_0 are the ordinary inradius and circumradius of c_0.

(3.4) THEOREM. *Let c_0 and c_1 be closed convex curves with areas A_0 and A_1 respectively and mixed area A_{01}. Let r_0, resp. R_0, be the inradius, resp. circumradius, of c_0 relative to c_1. Then*

$$A_{01}^2 - A_0 A_1 \geqq \frac{A_1^2}{4} (R_0 - r_0)^2.$$

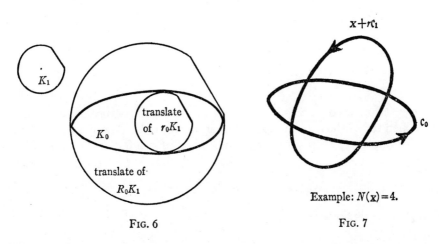

FIG. 6   FIG. 7

Example: $N(\mathbf{x}) = 4$.

4. The proof. We fix a number r with $r_0 \leqq r \leqq R_0$. For each point $\mathbf{x} = (x, y)$ of the plane we consider the translate $\mathbf{x} + rc_1$ of rc_1. We are interested in the number $N(\mathbf{x})$ of points of intersection of this translate with c_0. (See Fig. 7.) Thus

$$(4.1) \qquad N(\mathbf{x}) = \left| (\mathbf{x} + rc_1) \cap c_0 \right|.$$

We shall average this. Precisely, we set

$$(4.2) \qquad I = \iint_{E^2} N(\mathbf{x}) dx dy.$$

The function $N(\mathbf{x})$ may be infinite for certain values of $\mathbf{x}$ such that the curves c_0 and $\mathbf{x} + rc_1$ have a point of tangency. Now the points of $\mathbf{x}$ for which there is such a point of tangency lie on two curves which constitute a set of zero area. Consequently it makes no difference what $N(\mathbf{x})$ is equal to on these curves, the integral I is insensitive to these values.

The outside curve (Fig. 8) is precisely the curve c_2 we studied in Section 2 (see Fig. 5) but with c_1 replaced by rc_1. If x lies outside of this curve then $N(x)$ $=0$. But if x lies inside of this curve, $N(x) > 0$. For otherwise either $x+rc_1$ entirely surrounds c_0, a contradiction to $r \leq R_0$, or c_0 entirely surrounds $x+rc_1$, a contradiction to $r_0 \leq r$.

Actually if we ignore the boundary c_2 and the other curve (c_0 and $x+rc_1$ tangent internally) where $N(x)$ may have nasty values we may assert that $N(x) \geq 2$.

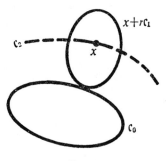

FIG. 8

For when two closed curves intersect at all (and really cross, no common tangents) then they intersect an even number of times. What comes in must go out.

According to (2.17) applied to c_0 and rc_1 we have for the area of K_2 where $c_2 = \partial K_2$,

$$A_2 = A_0 + 2rA_{01}^* + r^2 A_1.$$

From these remarks we have

$$I = \iint_{E^2} N(x)dxdy = \iint_{K_2} N(x)dxdy \geq \iint_{K_2} 2dxdy \geq 2A_2,$$

(4.3) $$I \geq 2(A_0 + 2rA_{01}^* + r^2 A_1).$$

One should worry why the discrete valued function $N(x)$ has an integral. It is easiest to ignore it, but a word to the wise in the ways of measure theory will be sufficient: the sets where $n(x) = 2m$ are open and E^2 is the join of these and a couple of piecewise smooth curves.

The relation (4.3) gives a lower bound for I. We now go after a precise evaluation of I. This is done in three steps. Step 1: we find all translates of rc_1 passing through a fixed point. Step 2: we find all translates of rc_1 which intersect a fixed short segment. Step 3: we break up c_0 into many short segments (or use an inscribed polygon approximation) and sum. Now the details!

If x_0 is a fixed point, the moving origin of rc_1 traces the oval $x_0 - rc_1$ as rc_1 is translated to all positions such that it passes through x_0, i.e., as $x_1(\theta)$ is translated

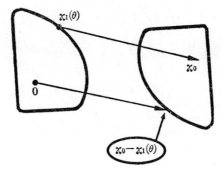

FIG. 9

to x_0, the origin gets translated to $x_0 - x_1(\theta)$. See Fig. 9.

Now consider a short segment of length ds_0. Think of this as a part of the fixed curve c_0. Which translates of rc_1 intersect this segment? Those for which the center of the translate lies in the thin strip swept out by $x_0 - rc_1$ as x_0 moves its distance ds_0 along the segment (Fig. 10). Ignoring the small shaded area at

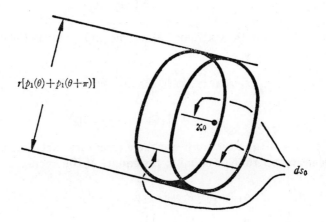

FIG. 10

the end, which is of smaller order of magnitude, the area of each half of this strip is obtained by multiplying its base $r[p_1(\theta) + p_1(\theta + \pi)]$ by its constant height ds_0. Now except for the shaded areas, each point of this strip is the center of a translate of rc_1 which intersects the segment once. Thus the contribution of this segment to the integral I is

$$2r[p_1(\theta) + p_1(\theta + \pi)]ds_0 + O(ds_0)^2.$$

Summing and passing to the limit we have

(4.4)
$$I = 2r \int_0^{2\pi} [p_1(\theta) + p_1(\theta + \pi)] ds_0.$$

By (2.16) and (2.17) we may write this as

(4.5)
$$I = 4r(A_{01} + A_{01}^*),$$

which completes our exact evaluation of I.

By (4.3) and (4.5) we have

(4.6)
$$2rA_{01} \geqq A_0 + r^2 A_1$$

for all r such that $r_0 \leqq r \leqq R_0$.

Now the existence of a single real number r such that the polynomial $x^2 A_1 - 2x A_{01} + A_0$ is less than or equal to zero at $x = r$ implies this polynomial has real roots and hence positive discriminant $A_{01}^2 - A_0 A_1 \geqq 0$. This is the Minkowski inequality which is good enough. However, we learn more by exploiting the fact that the polynomial is nonpositive on the whole interval $r_0 \leqq r \leqq R_0$. Simplest is to complete the square to rewrite (4.6) as

$$A_{01}^2 - A_0 A_1 \geqq (rA_1 - A_{01})^2.$$

We substitute for r the two extreme values and slyly change the sign inside the square:

$$A_{01}^2 - A_0 A_1 \geqq (R_0 A_1 - A_{01})^2,$$

$$A_{01}^2 - A_0 A_1 \geqq (A_{01} - r_0 A_1)^2.$$

We average these inequalities and use the elementary fact that the average of the squares is greater than or equal to the square of the average:

$$A_{01}^2 - A_0 A_1 \geqq \{\tfrac{1}{2}[(R_0 A_1 - A_{01}) + (A_{01} - r_0 A_1)]\}^2,$$

so finally $A_{01}^2 - A_0 A_1 \geqq A_1^2 (R_0 - r_0)^2 / 4$.

5. Remarks.

1. If c_1 is a circle of radius r then $L_1 = 2\pi r$, $A_1 = \pi r^2$, $A_{01} = \tfrac{1}{2} r L_0$. The curve c of (2.13) is the oval parallel to c_0 at distance r and (2.13) yields Steiner's formulas

$$L = L_0 + 2\pi r, \qquad A = A_0 + r L_0 + \pi r^2.$$

2. If c_1 is the unit circle, then (3.4) specializes to

$$L_0^2 - 4\pi A_0 \geqq \pi^2 (R_0 - r_0)^2$$

which is the inequality of Bonnesen mentioned in the Introduction. This is a striking relation between the length, area, circumradius and inradius of an oval.

3. Of course (3.4) implies $A_{01}^2 \geqq A_0 A_1$. It also implies that if $A_{01}^2 = A_0 A_1$, then $R_0 = r_0$ and (3.3) follows.

4. The estimate (4.3) may be replaced by the more precise

$$I = 2(A_0 + 2r A_{01}^* + r^2 A_1) + 2 \sum_{m=2}^{\infty} (m - 1) f_m,$$

where f_m is the area of the set on which $N(x) = 2m$. This follows from

$$I = \sum_1^{\infty} 2m f_m \quad \text{and} \quad \sum_1^{\infty} f_m = A_2.$$

It may be used to improve (4.6) and the subsequent inequalities.

6. The situation in space. Let K be a compact convex body in E^3 whose boundary $\mathfrak{S} = \partial K$ is a smooth closed convex surface with positive total curvature at each point. (This means that if n is the outward unit normal at x, then $x \rightarrow n$ is a one-one smooth mapping of $\mathfrak{S}$ onto the unit sphere whose inverse map is also smooth.) In addition to the volume V of K and the surface area A of $\mathfrak{S}$ there is another invariant,

(6.1)
$$M = \iint_S H \, dA,$$

where H is the mean curvature. The isoperimetric inequality in E^3, due to H. A. Schwarz, is

(6.2)
$$A^3 \geqq 36\pi V^2.$$

This may be sharpened to

(6.3)
$$A^3 - 36\pi V^2 \geqq [\sqrt{A} - \sqrt{4\pi}\, r_0]^6,$$

where r_0 is the inradius. This is enough to show that equality holds in (6.2) only for a sphere. Another refinement of (6.2) consists of the pair of inequalities

(6.4)
$$A^2 \geqq 3MV,$$

(6.5)
$$M^2 \geqq 4\pi A.$$

These imply both (6.2) and the further result

(6.6)
$$M^3 \geqq 48\pi^2 V.$$

These and (6.3) may all be summarized in the assertion that the function

(6.7)
$$f(t) = (V + At + Mt^2 + \tfrac{4}{3}\pi t^3)^{1/3}$$

is concave ($f'' \leqq 0$).

Now let K_0, K_1 be two convex bodies of the type considered. The first remarkable fact is that the volume of a linear combination $\lambda K_0 + \mu K_1 (\lambda, \mu \geqq 0)$ is given by a polynomial

(6.8) $$\left| \lambda K_0 + \mu K_1 \right| = V_0 \lambda^3 + 3V_{001}\lambda^2\mu + 3V_{011}\lambda\mu^2 + V_1\mu^3.$$

This defines the *mixed* volumes V_{001}, V_{011}. The function $f(t)^3$ from (6.7) is the special case $K_0 = K$, $K_1 =$ sphere of radius t. The Brunn-Minkowski inequality is the assertion that

(6.9) $$\left| K_0 + tK_1 \right|^{1/3}$$

is a concave function of t, $0 \leq t$. This has the consequences

(6.10) $$V_{001}^2 \geq V_0 V_{011},$$

(6.11) $$V_{011}^2 \geq V_{001}V_1,$$

which generalize (6.4) and (6.5) and furthermore

(6.12) $$V_{001}^3 \geq V_0^2 V_1,$$

(6.13) $$V_{011}^3 \geq V_0 V_1^2$$

which generalize (6.4) and (6.5).

This provides a glimpse into a large and fascinating part of geometry.

Presented to the Indiana Section, Nov. 11, 1967 at Marian College, Indianapolis, with the title, "Some ovals I have known." The author is indebted to Prof. G. D. Chakerian for reading his manuscript and making several helpful suggestions.

References

1. W. Blaschke, Kreis und Kugel, Veit, Leipzig, 1916 and Chelsea paperback reprint, New York 1949.

2. ———, Vorlesungen über Integralgeometrie I, II, Hamb. Math. Einzel. 20, 22, Teubner, Leipzig, 1936/37 and Chelsea reprint, New York, 1949.

3. T. Bonnesen, W. Fenchel, Theorie der konvexen Körper, Ergeb. der Math., 3 (1934) Springer, Berlin, and Chelsea reprint, New York, 1948.

4. H. Busemann, Convex Surfaces, Interscience, New York, 1958.

5. H. Hadwiger, Vorlesungen über Inhalt, Oberfläche und Isoperimetrie, Springer, Berlin, 1957.

6. ———, Altes und Neues über konvexe Körper, Birkhäuser, Basel, 1955.

7. L. A. Santaló, Introduction to integral geometry, A.S.I. 1198, Hermann, Paris, 1953.

8. ———, Integral geometry, M.A.A. Studies in Math., S.S. Chern ed., Studies in global geometry and analysis, 4 (1967) 147–193.

AN AREA-WIDTH INEQUALITY FOR CONVEX CURVES*

PAUL R. CHERNOFF, University of California, Berkeley

THEOREM. *Let C be a closed convex plane curve, bounding a region of area A. Let $w(\theta)$ denote the width of C in the θ-direction. Then*

$$(1) \qquad A \leqq \frac{1}{2} \int_0^{\pi/2} w(\theta) w\left(\theta + \frac{\pi}{2}\right) d\theta$$

with equality if and only if C is a circle.

Proof. We shall employ a method used by A. Hurwitz (see Courant [1] p. 213) to prove the isoperimetric inequality. It clearly suffices to establish the theorem for the case of C^2 curves such that for each $\theta, 0 \leqq \theta < 2\pi$, there is exactly one point, say $(x(\theta), y(\theta))$, at which the normal to C makes an angle θ with the X-axis. Then, as Courant indicates, there is a C^2 periodic function $p(\theta)$ such that

$$(2) \qquad \begin{aligned} x(\theta) &= p(\theta) \cos\theta - p'(\theta) \sin\theta \\ y(\theta) &= p(\theta) \sin\theta + p'(\theta) \cos\theta. \end{aligned}$$

In fact, if the origin 0 is chosen to be interior to C then $p(\theta)$ is the perpendicular distance from 0 to the tangent to C at $(x(\theta), y(\theta))$. We then have $w(\theta) = p(\theta) + p(\theta + \pi)$. Hence

$$I = \frac{1}{2} \int_0^{\pi/2} w(\theta) w\left(\theta + \frac{\pi}{2}\right) d\theta = \frac{1}{2} \int_0^{2\pi} p(\theta) p\left(\theta + \frac{\pi}{2}\right) d\theta,$$

while

$$A = \frac{1}{2} \int_0^{2\pi} (xy' - yx') d\theta = \frac{1}{2} \int_0^{2\pi} (p^2 - p'^2) d\theta.$$

We now express I and A in terms of the Fourier coefficients of $p(\theta)$. Let us write

$$p(\theta) = \sum_{k=-\infty}^{\infty} a_k e^{ik\theta}$$

where $\bar{a}_k = a_{-k}$ since p is real. We then have

$$(3) \qquad A = \pi \sum_{k=-\infty}^{\infty} |a_k|^2 (1 - k^2) = \pi |a_0|^2 + 2\pi \sum_{k=2}^{\infty} |a_k|^2 (1 - k^2)$$

and

$$I = \frac{1}{2} \sum_{k,l} a_k a_l \int_0^{2\pi} e^{ik\theta} e^{il(\theta + \pi/2)} d\theta$$

$$= \pi \sum_{k=-\infty}^{\infty} |a_k|^2 e^{ik(\pi/2)};$$

---

* From AMERICAN MATHEMATICAL MONTHLY, vol. 76 (1969), pp. 34–35.

that is,

$$(4) \qquad I = \pi \left| a_0 \right|^2 + 2\pi \sum_{k=2}^{\infty} \left| a_k \right|^2 \cos k \frac{\pi}{2} .$$

Comparing (3) and (4) we see that $A \leq I$. Equality holds if and only if $a_k = 0$ for $|k| > 1$, i.e., $p(\theta) = a_0 + 2Re(a_1 e^{i\theta})$, which corresponds to a circle.

As a corollary one has the well-known result that a convex region of area $> \pi/4$ must have width > 1 in some direction. In particular, such a region must contain two interior points a unit distance apart. (The discovery of (1) was inspired by the appearance of the latter statement as a problem in the 1967 Putnam competition.)

References

1. R. Courant, Differential and Integral Calculus, vol. II, E. J. McShane transl., Interscience, New York, 1936.

Further references.* A recent paper by Chakerian [2] deals with higher dimensional generalizations of the inequality given here. This article contains several related references, in particular to [3] in which this inequality is proved from other known inequalities rather than by Fourier analysis. Radziszewski mentions that (1) was conjectured by M. Biernacki.

2. G. D. Chakerian, The mean volume of boxes and cylinders circumscribed about a convex body, Israel J. of Math., 12 (1972), 249–56.

3. K. Radziszewski, Sur une fonctionnelle définie sur les ovales, Ann. Univ. Mariae Curie-Sklodowska, Lublin, Sect. A, 10 (1956), 57–9.

THE EARLY DEVELOPMENT OF ALGEBRAIC GEOMETRY †

SOLOMON LEFSCHETZ, Brown University and Princeton University

1. As I am neither a historian nor an archeologist of mathematics, I shall merely present my idea of the major contributions to the subject, say by the end of the last century.

By "algebraic geometry" one really understands what I have described as *bi-rational geometry*. A simple example will clarify this point. Take the irreducible curve

$$C : f(x,y) = y^2 - (x^3 - 1) = 0.$$

Let any complex rational function

$$R(x,y) = P(x,y)/Q(x,y),$$

---

* Provided by the author for the present volume.—ED.

† From AMERICAN MATHEMATICAL MONTHLY, vol. 76 (1969), pp. 451–460.

P and Q polynomials, be declared as null: $R = 0$, whenever P but not Q is divisible by f. Place in one class S^* all the rational functions which differ by zero from a given one. The collection $\{S^*\}$ is a *field*, an overfield of the field K of all complex numbers. This is the *function field* $K(C)$ of the curve C. The *algebraic geometry* of C is the study of the properties of C which depend solely upon the function field $K(C)$.

All this applies, of course, to any plane curve

$$C: f(x, y) = 0,$$

where f is a complex irreducible polynomial. It applies in fact also to higher dimensional varieties (surfaces, $\cdots$) but I shall restrict my discussion to mere plane curves.

The particular choice of the complex field K (as field of constants) rather than any other field, is simply because with K one may freely utilize complex analysis and topology.

The justification of "birational" is this: Let $D: g(u, v) = 0$ be a second curve like C and suppose that $K(C)$ is isomorphic over K with $K(D)$. One may identify the two fields under some fixed isomorphism T. As a consequence, say

$$Tx = S_1(u, v), \quad Ty = S_2(u, v); \qquad S_1, S_2 \in K(D)$$

and similarly

$$T^{-1}u = S_1'(x, y), \quad T^{-1}v = S_2'(x, y); \qquad S_1', S_2' \in K(C).$$

Thus T defines a *birational* transformation $D \to C$, which is almost a 1-1 correspondence between the points of the two curves.

Example: Let $f_h(x, y)$ denote a complex homogeneous polynomial of degree h. Let C be an irreducible curve with the origin as point of multiplicity h, so that its equation is

$$f(x, y) \equiv f_h(x, y) + f_{h+1}(x, y) = 0,$$

where f_h and f_{h+1} have no common factor. Let D be the line $v = 0$ of the u, v plane. Then $T^{-1}: u = y/x, v = 0$,

$$T: x = \frac{-f_h(1, u)}{f_{h+1}(1, u)}, \qquad y = \frac{-u f_h(1, u)}{f_{h+1}(1, u)}.$$

Thus all the curves such as C are birationally equivalent to the u line, and hence to one another (see Fig. 1).

Incidentally this underscores the vast richness of birational as compared with projective or Euclidean geometries. For it has enabled us to place in one class a seemingly vastly dissimilar collection of curves.

It has always tacitly been understood by the savants of the 19th century dedicated to our subject that the curves considered were inhabitants of a complex projective plane. However, for convenience in the utilization of complex analysis, they were placed, as here, in a complex Euclidean plane complemented

TRIPLE POINT (h=3)

Fig. 1

for "infinity" by some such device as a projective transformation

$$x = \frac{1}{x'}, \qquad y = \frac{y'}{x'},$$

replacing the line at infinity by the line $x'=0$. This is to be understood in what follows.

2. The first vestige of algebraic geometry is found in the extensive theory of elliptic functions by Legendre (before 1830). The first step was the natural extension of trigonometric integrals to the type

$$u = \int_0^x \frac{dx}{\sqrt{(1 - x^2)(1 - k^2x^2)}} = \int_0^x \frac{dx}{y},$$

where $k^2 \neq 0$, 1 and $y^2 = (1-x^2)(1-k^2x^2)$.

Legendre confined his attention to the real domain. He defined the inverse function $x = \operatorname{sn} u$ as *sine amplitude*, introduced other functions which I shall not describe, and was primarily interested in the applications, notably to the arc of an ellipse (hence elliptic functions) and to large oscillations of a pendulum.

The integral u, finite for all x, has evidently birational character. However, perhaps not as yet influenced by the fundamental work of Cauchy, Legendre failed to catch the double periodicity of sn u. It was brought out (around 1830) by Abel, Jacobi, and Gauss. A brilliant proof of the fact that both distinct periods of sn u could not be real is due to Jacobi. The proof is so simple that I cannot escape the temptation of outlining it. Suppose that there exist two rationally independent periods ω_1, ω_2, both real. Then for any $\epsilon > 0$ there exist integers m, n such that $|m\omega_1 + n\omega_2| < \epsilon$. Hence for real u we have $|\operatorname{sn} u| < \epsilon$, which is untrue.

3. By 1850 these basic steps had been taken:

Abel. Generalization of elliptic integrals as follows: Given C, let $R \in K(C)$ be expressed in acceptable cartesian coordinates x, y. Then

$$u = \int_{(x_0, y_0)}^{(x, y)} R dx,$$

limits and path in C, is known as an *abelian* integral. There are 3 types: first kind $|u|$—bounded; 2nd kind—u has poles only; 3rd kind—remaining type.

ABEL'S THEOREM. Let the variable curve $\phi = \sum c_h \phi_h(x, y) = 0$, (the ϕ_h are linearly independent modulo f and are polynomials of the same degree) cut C in points $P_1, \cdots, P_n$. Then

$$\sum_k \int_A^{P_k} du = v$$

is a constant (independent of the c_h).

JACOBI. Let $u_1, \cdots, u_p$ be a maximal set of linearly independent integrals of the first kind modulo constants. Given p general values $v_1, \cdots, v_p$, the system in the unknowns P_k, $k \leq p$

$$\sum_k \int_A^{P_k} du_h = v_h$$

has a unique solution.

WEIERSTRASS. The maximal number of linearly independent integrals of the second kind modulo $K(C)$ is $2p$. (That is K-linearly independent modulo elements of the function field $K(C)$.)

PUISEUX. Let C be *any* algebraic curve and $P(x_0, y_0)$ any (finite) point of C. The complete neighborhood of the point on C is represented by a finite number of fractional power series

$$y - y_0 = a_1(x - x_0)^{q_1/q_0} + a_2(x - x_0)^{q_2/q_0} + \cdots,$$

where $|x - x_0| < \sigma$, and the q_h have no common factor. These series come in collections of q_0, forming a system of roots $y_1, y_2, \cdots, y_{q_0}$ of $f(x, y) = 0$, which are circularly permuted as x turns around x_0 in its complex plane. This set of solutions is jointly represented parametrically in the form

$$x = x_0 + t^{q_0}, \quad y = y_0 + a_1 t^{q_1} + a_2 t^{q_2} + \cdots, \qquad 0 \leq |t| < \sigma^{1/q_0}.$$

Such a pair of series defines what is now known as a *place of center P*. There are simple analogues for the points "at infinity" which I shall not describe.

One may therefore summarize Puiseux's fundamental result as follows: *Each point P of the curve C has a neighborhood on C consisting of a finite number of places of which P is the center.* (I am anticipating in this statement a formulation really due to Hermann Weyl.)

General observation. All the results described so far have strict birational character.

4. I come now to one of the greatest contributions ever made to mathematics. The author is Riemann, the period around 1860 and the topic:

RIEMANN SURFACES. The representation of an algebraic curve by a truly geometric, and possibly simple model was assuredly most desirable. This is exactly what Riemann accomplished.

Let C be our usual curve and let m be the degree of f in y. A *branch point* of y is a value $x = a$, marked on the sphere S of the complex variable x, where two or more roots $y(x)$ of $f = 0$ are permuted as x turns around a. For instance if C is the curve

$$y^2 = (x - a_1)(x - a_2) \cdots (x - a_{2n})$$

the a_h (assumed all distinct) are the branch points.

Mark the positions a_j of the branch points on S and let α, β be two diametral points on the sphere such that no two branch points are on a great circle through α and β. Draw arcs of great circles αa_j; none containing β. Cut S open along these arcs and let Ω be their complement in S. In Ω each of the roots $y_j(x)$, $j \leq m$, of $f(x, y) = 0$ is uniquely determined by the value $y_j(\beta)$.

Now choose for each k a copy S_k of S corresponding to $y_k(x)$, save that in S_k we only draw the a_j and cuts αa_j which permute $y_k(x)$. The complement Ω_k of the cuts in S_k is a 2-cell, i.e., a simply connected region of the sphere.

To simplify matters suppose that the cut αa_1 permutes y_1 and y_2. Thus this cut will be found in S_1 and S_2. The situation in both spheres is shown in Fig. 2 with proper orientations.

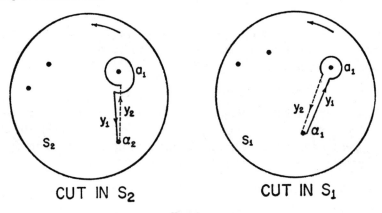

CUT IN S_2 CUT IN S_1

FIG. 2

Now match the two sides labelled y_1 and those labelled y_2 and repeat the process throughout all the spheres S_k. The result is a configuration consisting of m polygons which are all oriented in such a way that where a side belongs to two polygons (never more than two) the side is oppositely oriented relatively to the two polygons. (See Fig. 3.)

This configuration is precisely the Riemann surface $\Phi(C)$ of the curve C.

5. We now state a certain number of the major properties of $\Phi(C)$.

(a) Φ is a smooth surface which is decomposed into m simple polygons (closed

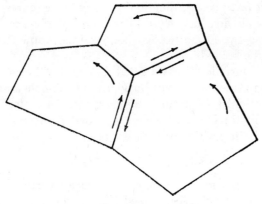

FIG. 3

2-cells). This is technically an orientable and oriented 2-manifold. ("2-manifold" $=2$ dimensional manifold—less precisely a figure as smooth as a plane around each point. It is often called a surface.)

(b) Here I fall back upon well-known elementary properties of topology. If $\alpha_0, \alpha_1, \alpha_2$ designate the number of vertices, arcs, and polygons in a decomposition of a surface like Φ, then the expression

$$\chi(\Phi) = \alpha_0 - \alpha_1 + \alpha_2 = 2 - 2p$$

is the *characteristic* of Φ. The integer p is the *genus* of the surface and is independent of the mode of decomposition of Φ. Its calculation is very simple. Let a place π_j result from the cyclic permutation of q_j roots. Set $N = \sum (q_j - 1)$. (When C has only multiple points with distinct tangents, N is the class of C: number of tangents issued from a general point of the plane.) The calculation of the number χ yields easily this result: If $\beta_0, \beta_1, \beta_2$ are the α_j for a sphere then it is known that

$$\beta_0 - \beta_1 + \beta_2 = 2.$$

Here $\alpha_1 = m\beta_1$, $\alpha_2 = m\beta_2$, $\alpha_0 = m\beta_0 - N$, hence

$$\chi(\Phi) = 2 - 2p = 2m - N.$$

This yields the following formula, due to Riemann:

$$N = 2(p + m - 1).$$

Since N and m are readily obtained from the equation of C, this expression yields the determination of p, the *genus* of C, from the equation of C.

I believe that p had already been found earlier by Plücker and even shown to be birationally invariant (at least under certain simple transformations).

I must underscore that essentially Riemann had constructed his surface as a *topological image* of the "space of places." In other words he had come to realize that if C has, say, a point A of multiplicity k with k distinct tangents (hence A is the center of k *distinct* places), then in $\Phi(C)$ it is to be represented by k *distinct points*. That is, he already had the intuitive concept of *place*.

Of the following two properties Riemann did not know the first but in some manner knew of the second.

(c) Φ is homeomorphic to a two sided disk with p holes.

(d) Let a_h be an oriented circuit on Φ around the hole h and b_h one through the same hole, both initiated from some fixed point A of Φ. Then Φ may be reconstructed as follows: Draw a regular $4p$-sided plane polygon Π with sides labeled successively in positive orientation as

$$a_1, b_1, a_1^{-1}, b_1^{-1}, \cdots, b_p^{-1}$$

(a_h^{-1}, b_h^{-1} are a_h, b_h oppositely oriented). Match all vertices with one point A, then a_h with a_h^{-1}, b_h with b_h^{-1}. The closed 2-manifold thus obtained is Φ.

(e) On the preceding manifold one may operate upon the functions on Φ, locally analytic in terms of the local place variables t, as with complex analytic functions on a complex plane. This means that the basic results of Cauchy extend to the Riemann surface.

RIEMANN'S EQUALITY AND INEQUALITY. These are two results whose importance could not be exaggerated. Let u, v be two integrals of the first kind and define their basic *periods* as

$$\int_{a_\mu} du = \omega_\mu, \quad \int_{b_\mu} du = \omega_{p+\mu}, \quad 1 \leq \mu \leq p,$$

and let ω_μ', $\omega_{p+\mu}'$ be the same for v. Then by integrating $\int u\, dv$ along the polygon Π and applying Cauchy's result for holomorphic functions, we obtain

RIEMANN'S EQUALITY: $\displaystyle\sum_{\mu=1}^{p} \begin{vmatrix} \omega_\mu & \omega_{p+\mu} \\ \omega_\mu' & \omega_{p+\mu}' \end{vmatrix} = 0.$

Let now $u = u' + iu''$ and $\omega_\mu = a_\mu + ib_\mu$, $\mu \leq 2p$. There follows from Cauchy's inequality and again from $\int_\pi u'du'' > 0$

RIEMANN'S INEQUALITY: $\displaystyle\sum_{\mu=1}^{p} \begin{vmatrix} a_\mu & a_{p+\mu} \\ b_\mu & b_{p+\mu} \end{vmatrix} > 0,$

(Jacobi for $p = 1$). By means of these properties plus difficult Dirichlet principle analysis Riemann proved:

THEOREM 1. *There is exactly a maximum p of linearly independent differentials of the first kind and similarly a maximum of $2p$ linearly independent differentials of the second kind modulo $dK(C)$.*

Noteworthy also is:

THEOREM 2. (Part of the Theorem of Riemann-Roch.) *Let $G = n_1\pi_1 + \cdots + n_s\pi_s$ be a set of places, with n_h as the multiplicity of π_h. Set $n = \sum n_h$. Then the* *elements of the function field $K(C)$ with G as set of poles and $n > 2p$ form a linear system of dimension $n - p$.*

Obvious implications: The Riemann surface as topological structure is a birational invariant of the curve C. This holds also for the genus p and for the space of places (homeomorphic to $\Phi(C)$).

I hope that the little I have said about Riemann's contribution will at least make clear its enormous importance for algebraic geometry.

6. If one had asked Riemann or any one of his great predecessors where he would place his contribution he would certainly have declared "in functions of a complex variable." The first, and all important, deviation towards more algebra was made in 1870 by *Max Noether*. His work marks definitely the beginning of a new epoch in algebraic geometry.

The starting point is a famous theorem due to Noether and usually referred to as the $A\phi + B\psi$ *Theorem*. Given two algebraic curves $\phi(x, y) = 0$ and $\psi(x, y) = 0$ intersecting only in isolated points, find n.a.s.c. in order that a polynomial $f(x, y)$ be representable in the form

$$f = A\phi + B\psi; \quad A, B \text{ polynomials.}$$

That is, under what condition is f an element of the ideal (ϕ, ψ)?

This question was completely solved by Noether (around 1870). When ϕ, ψ have no common tangent where they intersect, a sufficient condition is this: if an intersection P of $\phi = 0$ and $\psi = 0$ is of multiplicity p and q for the curves then it is sufficient that $f = 0$ be of multiplicity $\geq p + q - 1$ at P.

Applications were made a couple of years later in a fundamental memoir of Brill and Noether. The basic theorem there is the *Remainder Theorem*.

Recall that a *Cremona transformation* of a projective plane P^2 is the result of a finite succession of quadratic and projective transformations of P^2. This is a general birational transformation of P^2 into itself.

Let me admit this very important result due to Noether: By a suitable Cremona transformation, an irreducible curve C is transformable (implicit: birationally) into a curve still called C, which has only multiple points with distinct tangents. An *adjoint* to C is a curve having at any p-tuple point of C at least multiplicity $p - 1$. The Remainder Theorem asserts this: Let an adjoint of degree n to C intersect it besides the imposed intersections in two sets Q and R. Let two adjoints ϕ, ψ of degree n intersect C respectively in $R + Q'$ and $Q + R'$. Then there is an adjoint ω of degree n intersecting C in $R' + Q'$ (in all cases intersections besides those imposed).

The goal of Brill and Noether was this: First let the linear system $\psi = c_0\phi_0 + \cdots + c_r\phi_r = 0$ (where the ϕ_h are of equal degree and linearly independent modulo f) intersect C in n points, some of which may be fixed. The collection of all such sets of n points is called a *linear series* of *degree n* and *dimension r*; the series is *complete* if not amplifiable with the same n but larger r. General notation g_n^r (usually complete series).

Let the curve C have only ordinary singularities, that is multiple points with distinct tangents.

I. Every complete g_n^r may be cut out by all the adjoints ϕ_μ of a suitable degree μ passing through a certain fixed set of points of C.

II. Let m be the degree of C. The adjoints ϕ_{m-3} of degree $m-3$ are called *canonical*. Their complete series is a unique g_{2p-2}^{p-1} without fixed points: the *canonical series*. The complete set of differentials

$$\left\{ \frac{\phi_{m-3}dx}{f_y'} \right\}$$

is a complete set of differentials of the first kind. This implies the birational invariance of p and of the canonical series (by purely algebraic methods).

III. THEOREM OF RIEMANN-ROCH. *Let g_n^r be complete and let σ be the maximal number of linearly independent canonical curves through an element of g_n^r. Then $n-r=p-\sigma$. Hence if no canonical curve passes through any element of g_n^r then $n-r=p$.* (Strict "Riemann part.")

IV. If the complete g_n^r is generated by the linear system

$$\sum_0^r c_h\psi_h(x) = 0$$

then $\{\psi_h/\psi_0\}$ is a linear base for the complete system of rational functions on C (elements of $K(C)$) whose poles consist of the element of g_n^r cut out by $\psi_0 = 0$. This identifies the "rational function" problem and the complete g_n^r problem.

A constant tool of Brill and Noether was addition and subtraction of complete series $g \pm g'$.

Important observation: The Noether-Brill results are applicable to algebraic curves over any algebraically closed field of characteristic zero. This was in no sense underscored by Noether and Brill but is implicit in their results and methods.

7. The resonance from the Brill-Noether memoir was scarcely felt in Germany, but had very great effect in Italy. For assuredly the work of the very brilliant school of Italian geometers was a direct consequence of the Brill-Noether stimulus. It began in 1882 with Castelnuovo, associated quite soon with Enriques, and greatly enriched around 1900 by the addition of Severi. Briefly speaking, while Brill and Noether studied the effect of the addition of complete linear series on a curve, the Italian group dedicated much effort upon the far more complicated addition of algebraic curves on a surface.

During the same period Émile Picard in France developed entirely alone, and practically without topology, a considerable portion of Riemann's work for algebraic surfaces.

With these few observations I must stop, since beyond it would decidedly take me out of "early development."

Bibliography

The reader desiring fuller information upon the topics discussed in my lecture may profitably consult the following sources (chosen almost at random among the vast literature on the subject):

For the contributions up to and including the work of Riemann:

Émile Picard, Traité d'Analyse, 3rd edition, Vol. 2, Gauthier-Villars, Paris, 1922–28. Chs. 13–18.

Hermann Weyl, Die Idee der Riemannschen Fläche, Teubner, Leipzig, 1913. English translation (3rd ed.) by G. R. McLane, Addison-Wesley, Reading, Mass. 1955.

For the work of Max Noether and Noether-Brill:

Émile Picard and George Simart, Théorie des fonctions algébriques de deux variables, Vol. 2, Gauthier-Villars, Paris, 1906.

For elementary algebraic topology:

See, for example, my monograph: Introduction to Topology, Princeton University Press, 1949.

Editorial Note: See also "The historical development of algebraic geometry" by Jean Dieudonné, reprinted on pp. 248–288 of this volume, and "Historical ramblings in algebraic geometry and related algebra" by S. S. Abhyankar, this MONTHLY, vol. 83 (1976), 409–48.

THE KISS PRECISE*

W. S. BROWN, Bell Telephone Laboratories, Murray Hill

In a poem whose title we have borrowed for this paper, Soddy [1] popularized a theorem relating the curvatures of four mutually tangent spheres. Gosset [2] extended the poem, and the theorem, to the case of n mutually tangent $(n-3)$-dimensional spheres in $(n-2)$-dimensional space. Mauldon [3] proved the extended theorem and its converse as special cases of a more general theorem about sets of equally inclined spheres. Pedoe [4] traced the circle theorem back to Descartes, and gave three proofs of the extended theorem, including an algebraic proof by A. Aeppli.

This paper presents brief, elementary, self-contained proofs of the extended theorem and its converse, and also an interpretation of the basic formula. Although our proof of the extended theorem is similar to Aeppli's, the starting point is more familiar.

THEOREM 1. *Let $S_1, \cdots, S_n$ be $(n-3)$-dimensional spheres in $(n-2)$-dimensional Euclidean space $(n \geqq 3)$, such that each is tangent to all of the others at distinct points. (Our n is the same as the $n+2$ of $[2]-[4]$.) Let $C_1, \cdots, C_n$ be the centers of these spheres; $r_1, \cdots, r_n$ their radii; and $\epsilon_1, \cdots, \epsilon_n$ their curvatures (that is, the reciprocals of the radii). Then*

    (a) *either all of the spheres touch each other externally, or one of them contains the others;*

---

* From AMERICAN MATHEMATICAL MONTHLY, vol. 76 (1969), pp. 661–663.

(b) *the curvatures satisfy the formula*

$$\left(\sum_{i=1}^{n} \epsilon_i \right)^2 = (n-2) \sum_{i=1}^{n} \epsilon_i^2,$$

provided that the radius and curvature of the containing sphere, if any, are taken to be negative.

Proof of (a). If one sphere S contains any other, then it must contain all of the others, since a sphere inside S and a sphere outside S, touching S at distinct points, could not touch each other.

Proof of (b). Let $x_1, \cdots, x_{n-1}$ be vectors from C_n to $C_1, \cdots, C_{n-1}$ respectively. Since the spheres are mutually tangent, we have

$$x_i^2 = (r_i + r_n)^2 \qquad (i = 1, \cdots, n-1),$$
$$(x_i - x_j)^2 = (r_i + r_j)^2 \qquad (i \neq j).$$

On the other hand, since the vectors lie in a space of dimension $n-2$, they are linearly dependent, and therefore

$$\det[x_i \cdot x_j] = 0 \qquad (i, j = 1, \cdots, n-1).$$

From the above tangency conditions, we find

$$(1) \qquad x_i^2 = \frac{(\epsilon_i + \epsilon_n)^2}{\epsilon_i^2 \epsilon_n^2} \qquad (i = 1, \cdots, n-1),$$

$$(2) \qquad x_i \cdot x_j = \frac{(\epsilon_i + \epsilon_n)(\epsilon_j + \epsilon_n) - 2\epsilon_n^2}{\epsilon_i \epsilon_j \epsilon_n^2} \qquad (i \neq j).$$

The determinant can be reduced to

$$(3) \qquad \det[x_i \cdot x_j] = \frac{2^{n-2}}{\epsilon_1^2 \cdots \epsilon_n^2} \left[\left(\sum_{i=1}^{n} \epsilon_i \right)^2 - (n-2) \sum_{i=1}^{n} \epsilon_i^2 \right],$$

and our conclusion follows immediately.

THEOREM 2. *Let $\epsilon_1, \cdots, \epsilon_n$ be nonzero real numbers, all or all but one positive, such that* (b) *is satisfied, and let $S_1, \cdots, S_n$ be $(n-3)$-dimensional spheres in $(n-2)$-dimensional space, with curvatures $\epsilon_1, \cdots, \epsilon_n$ respectively. Then $S_1, \cdots, S_n$ can be located in such a way that each is tangent to all of the others at distinct points.*

LEMMA. *Let K be any subset of k of the integers $1, \cdots, n$ $(k \geqq 2)$. If* (b) *holds, then*

$$\left(\sum_{i \in K} \epsilon_i \right)^2 - (k-2) \sum_{i \in K} \epsilon_i^2 \geqq 0.$$

Proof of the Lemma. For $k = 2$ and $k = n$ the lemma is trivial. Let $E(K)$ denote the left hand side of the inequality, and suppose the lemma is false. Then there is a maximal set K with $E(K) < 0$, and we have $2 < k < n$. Adjoining to K an integer j not in K, we obtain a set K', with $E(K') \geqq 0$ since K is maximal. But

$$(k - 2)E(K') - (k - 1)E(K) = -\left[(k - 2)\epsilon_j - \sum_{i \in K} \epsilon_i \right]^2 \leqq 0,$$

so $E(K') < 0$, which is a contradiction.

Proof of Theorem 2. It is sufficient to construct vectors $x_1, \cdots, x_{n-1}$ satisfying (1) and (2). Let M be the real symmetric matrix of these right hand sides. By (b) the determinant of M is zero, and by the lemma its principal minors are nonnegative. Therefore it is positive semidefinite and has a decomposition ([5], page 256) of the form

$$XX^T = M$$

with X real. Taking the i-th row of X as the vector x_i, for $i = 1, \cdots, n-1$, we are finished.

Interpretation. Let $\bar\epsilon$ and σ be the root mean square and the standard deviation, respectively, of the curvatures $\epsilon_1, \cdots, \epsilon_n$. Then (b) can be rewritten as

$$\sigma = \bar\epsilon\sqrt{2/n}; \qquad n \geqq 3.$$

This formula specifies the degree of nonuniformity among the curvatures, and shows that this nonuniformity decreases with increasing n.

The author thanks Prof. H. S. M. Coxeter for calling his attention to [3].

References

1. Frederick Soddy, The kiss precise, Nature, 137 (1936) 1021.
2. Thorold Gosset, The kiss precise, Nature, 139 (1937) 62.
3. J. G. Mauldon, Sets of equally inclined spheres, Canad. J. Math., 14 (1962) 509–516.
4. Daniel Pedoe, On a theorem in geometry, this MONTHLY, 74 (1967) 627–640.
5. G. Birkhoff and S. MacLane, A Survey of Modern Algebra, 3rd ed., Macmillan, New York, 1965.

Appendix.* The following proof of equation (3) is due to A. J. Goldstein of the Bell Telephone Laboratories.

After forming the determinant from the right hand sides of (1) and (2), we proceed as follows:

    1. For $i, j = 1, \ldots, n - 1$, multiply row (i) by $\varepsilon_i \varepsilon_n$ and col(j) by $\varepsilon_j \varepsilon_n$.

---

* Added for the present volume.—ED.

2. Border the determinant with

$$\text{row}(n) = (0, \ldots, 0, 1, 0)$$
$$\text{row}(n+1) = (0, \ldots, 0, 0, 1)$$
$$\text{col}(n) = (\varepsilon_1 + \varepsilon_n, \ldots, \varepsilon_{n-1} + \varepsilon_n, 1, 0)$$
$$\text{col}(n+1) = (1, \ldots, 1, 0, 1).$$

3. For $j = 1, \ldots, n-1$, replace $\text{col}(j)$ by $\text{col}(j) - (\varepsilon_j + \varepsilon_n)\text{col}(n) + 2\varepsilon_n^2\text{col}(n+1)$.

4. Replace $\text{row}(n)$ by $\text{row}(n) + \dfrac{1}{2\varepsilon_n}\,\text{row}(n+1)$.

5. Replace $\text{col}(n)$ by $\text{col}(n) - \varepsilon_n\text{col}(n+1)$.

6. Replace $\text{row}(n+1)$ by $\text{row}(n+1) - \sum_{i=1}^{n-1} \text{row}(i)$.

7. Replace $\text{row}(n)$ by $\text{row}(n) + \dfrac{1}{2\varepsilon_n^2} \sum_{i=1}^{n-1} \varepsilon_i\text{row}(i)$.

8. For $j = 1, \ldots, n-1$, divide $\text{col}(j)$ by $2\varepsilon_n^2$.

9. Multiply $\text{row}(n)$ by $2\varepsilon_n^2$.

10. Evaluate the lower right 2×2 minor.

A PROBLEM IN CARTOGRAPHY*

JOHN MILNOR, Massachusetts Institute of Technology†

1. Introduction. The central problem of mathematical cartography is the problem of representing a portion of the curved surface of the earth on a flat piece of paper without introducing any more distortion than is absolutely necessary. This note will propose a quantitative definition for the term "distortion," and then study the mathematical problem of choosing a method of mapping which minimizes distortion.

To simplify the problem we first replace the rather irregular surface of the earth by a perfect sphere.

DEFINITIONS. Let S be the sphere of radius r consisting of all points x in the 3-dimensional euclidean space with distance r from the origin, and let U be any nondegenerate subset of S. (By "nondegenerate" we mean that U must contain at least two distinct points.)

A map projection f on the domain U will mean a function which assigns to each point x of U some point $f(x)$ of the euclidean plane E.

Let $d_S(x, y)$ denote the geodesic distance between two points x and y of the sphere S. By definition, this is equal to the length of the shorter great circle arc

---

* From AMERICAN MATHEMATICAL MONTHLY, vol. 76 (1969), pp. 1101–1112.

† Prof. Milnor is now at the Institute for Advanced Study.

joining x to y. The euclidean distance between two points a and b of the plane E will be denoted analogously by $d_E(a, b)$.

The scale of a map projection f with respect to a pair of distinct points x and y in the domain U is defined to be the ratio

$$d_E(f(x), f(y))/d_S(x, y).$$

Ideally we would like this scale to be the same for all pairs of points x and y in U, but this is not usually possible. So we must introduce the *minimum scale* σ_1, defined to be the infimum of the ratio $d_E(f(x), f(y))/d_S(x, y)$ as x and y vary over all pairs of distinct points in U, and the *maximum scale σ_2*, defined to be the supremum of the ratio $d_E(f(x), f(y))/d_S(x, y)$. In other words σ_1 and σ_2 are the "best" possible constants such that the inequality

$$\sigma_1 d_S(x, y) \leqq d_E(f(x), f(y)) \leqq \sigma_2 d_S(x, y)$$

holds for all points x and y in U.

To measure the extent to which scale fails to be constant we propose the following:

DEFINITION. *The distortion of the map projection f is the natural logarithm*

$$\delta = \log(\sigma_2/\sigma_1)$$

of the ratio of maximum scale to minimum scale.

Thus $0 \leqq \delta \leqq \infty$, where δ is finite if and only if both σ_1 and σ_2 are positive and finite numbers. If δ is finite, notice that the function f is continuous and one-to-one.

We would like to find a map projection f with no distortion at all $(\delta = 0)$. Since this is not possible except in a few special and uninteresting cases (e.g., the case of a domain U consisting of only three points), the best we can actually do is to try to find a map projection for which δ is as small as possible.

DEFINITION. *A minimum distortion map projection f_0 on U will mean a map projection whose distortion δ_0 is less than or equal to the distortion of every other map projection on U.*

PRELIMINARY THEOREM. *For every nondegenerate set of points U on the sphere there exists a minimum distortion map projection f_0 with domain U.*

The proof of this theorem, which is quite elementary, will be deferred until Appendix A.

Unfortunately the proof will fail to suggest answers to a number of relevant questions: Is this minimum distortion map f_0 unique in some sense? Is f_0 differentiable (assuming that U is a nice enough set so that differentiability makes sense)? How can one actually construct f_0, or even a reasonable approximation to f_0? How can one estimate the minimum possible distortion δ_0 associated with a given set U?

This note will succeed in answering these questions only in one very special case, namely, the case of the region bounded by a circle on S.

Given a fixed point x_0 of S, let D_α denote the closed *disk* of geodesic radius $r\alpha$, consisting of all points x in S for which $d_S(x, x_0) \leqq r\alpha$. Here α can be any number in the interval $0 < \alpha < \pi$.

MAIN THEOREM. *There is one and, up to similarity transformations of the plane, only one minimum distortion map projection f_0 on the domain D_α. This map projection is infinitely differentiable, and has distortion δ_0 equal to $\log(\alpha/\sin \alpha)$.*

This minimum distortion projection f_0, known to cartographers as the "azimuthal equidistant projection," can be characterized by the fact that it preserves both distances and directions from the central point x_0. The explicit formula $\delta_0 = \log(\alpha/\sin \alpha)$ shows that the distortion δ_0 is small for small values of α, being asymptotically equal to

$$\alpha^2/6 \sim \tfrac{2}{3} \text{ area } D_\alpha/\text{area } S$$

as $\alpha \to 0$. However δ_0 tends to infinity as $\alpha \to \pi$.

This theorem will be proved in Section 2. The problem of estimating the δ_0 associated with a more general domain U is discussed in Section 3. There are two appendices, one proving that minimum distortion map projections exist, and a second discussing a corresponding problem for conformal map projections, following Chebyshef.

2. The azimuthal equidistant projection. Again let D_α denote a spherical disk of geodesic radius $r\alpha$ centered at x_0.

LEMMA 1. *The distortion δ for any map projection f with domain D_α satisfies $\delta \geqq \log(\alpha/\sin \alpha)$.*

Proof. We may assume that f has finite distortion. Hence the "Lipschitz inequality"

$$(1) \qquad\qquad d_E(f(x), f(y)) \leqq \sigma_2 d_S(x, y)$$

is satisfied, where σ_2 is a finite constant, and it follows that f is continuous. Furthermore f is one-to-one.

Let C_α denote the boundary of the disk D_α. Clearly the image $f(C_\alpha)$ is a simple closed curve in the plane. We shall first prove:

ASSERTION A. *Every half-line emanating from the point $f(x_0)$ in the plane must intersect the simple closed curve $f(C_\alpha)$ at least once.*

Proof. The Jordan Curve Theorem asserts that the simple closed curve $f(C_\alpha)$ cuts the plane into two components

$$E - f(C_\alpha) = E_1 \cup E_2,$$

one of these components, say E_1, being bounded, and the second unbounded. But the bounded component E_1 is just the image, under the continuous one-to-one function f, of the interior of the disk D_α. This is proved, for example, in Newman [10, Theorem 12.2, p. 121]. In particular it follows that the point $f(x_0)$ must belong to the bounded component E_1. Hence every half-line emanating from x_0 must cross $f(C_\alpha)$, since otherwise it would lie completely within the bounded set E_1 which is impossible. This proves Assertion A.

Since the curve C_α on S has finite length $2\pi r \sin \alpha$, it follows easily from the Lipschitz inequality (1) that $f(C_\alpha)$ also has finite length L, where

$$(2) \qquad\qquad L \leq 2\pi\sigma_2 r \sin \alpha.$$

(The *length* of a not necessarily smooth curve is defined for example in [6, p. 36].)
    Now let us make use of the inequality

$$(3) \qquad\qquad d_E(f(x), f(y)) \geq \sigma_1 d_S(x, y).$$

Since every point of C_α has geodesic distance exactly $r\alpha$ from x_0 it follows that every point of $f(C_\alpha)$ has euclidean distance $\geq \sigma_1 r\alpha$ from $f(x_0)$.

Thus $f(C_\alpha)$ is a simple closed curve of finite length L which lies outside an open disk D^* of radius $\sigma_1 r\alpha$ in the plane, and cuts every half-line through the center of this disk.

ASSERTION B. *This implies that $L \geq 2\pi\sigma_1 r\alpha$, where equality holds if and only if $f(C_\alpha)$ is precisely equal to the boundary of D^*.*

Proof. Cut $f(C_\alpha)$ by a straight line through the center of D^* and choose intersection points, say a and b, which lie on opposite sides of D^*. Let A be either one of the two arcs of $f(C_\alpha)$ from a to b. Introducing polar coordinates ρ and θ about the center of D^*, first assume that the arc A can be described, in terms of a parameter t, by piecewise smooth functions

$$\rho = \rho(t), \qquad \theta = \theta(t).$$

Then

$$\text{length } A = \int (\dot{\rho}^2 + \rho^2\dot{\theta}^2)^{1/2} dt \geq \int \rho |\dot{\theta}| \, dt,$$

where the dot denotes differentiation. Since

$$\rho \geq \sigma_1 r\alpha \quad \text{and} \quad \int |\dot{\theta}| \, dt \geq \left| \int \dot{\theta} dt \right| \geq \pi,$$

this proves that length $A \geq \pi\sigma_1 r\alpha$, and therefore $L \geq 2\pi\sigma_1 r\alpha$, as required.

If A is not piecewise smooth, then an extra step is needed. For each $\epsilon > 0$ it is possible to approximate A by a polygonal path A_ϵ' from a to b which lies outside the disk of radius $\sigma_1 r\alpha - \epsilon$ and satisfies

$$\text{length } A \geq \text{length } A_\epsilon' \geq \pi(\sigma_1 r\alpha - \epsilon).$$

Letting $\epsilon \rightarrow 0$, we obtain length $A \geq \pi\sigma_1 r\alpha$, as before.

Now suppose that the length of A is precisely equal to $\pi\sigma_1 r\alpha$. Then any portion of A which has distance greater than $\sigma_1 r\alpha$ from the center of D^* must be a straight line segment. Otherwise, replacing some small portion of A by a straight line segment we could decrease its length, which is impossible.

Any maximal line segment A_0 which forms a part of A must lead from one of the end points a or b of A to a point of the circle bounding D^*. The only other possibility would be that both end points of A_0 lie on the circle, which is impossible. Thus A consists of a line segment (possibly degenerate) from a to the circle, followed by a circle arc, followed by a line segment to b. Elementary geometry now shows that the minimal length $\pi\sigma_1 r\alpha$ is achieved only if A is the semicircle. Hence L can equal $2\pi\sigma_1 r\alpha$ only if $f(C_\alpha)$ is the full circle. This completes the proof of Assertion B.

Combining Assertion B with the inequality (2) we obtain

$$2\pi\sigma_1 r\alpha \leqq 2\pi\sigma_2 r \sin \alpha$$

or

$$\alpha/\sin \alpha \leqq \sigma_2/\sigma_1$$

and hence $\log(\alpha/\sin \alpha) \leqq \delta$, which completes the proof of Lemma 1.

LEMMA 2. *If the distortion of f is precisely equal to $\log(\alpha/\sin \alpha)$, then f is an azimuthal equidistant projection.*

By definition this means that f carries each great circle passing through x_0 into a straight line in the plane, the angle between two great circles being equal to the angle between the corresponding straight lines, and that f carries each circle C centered at x_0 to a circle $f(C)$ centered at $f(x_0)$, the radius of $f(C)$ being proportional to the geodesic radius of C.

To differential geometers, this means that f is the inverse of the so called exponential map. It follows that f is infinitely differentiable, even at x_0. See for example [9, p. 147].

Proof of Lemma 2. If $\delta = \log(\alpha/\sin \alpha)$, then according to Assertion B the image $f(C_\alpha)$ must be precisely equal to the circle of radius

$$\sigma_1 r\alpha = \sigma_2 r \sin \alpha$$

centered at $f(x_0)$. Hence the image $f(D_\alpha)$ must be precisely the closed disk bounded by this circle. (Compare the proof of Assertion A.)

Now consider an arbitrary point x of D_α. Construct a great circle segment from x_0 through x to a point $\bar{x}$ on the boundary C_α of D_α. If c denotes the geodesic distance $d_S(x_0, x)$, note that x has geodesic distance precisely $r\alpha - c$ from $\bar{x}$, and geodesic distance strictly greater than $r\alpha - c$ from every other point of C_α. Hence, using inequality (3), the image $f(x)$ must

(a) have distance at least $\sigma_1 c$ from $f(x_0)$,
(b) have distance at least $\sigma_1(r\alpha - c)$ from $f(\bar{x})$, and
(c) have distance greater than $\sigma_1(r\alpha - c)$ from every other point of $f(C_\alpha)$.

Clearly there is one and only one point in the disk $f(D_\alpha)$ which satisfies these three conditions: namely, the point which lies at distance $\sigma_1 c$ along the line segment from $f(x_0)$ to $f(\bar{x})$. Thus the map projection f on D_α is completely determined by what is does to boundary points of D_α.

To complete the proof of Lemma 2 we need only verify that f carries the circle C_α to the circle $f(C_\alpha)$ by a similarity transformation which multiplies all lengths by the constant factor σ_2. Suppose that we cut C_α into two arcs A and A', so that

$$\text{length } A + \text{length } A' = \text{length } C_\alpha = 2\pi r \sin \alpha.$$

The Lipschitz inequality (1) implies that

(4) $\text{length } f(A) \leqq \sigma_2 \text{ length } A, \qquad \text{length } f(A') \leqq \sigma_2 \text{ length } A'.$

But

$$\text{length } f(A) + \text{length } f(A') = \text{length } f(C_\alpha)$$

is precisely equal to σ_2 times the length $2\pi r \sin \alpha$ of C_α. So both of the inequalities (4) must actually be equalities. This proves Lemma 2.

Now we must prove the converse.

LEMMA 3. *The azimuthal equidistant projection on the disk D_α has distortion δ precisely equal to $\log(\alpha/\sin \alpha)$.*

Proof. Centering D_α at the north pole, we will use the longitude $0 \leqq \theta \leqq 2\pi$ and the colatitude $0 \leqq \gamma \leqq \alpha$ as coordinates. Suppose that f maps the point with colatitude γ and longitude θ to the point with cartesian coordinates $(r\gamma \cos \theta, r\gamma \sin \theta)$ in the plane. The length of any smooth curve $\gamma = \gamma(t)$, $\theta = \theta(t)$ in D_α is given by the integral

$$L = r \int (\dot{\gamma}^2 + \dot{\theta}^2 \sin^2 \gamma)^{1/2} dt,$$

and the length of the corresponding curve in $f(D_\alpha)$ is

$$L' = r \int (\dot{\gamma}^2 + \dot{\theta}^2 \gamma^2)^{1/2} dt.$$

But, since $\gamma/\sin \gamma$ is a monotone increasing function of γ, we have

$$\sin \gamma \leqq \gamma \leqq (\alpha/\sin \alpha) \sin \gamma,$$

from which it follows easily that

(5) $L \leqq L' \leqq (\alpha/\sin \alpha) L.$

Starting from this inequality (5) we will prove that

$$d_S(x, y) \leqq d_E(f(x), f(y)) \leqq (\alpha/\sin \alpha) d_S(x, y)$$

for every x and y in D_α. Clearly this will imply that $\delta \leqq \log(\alpha/\sin \alpha)$ and hence, by Lemma 1, that $\delta = \log(\alpha/\sin \alpha)$.

Proof that $d_S(x, y) \leq d_E(f(x), f(y))$. Join $f(x)$ to $f(y)$ within the convex set $f(D_\alpha)$ by a line segment of length L' precisely equal to $d_E(f(x), f(y))$. The corresponding curve in D_α will have length $L \geq d_S(x, y)$. Since $L \leq L'$, we obtain $d_S(x, y) \leq d_E(f(x), f(y))$, as required.

Proof that $d_E(f(x), f(y)) \leq (\alpha/\sin \alpha)d_S(x, y)$. First suppose that $\alpha \leq \pi/2$, so that the disk D_α is "geodesically convex." Then the proof is quite analogous. Join x to y, *within* D_α, by a great circle segment A of length $L = d_S(x, y)$. Then $f(A)$ has length $L' \geq d_E(f(x), f(y))$, so the inequality $L' \leq (\alpha/\sin \alpha)L$ implies that $d_E(f(x), f(y)) \leq (\alpha/\sin \alpha)d_S(x, y)$, as required.

If $\alpha > \pi/2$, so that the disk D_α is not geodesically convex, then a more complicated argument is necessary. Suppose that the shortest great circle arc from x to y does not lie completely within D_α, but rather crosses out of D_α at a boundary point $\bar{x}$, and then crosses back in at another boundary point $\bar{y}$. We shall show that

$$(6) \qquad\qquad d_E(f(x), f(\bar{x})) \leq (\alpha/\sin \alpha)d_S(x, \bar{x}),$$

$$(7) \qquad\qquad d_E(f(\bar{x}), f(\bar{y})) \leq (\alpha/\sin \alpha)d_S(\bar{x}, \bar{y}),$$

$$(8) \qquad\qquad d_E(f(\bar{y}), f(y)) \leq (\alpha/\sin \alpha)d_S(\bar{y}, y).$$

Adding these three inequalities, we shall clearly obtain the required inequality.

But (6) and (8) can be proved by the argument above. To prove (7) we introduce an auxiliary azimuthal equidistant projection g whose domain is the complementary disk $D'_{\pi-\alpha}$ centered at the south pole. Since $\pi - \alpha \leq \pi/2$ we have

$$d_E(g(\bar{x}), g(\bar{y})) \leq ((\pi - \alpha)/\sin(\pi - \alpha))d_S(\bar{x}, \bar{y}).$$

Multiplying this by $\alpha/(\pi - \alpha)$ we obtain the required inequality (7). This completes the proof of Lemma 3.

Clearly Lemmas 1, 2, and 3 imply the "Main Theorem" of Section 1.

3. Discussion. How can one estimate the minimum possible distortion δ_0 for map projections on a given set U? Here is a crude estimate. Define the *angular width* w of a set U as follows. Choose a smallest possible "lune" (figure bounded by two great semicircles) containing U, and let w be the angle at the vertex of this lune.

ASSERTION. *Any set with angular width* $w < \pi$ *possesses a map projection with distortion* $\delta \leq \log \sec(w/2)$.

This is proved by rotating so that the lune is centered on the equator, and then using the latitude and longitude of x as the cartesian coordinates of $f(x)$. The computations are similar to those in the proof of Lemma 3.

It is conjectured that this estimate gives the right order of magnitude in the case of a small geodesically convex region, in the sense that δ_0 is greater than say one sixth of $\log \sec(w/2)$. But $\log \sec(w/2)$ is not a really good estimate for δ_0, except perhaps in the case of a long narrow region.

It would be more interesting to find a relation between δ_0 and area.

PROBLEM. *Among all geodesically convex regions of given area, does the disk D_α require the largest distortion?*

In other words, if area(U) = area(D_α) does it follow that U has a map projection with distortion $\delta \leqq \log(\alpha/\sin \alpha)$? If true this would imply the existence of map projection with smaller distortion than any which are actually known for many regions on the sphere. A test case which would be particularly interesting would be that of a small "rectangular" region on the sphere.

Slightly cruder is the following possible estimate.

PROBLEM. *Does every geodesically convex region U possess a map projection with distortion less than the normalized area,*

$$\delta < \text{area } U/\text{area } S?$$

As an example, for the continental United States with about 1.5 percent of the earth's area, does there exist a map projection with scale errors of no more than 1.5 percent (or perhaps $1.5 + \epsilon$ to allow for the lack of geodesic convexity)? All standard map projections for the continental United States seem to have scale errors of at least 2.2 percent.

Appendix A. Minimum distortion projections always exist. We shall first prove the following. Let U be a subset of the sphere S and let $\overline{U}$ denote the topological closure of U.

LEMMA 4. *Any map projection f on U with distortion $\delta < \infty$ extends uniquely to a map projection $\overline{f}$ on $\overline{U}$ having the same distortion δ.*

Proof. The inequalities

$$\sigma_1 d_S(x, y) \leqq d_E(f(x), f(y)) \leqq \sigma_2 d_S(x, y)$$

show that f is uniformly continuous, and hence extends uniquely to a continuous function $\overline{f}$ on $\overline{U}$. (See [3, p. 55].) Clearly $\overline{f}$ will also satisfy these inequalities.

Now, given some fixed set U, consider all possible map projections f with domain U, and let δ_0 denote the infimum of the corresponding distortions $\delta(f)$. We must construct a map projection f_0 whose distortion is precisely equal to δ_0. We may assume that $\delta_0 < \infty$, since otherwise there is nothing to prove.

REMARK. Note that there exists a map projection with finite distortion on U if and only if the closure $\overline{U}$ is not the entire sphere. For if U is not everywhere dense on S then U is contained in some disk $D_{\pi-\epsilon}$ and hence possesses a map projection with distortion $\delta \leqq \log((\pi-\epsilon)/\sin(\pi-\epsilon)) < \infty$. But if $\overline{U} = S$ then a map projection with finite distortion on U would extend to a map projection with finite distortion on S, which is impossible since $S \supset D_\alpha$ for all α, or since S is not homeomorphic to any subset of E. (See for example [10, p. 122].)

Choose a sequence of map projections $\{f_1, f_2, f_3, \cdots\}$ on U so that the corresponding sequence $\{\delta_1, \delta_2, \delta_3, \cdots\}$ of distortions tends to the limit δ_0. We may assume that each f_i has been chosen so as to have maximum scale equal to 1, and so that the image $f_i(U)$ contains the origin.

Choose a countable dense subset

$$U' = \{x_1, x_2, x_3, \cdots\}$$

of U. Since the points $f_1(x_1), f_2(x_1), \cdots$ all have distance $\leq \pi r$ from the origin, we can choose a convergent subsequence. That is there exists an infinite set I_1 of positive integers so that the sequence of points $f_i(x_1)$, where i tends to infinity through the set I_1, converges to some limit a_1 in E. Similarly we can find an infinite set $I_2 \subset I_1$ so that the limit

$$\lim\{f_i(x_2) \mid i \to \infty, i \in I_2\}$$

exists. Call this limit a_2. Continuing inductively we can define a function f from U' to the plane by $f(x_j) = a_j = \lim\{f_i(x_j) \mid i \to \infty, i \in I_j\}$. Since the inequalities

$$e^{-\delta_i}d_S(x, y) \leq d_E(f_i(x), f_i(y)) \leq d_S(x, y)$$

hold for all i, it follows, taking the limit as i tends to infinity through an appropriate I_j, that

$$e^{-\delta_0}d_S(x, y) \leq d_E(f(x), f(y)) \leq d_S(x, y)$$

for all x and y in U'. Thus f is a map projection on U' with distortion δ_0.

Now applying Lemma 4 we obtain the required map projection on U with distortion δ_0.

Appendix B. Conformal map projections. Recall that a map projection f, defined on an open set U, is called *conformal* (cartographers prefer the term "orthomorphic") if it is differentiable and preserves angles. (That is, f transforms any pair of curves in U, whose tangent vectors at a point of intersection span the angle α into a pair of curves in E, whose tangent vectors at the corresponding intersection point span the same angle α.)

It follows that f has a well defined *infinitesimal-scale* $\sigma(x)$ at each point x of U. By definition $\sigma(x)$ is the limit of the ratio $d_E(f(x), f(y))/d_S(x, y)$ as y tends to the limit x. (Compare [1, p. 74].)

We shall make use of the *Laplace-Beltrami operator* Δ, a second order partial differential operator which assigns to each twice differentiable real valued function g on a Riemannian manifold a new real valued function Δg. In euclidean space this is the familiar Laplace operator. We shall use Δ only on the sphere S of radius r. Using latitude λ and longitude θ as coordinates, the operator Δ on the sphere takes the form

$$r^2 \Delta g = g_{\lambda\lambda} - g_\lambda \tan \lambda + g_{\theta\theta} \sec^2 \lambda.$$

(Compare [14, p. 160]. The subscripts denote partial derivatives.)

Suppose now that U is a simply connected open subset of the sphere S.

LEMMA 5. *The infinitesimal-scale function $\sigma(x)$ associated with a conformal map projection f on U determines f up to an (orientation preserving or reversing) rigid motion of the plane. A given positive real valued function σ on U is the infinitesimal-scale function associated with some conformal f if and only if σ is twice differentiable and satisfies the differential equation $r^2 \Delta \log \sigma = 1$.*

As an example, the function $\sigma(x) = \sec(\text{latitude } x)$ provides a solution to this equation $r^2 \Delta \log \sigma = 1$, except at the north and south poles. The corresponding f turns out to be the familiar Mercator projection.

(Note that our differential equation cannot have any solution which is defined and smooth throughout the entire sphere, since the condition $\Delta \log \sigma > 0$ implies easily that σ cannot have any local maximum.)

Proof of Lemma 5. More generally, consider a smooth surface M provided with a Riemannian metric, expressed in terms of local coordinates u and v as $ds^2 = E du^2 + 2F du dv + G dv^2$. Let Δ denote the associated Laplace-Beltrami operator, and let K denote the Gaussian curvature of M. Consider a second Riemannian metric of the form $\sigma^2 ds^2$ on M, where σ is a positive twice differentiable function. Computation (using for example [14, pp. 113, 160]) shows that the Gaussian curvature K' associated with this new Riemannian metric is given by the formula $K' = (K - \Delta \log \sigma)/\sigma^2$.

If σ is the infinitesimal-scale function associated with a conformal mapping f from M to M', then clearly $K'(x)$ is just the Gaussian curvature of M' at $f(x)$. Thus if M' is the euclidean plane, with $K' \equiv 0$, we see that the differential equation

$$\Delta \log \sigma = K$$

must be satisfied. In particular, taking M to be the open subset U of S, with $K \equiv 1/r^2$, we obtain the required equation

$$r^2 \Delta \log \sigma = 1.$$

Conversely, given any solution σ to the differential equation $\Delta \log \sigma = K$, the Riemannian metric $\sigma^2 ds^2$ has curvature K' identically zero. Hence any sufficiently small connected open subset of M, with the metric $\sigma^2 ds^2$, can be mapped isometrically onto an open subset of the plane ([14, p. 145]). This isometry is unique up to rigid motions of the plane, since any isometry ϕ from one connected open subset of the plane to another extends to an isometry of the entire plane. (Assuming that ϕ preserves orientation, we can think of ϕ as a complex analytic function [1, p. 74] with $|d\phi/dz| \equiv 1$. Hence $d\phi/dz$ is constant and $\phi(z) = cz + c'$ with $|c| = 1$.)

Now if M is simply connected then a monodromy argument shows that these local isometries can be chosen so as to fit together to yield a smooth mapping f from all of M to E.

(Compare [8, p. 1297]. The "Monodromy Theorem" says that if we are given connected open sets U_α covering a simply connected manifold M, and for each

U_α a collection F_α of functions from U_α to Y satisfying the following condition, then there exists a function from M to Y whose restriction to each U_α belongs to F_α. The condition is that for each f_α in F_α and each x in $U_\alpha \cap U_\beta$ there should exist one and only one f_β in F_β which coincides with f_α throughout some neighborhood of x. Compare [1, p. 285], [12, p. 67].)

In the large, this mapping f from M to E may not be one-to-one, but locally it carries M, with the metric $\sigma^2 ds^2$, isometrically to E. Hence it carries M with the original metric ds^2 conformally to E, the infinitesimal-scale function of f being precisely equal to σ. This completes the proof of Lemma 5.

Chebyshef [2] studied conformal map projections, using the ratio sup $\sigma(x)/$ inf $\sigma(x)$ of maximum infinitesimal-scale to minimum infinitesimal-scale as a measure of distortion.

REMARK. If the domain U is geodesically convex, note that the maximum infinitesimal-scale sup $\sigma(x)$ is equal to the maximum scale σ_2 of Section 1. (Compare the proof of Lemma 3.) Similarly, if f is one-to-one and $f(U)$ is convex, then inf $\sigma(x) = \sigma_1$.

CHEBYSHEF THEOREM. *If U is a simply connected region bounded by a twice differentiable curve, then there exists one and, up to a similarity transformation of E, only one conformal map projection which minimizes this ratio* sup $\sigma/$inf σ. *This "best possible" conformal map projection is characterized by the property that its infinitesimal-scale function $\sigma(x)$ is constant along the boundary of U.*

This result has been available for more than a hundred years, but to my knowledge it has never been used by actual map makers.

Proof. Setting $g(x) = \log \sigma(x)$, first note that the differential equation $r^2 \Delta g = 1$ has a unique solution satisfying the boundary condition $g(x) = 0$ for $x \in bd(U)$. See for example [5, p. 288]. If h is any other function which is twice differentiable and satisfies the equation $r^2 \Delta h = 1$ throughout the interior of U, then we shall show that

(9)   sup $h -$ inf $h \geqq$ sup $g -$ inf g,

where equality holds if and only if

$$h = g + \text{constant.}$$

(Note that sup $g -$ inf g is just the logarithm of the ratio sup $\sigma/$inf σ which we want to minimize.) Clearly this will complete the proof.

Since $\Delta g > 0$, an easy argument shows that the function g cannot attain its maximum at any interior point of U. Since g must achieve a maximum at some point of the compact set $\overline{U}$, it follows that the maximum must be attained on $bd(U)$. Thus sup $g(x) = 0$.

The difference $h - g$ satisfies the homogeneous equation $\Delta(h - g) = 0$, and so cannot achieve its maximum at an interior point of U unless $h - g = $ constant. (See [5, p. 232].) Hence any sequence of points $x_1, x_2, \cdots$ for which

$$\lim_{i \to \infty} (h(x_i) - g(x_i)) = \sup(h - g)$$

must be a sequence tending to the boundary of U, unless $h - g =$ constant.

Setting $c = \sup h$ (we may assume that c is finite since otherwise (9) would trivially be satisfied), we have

$$g(x_i) \to 0, \qquad h(x_i) \leqq c,$$

hence

$$\sup(h - g) = \lim(h(x_i) - g(x_i)) \leqq c,$$

or in other words

$$h(x) \leqq g(x) + c$$

for all x. Therefore $\inf h \leqq \inf g + c$, which proves (9).

If equality holds, then at the interior point x_0 of U where g achieves its minimum we have

$$h(x_0) = g(x_0) + c.$$

Thus $h - g$ achieves its maximum c at an interior point, and hence is constant. This completes the proof.

REMARK. The "best possible" conformal map projection f, although locally well behaved, may not be one-to-one in the large. However, if U is geodesically convex, then it can be shown that f is one-to-one and that $f(U)$ is also convex.

References

1. L. V. Ahlfors, Complex Analysis, 2nd ed., McGraw-Hill, New York, 1966.

2. P. L. Chebyshef (or Tchebychef), Sur la construction des cartes géographiques, Oeuvres I, Chelsea, New York, 1962, pp. 233–236 and 239–247.

3. J. Dieudonné, Foundations of Modern Analysis, Academic Press, New York, 1960.

4. L. Driencourt and J. Laborde, Traité des projections des cartes géographiques, 4 volumes, Hermann, Paris, 1932.

5. P. Garabedian, Partial Differential Equations, Wiley, New York, 1964.

6. E. Hille, Analytic Function Theory, vol. 1, Ginn, New York, 1959.

7. A. R. Hinks, Map Projections, Cambridge Univ. Press, New York, 1921.

8. T. Klotz, A complete R_Λ-harmonically immersed surface in E^3 on which $H \neq 0$, Proc. Amer. Math. Soc., 19 (1968) 1296–1298.

9. S. Kobayashi and K. Nomizu, Foundations of Differential Geometry, Interscience, New York, 1963.

10. M. H. A. Newman, Elements of the Topology of Plane Sets of Points, 2nd ed., Cambridge University Press, New York, 1951.

11. A. H. Robinson, Elements of Cartography, Wiley, New York, 1960.

12. N. E. Steenrod, The Topology of Fibre Bundles, Princeton University Press, Princeton, N. J., 1951.

13. J. A. Steers, An Introduction to the Study of Map Projections, University of London Press, London, 1957.

14. D. J. Struik, Lectures on Classical Differential Geometry, Addison-Wesley, Reading, Mass., 1950.

CRITICAL POINTS AND CURVATURE FOR EMBEDDED POLYHEDRAL SURFACES*

T. F. BANCHOFF, Brown University

The Gauss-Bonnet theorem for a surface M^2 in Euclidean 3-space E^3 and the Critical Point Theorem for height functions on an embedded surface are two of the earliest and most important theorems of "geometry in the large." Both theorems relate geometric properties of the embedded surface (the total curvature of the surface or the sum of a set of geometrically defined indices of singularity) to a topological property of the surface, the Euler-Poincaré characteristic $\chi(M^2)$. Both theorems are very geometric in character despite the fact that the standard definitions of total curvature and index of singularity appear to involve the use of differential calculus and the hypothesis that the surfaces are smooth. In fact both theorems have analogues for polyhedral surfaces embedded in E^3, and the proofs in the polyhedral case are entirely elementary.

Some of the most interesting results in global geometry have exploited the connection between total curvature and critical point theory, as in the work of Kuiper [4]. In this paper we shall follow this same procedure to prove the

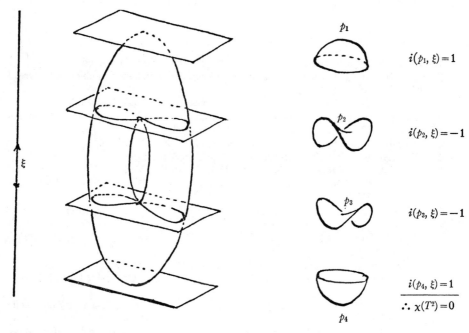

FIG. 1

---

* From AMERICAN MATHEMATICAL MONTHLY, vol. 77 (1970), pp. 475–485.

critical point theorem and use it to prove the Gauss-Bonnet theorem. In the polyhedral case, a new feature is an interpretation of the *Theorema Egregium* of Gauss which relates the extrinsic curvature to the intrinsic curvature on a surface.

All of the theorems in this paper have appeared in a generalized (and technical) form in the author's paper [2].

1. The Critical Point Theorem. Consider a closed smooth surface M^2 embedded in E^3 and consider a linear function $\xi^\dagger$ on E^3 given by projecting all of E^3 to the line determined by a unit vector ξ. A point p of M^2 is said to be a *critical point* for ξ if the tangent plane to M^2 at p is perpendicular to ξ; all other points of M^2 are called *ordinary points* for ξ. In the "standard" example of a height function on a torus of revolution held vertically there are just four critical points, a maximum, a minimum, and two (nondegenerate) saddle points.

The *Critical Point Theorem* for height functions states that if ξ has a finite number of critical points on M^2 and all are of the three types described above, then (number of local maxima) + (number of local minima) − (number of saddle points) = $\chi(M^2)$, where $\chi(M^2)$ is the Euler-Poincaré characteristic of M^2.

We express this theorem more succinctly by indexing each critical point by $i(p,\xi) = 1$ if p is a local maximum or minimum and $i(p,\xi) = -1$ if p is a (non-degenerate) saddle point. The theorem then states

$$\sum_{p \text{ critical for } \xi} i(p,\xi) = \chi(M^2).$$

In classical critical point theory (= Morse Theory) the index is given by considering the sign of the determinant of the matrix of second derivatives, as in [5], but since we are interested in developing the polyhedral analogue of the theorem, we proceed to give a more geometric presentation of this indexing procedure.

If a point q is ordinary for the height function ξ, then the tangent plane at q is not perpendicular to ξ. Thus the tangent plane divides a "small disc neighborhood" U of q on M^2 into exactly two pieces and it meets a "small circle" about q in precisely two points. This distinguishes an ordinary point from a local maximum or minimum (where a "small circle" about the critical point will not meet the tangent plane at all) and from a nondegenerate saddle point p (where the plane at p perpendicular to ξ meets a "small circle" about p on M^2 in four distinct points).

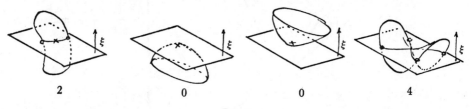

2   0   0   4

FIG. 2

---

† Called the height function for ξ.

We may then give an arithmetic definition of the *index* as follows:

$i(p, \xi) = 1 - \frac{1}{2}$ (number of points in which the plane through p perpendicular to ξ meets a "small circle" about p on M^2).

This definition agrees with the previous indexing procedure and has the additional property that $i(q, \xi) = 0$ if q is not a critical point. In the smooth case, however, the definition is somewhat unsatisfactory due to the difficulty of defining precisely the notion of a "small circle." In the polyhedral case, on the other hand, this is exactly the sort of definition which we want.

Consider a polyhedral surface M^2 in E^3 which is expressed as a union of V vertices, E edges, and T triangular faces. The *Euler characteristic* of M^2 is defined to be

$$\chi(M^2) = V - E + T.$$

A height function ξ on E^3 is said to be *general for the polyhedral surface M^2* if $\xi(v) \neq \xi(w)$ whenever v and w are distinct vertices of M^2. If ξ is general for M^2,

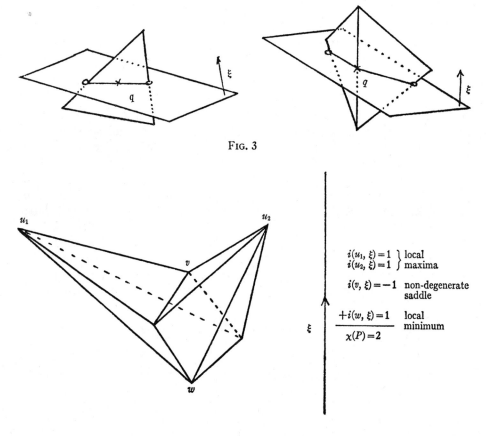

Fig. 3

Fig. 4

$i(u_1, \xi) = 1$ } local
$i(u_2, \xi) = 1$ } maxima

$i(v, \xi) = -1$ non-degenerate saddle

$+i(w, \xi) = 1$ local
                   minimum
——————
$\chi(P) = 2$

then the point q is said to be *ordinary* for ξ if the plane through q perpendicular to ξ cuts the disc neighborhood Star(q) into two pieces, where Star(q) is the union of all vertices, edges, and faces which include q. (When we say M^2 is a polyhedral surface, we mean that for each point q, Star(q) is the image of an open disc in the plane under a one-to-one continuous map.) With this definition, any point q in the interior of a face or an edge has to be ordinary since no face or edge can be perpendicular to the vector ξ if ξ is general for M^2.

For vertices, however, there are critical points corresponding to all the types presented for smooth functions, as, for example, in the indicated polyhedron. We may then use the indexing procedure developed for smooth surfaces, where instead of a "small circle" we use the embedded polygon which is the boundary of the star of the vertex v. The number of times the plane through v perpendicular to ξ meets this polygon is then equal to the number of triangles $\triangle$ in Star(v) such that one of the vertices of $\triangle$ lies above the plane and the other lies below.

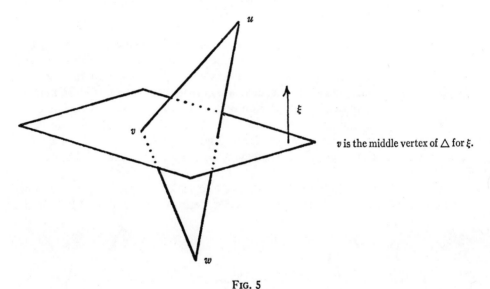

v is the middle vertex of $\triangle$ for ξ.

FIG. 5

In such a case we say that v is the middle vertex of $\triangle$ for ξ, and we may write the index as follows:

$$i(v, \xi) = 1 - \tfrac{1}{2} \text{ (number of } \triangle \text{ with } v \text{ middle for } \xi).$$

Again this definition corresponds to the definitions given in the smooth case and it gives index 0 for an ordinary point.

The critical point theorem then states:

THEOREM 1. *If ξ is general for M^2, then*

$$\sum_{v \in M^2} i(v, \xi) = \chi(M^2).$$

We require a lemma on polyhedral surfaces:

LEMMA. *For a polyhedral surface, $3T = 2E$.*

Proof of Lemma. Since an edge in a polyhedral surface has precisely two triangles in its star,

$$3T = \text{number of pairs } (\triangle, \text{ edge of } \triangle) = 2E.$$

Proof of Theorem. If ξ is general for M^2,

$$\sum_{v \in M} i(v, \xi) = \sum_{v \in M} (1 - \tfrac{1}{2}(\text{number of } \triangle \text{ with } v \text{ middle for } \xi)),$$

$$= V - \tfrac{1}{2} \sum_{v \in M} (\text{number of } \triangle \text{ with } v \text{ middle for } \xi),$$

$$= V - \tfrac{1}{2}T \text{ (since each } \triangle \text{ has exactly one middle vertex for } \xi),$$

$$= V - \tfrac{1}{2}(2E - 2T) \text{ (since } T = 2E - 2T \text{ by the lemma)},$$

$$= V - E + T.$$

REMARK. For a smooth surface M^2 embedded in E^3 it is a classical result that for almost every unit vector ξ on S^2 (i.e., except for a set of measure zero on S^2), the height function ξ has only finitely many critical points, and furthermore, for almost all ξ, this height function has as critical points only local maxima and minima, and nondegenerate saddle points. In the polyhedral case it is immediate that almost all ξ are general for M^2 (since the nongeneral ξ lie in the finite union of great circles $\{\xi \in S^2 \,|\, \xi(v) = \xi(w)\}$, one for each pair of distinct vertices v, w). The stronger result, however, is not correct in the polyhedral case. Consider an isolated critical point of a smooth surface which is degenerate—the "monkey saddle" (so called in Hilbert and Cohn-Vossen [3], p. 191, since a monkey riding a bicycle would need three depressions in his saddle, one for

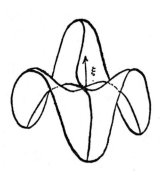

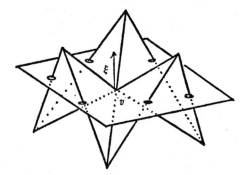

$$i(v, \xi) = -2$$

FIG. 6

each leg and one for his tail). Although on the smooth monkey saddle, height functions η near ξ on S^2 have only nondegenerate critical points, any η near ξ

has exactly six triangles with v middle, so for each such η, $i(v, \eta) = -2$.

However, in the proof of Theorem 1, we never required that the indices be only 0, 1 or -1. The proof goes through without change for any polyhedral surface M and ξ general for M, regardless of the complexity of the stars of the vertices of M.

2. Total Curvature and the Gauss-Bonnet Theorem. The total curvature or Gaussian curvature of a neighborhood U on a smooth surface M^2 in E^3 has several definitions which appear in texts in differential geometry. The definition which most easily leads to an extrinsic curvature theory for polyhedra is the one originally given by Gauss. We sketch his procedure in the smooth case and then develop the analogous theory for arbitrary embedded polyhedral surfaces in E^3.

Consider a "small" neighborhood U_1 on a convex surface M^2 in E^3. The Gauss map $g: U_1 \to S^2$ is defined by setting $g(p) =$ outward unit normal vector to M^2 at p. If the mapping g restricted to U_1 is one-to-one, and g is orientation-preserving (outward normals at corresponding points correspond) then U_1 is said to be *strictly convex*. The *total curvature* $\tilde{K}(U_1)$ of U_1 is then defined to be the area of the spherical image $g(U_1)$ on S^2.

If U_2 is a region of a nonconvex surface M^2 on which g is one-to-one and orientation-reversing, then $\tilde{K}(U_2)$ is defined to be the negative of the area of $g(U_2)$.

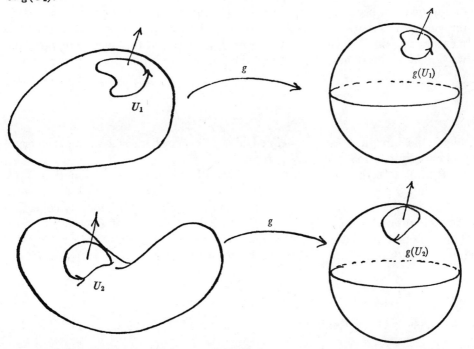

Fig. 7

We may use the index $i(p, \xi)$ to describe the total curvature of U_1 in these cases. Let $d\omega$ denote the area integrand for S^2, so

$$\int_{S^2} d\omega = 4\pi.$$

Then the area of $g(U_1)$ is the integral over S^2 of the characteristic function: $F_{g(U_1)}(\xi) = 1$ if ξ is in $g(U_1)$ and 0 otherwise. But if ξ is in $g(U_1)$, then the height function ξ has a critical point at a point p on U_1. In fact, $i(p, \xi) \neq 0$ for a point p of U_1 if and only if ξ is in $g(U_1)$ or $-\xi$ is in $g(U_1)$. We may assume that U_1 is small enough that $g(U_1)$ contains no pair of antipodal points. The total curvature may then be described as follows:

$$\tilde{K}(U_1) = \int_{S^2} F_{g(U_1)}(\xi) d\omega = \frac{1}{2} \int_{S^2} \sum_{p \in U_1} i(p, \xi) d\omega.$$

But the right-hand expression also serves to define $\tilde{K}(U_2)$, since this expression gives the negative of the area of $g(U_2)$.

The following two paragraphs explain the procedure of defining the total curvature in the smooth case, and this serves to motivate the definition for the polyhedral analogue. In the polyhedral case, however, the technical difficulties concerning the convergence of the integrand do not occur.

In order to define the total curvature of a neighborhood U on which g is not one-to-one, we begin by expressing U as a countable disjoint union of sets U_i on which g is one-to-one together with a set $V = U - \bigcup_{i=1}^{\infty} U_i$ such that $g(V)$ has measure zero on S^2. Then we set $\tilde{K}(U) = \sum_{i=1}^{\infty} \tilde{K}(U_i)$ if this sum converges, and we obtain a totally additive set function on M^2. This definition then coincides with

$$\tilde{K}(U) = \frac{1}{2} \int_{S^2} \sum_{p \in U} i(p, \xi) d\omega,$$

where the integrand is well defined almost everywhere since almost every ξ has only finitely many critical points.

REMARK. In the case that M^2 is sufficiently smooth, the set function $\tilde{K}(U)$ is absolutely continuous with respect to the area measure $A(U)$ on M^2, and we may define the point function $K(p)$ as the limit of $\tilde{K}(U)/A(U)$ for any collection of neighborhoods U_i of p with the limit of the diameters of U_i going to zero. This function is called the *Gaussian curvature* at p, and by integrating this function with respect to the area measure, we obtain

$$\tilde{K}(U) = \int_U K(p) dA.$$

The classical Gauss-Bonnet theorem for embedded closed surfaces states that

$$\tilde{K}(M^2) = \int_{M^2} K(p) dA = 2\pi \chi(M^2).$$

In the case of a polyhedral manifold M^2 in E^3, we may use the same definition as that developed for smooth surfaces. For any open set U of M^2, we set

$$K(U) = \frac{1}{2} \int_{S^2} \sum_{v \in U} i(v, \xi) d\omega.$$

When M^2 is a convex polyhedron, if U is a neighborhood containing only one vertex v, then $\tilde{K}(U)$ gives the measure of the exterior angle at v, i.e., the set of normals to support planes to M^2 at v, and this approach has been used in the classical theory of convex polyhedra, for example in [1].

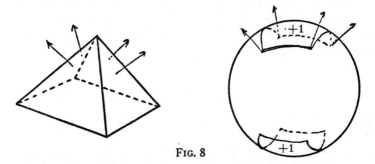

FIG. 8

The expression above also yields a definition of curvature for nonconvex vertices of saddle type as well as for the monkey saddles.

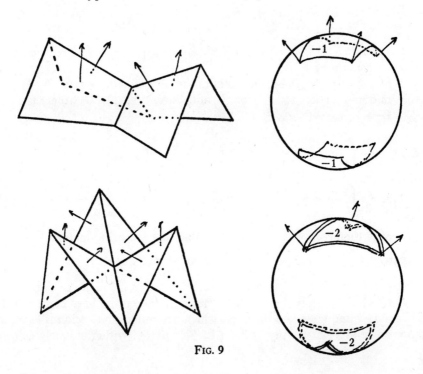

FIG. 9

The Gauss-Bonnet theorem for embedded surfaces, smooth or polyhedral, follows immediately:

THEOREM 2. $\tilde{K}(M^2) = 2\pi\chi(M^2)$.

Proof.

$$\tilde{K}(M^2) = \frac{1}{2}\int_{S^2}\sum_{p\in M^2} i(p,\xi)d\omega = \frac{1}{2}\int_{S^2}\chi(M^2)d\omega = \frac{1}{2}\chi(M^2)\int_{S^2}d\omega = 2\pi\chi(M^2).$$

3. The Theorema Egregium. Total curvature was defined for a set U on a smooth surface M in a way which involved the extrinsic properties of the surface, that is, the way M is situated in E^3. Gauss, however, proved that in fact $\tilde{K}(U)$ depends only on intrinsic properties of U, i.e., on properties that are determined by measurements made along the surface not taking into account the way the surface is situated in space, and Gauss called this result his *Theorema Egregium.*

We shall prove the analogous theorem for embedded polyhedral surfaces. Observe first of all that for an open set U of M,

$$\tilde{K}(U) = \frac{1}{2}\int_{S^2}\sum_{v\in U} i(v,\xi)d\omega = \sum_{v\in U}\frac{1}{2}\int_{S^2} i(v,\xi)d\omega$$

so we may set

$$\tilde{K}(U) = \sum_{v\in U}\tilde{K}(v), \quad \text{where} \quad \tilde{K}(v) = \frac{1}{2}\int_{S^2} i(v,\xi)d\omega.$$

THEOREM 3. $\tilde{K}(v)$ *is intrinsic, in fact,* $\tilde{K}(v) = 2\pi - $ *(sum of interior angles at* v *of triangles containing* v).

Proof. Let $m(v, \triangle, \xi)$ be the function on E^3 defined by $m(v, \triangle, \xi) = 1$ if v is the middle vertex of $\triangle$ for ξ and $m(v, \triangle, \xi) = 0$ otherwise. Then $i(v, \xi) = 1 - \frac{1}{2}\sum_{\triangle\in M} m(v, \triangle, \xi)$ and

$$\tilde{K}(v) = \frac{1}{2}\int_{S^2} i(v,\xi)d\omega = \frac{1}{2}\int_{S^2}\left(1 - \frac{1}{2}\sum_{\triangle\in M} m(v,\triangle,\xi)\right)d\omega$$

$$= \frac{1}{2}\int_{S^2} d\omega - \frac{1}{4}\int_{S^2}\sum_{\triangle\in M} m(v,\triangle,\xi)d\omega = 2\pi - \sum_{\triangle\in M}\frac{1}{4}\int_{S^2} m(v,\triangle,\xi)d\omega.$$

The proof is then complete once we establish the following lemma:

LEMMA. $\int_{S^2} m(v, \triangle, \xi)d\omega = 4$ *(interior angle of* $\triangle$ *at* v).

Proof. First of all, observe that for vectors **n** at v in the plane E^2 containing $\triangle$, $m(v, \triangle, \eta) = 1$ if and only if **n** lies in the region between the lines perpendicular to the edges $\mathbf{u} - \mathbf{v}$ and $\mathbf{w} - \mathbf{v}$ and the angle which determines this region is equal to the interior angle of $\triangle$ at v. Any vector ξ of S^2 may be written uniquely as $\xi = \eta + \zeta$

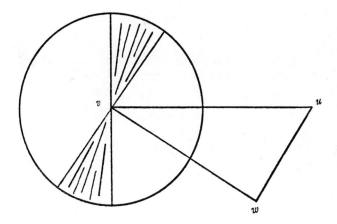

FIG. 10

where η lies in the plane E^2 containing $\triangle$ and ζ is perpendicular to E^2. Then $m(v, \triangle, \xi) = 1$ if and only if $\xi \cdot (u - v) > 0 > \xi \cdot (v - w)$ or $\xi \cdot (u - v) < 0 < \xi \cdot (v - w)$ so $m(v, \triangle, \xi) = 1$ if and only if $m(v, \triangle, \eta) = 1$. Thus the set of ξ on S^2 centered at v for which $m(v, \triangle, \xi) = 1$ forms a double lune with axis perpendicular to E^2 and with each angle equal to the interior angle of $\triangle$ at v. But the area of a lune is twice the angle of the lune, so

$$\int_{S^2} m(v, \triangle, \xi)\,d\omega = 4 \text{ (interior angle of } \triangle \text{ at } v).$$

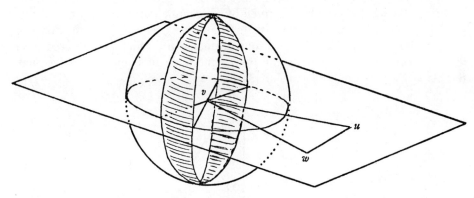

FIG. 11

This completes the proof of the *Theorema Egregium* for polyhedra, no matter how complicated the vertex stars are. Compare Hilbert and Cohn-Vossen ([3], p. 195) for a similar argument for vertices corresponding to nondegenerate saddle points.

REMARK. The theorems of section 2 may be considered as a generalization of the approach of G. Pólya for embedded convex polyhedral discs [6].

References

1. A. D. Alexandroff, Konvexe Polyeder, Akademie Verlag, Berlin, 1958 (Russian original: Moscow, 1950).

2. T. F. Banchoff, Critical Points and Curvature for Embedded Polyhedra, J. of Differential Geometry, number 3, 1 (1967) 257–268.

3. D. Hilbert and S. Cohn-Vossen, Geometry and the Imagination, Chelsea, New York, 1952 (German original: Springer, Berlin, 1932).

4. N. H. Kuiper, Der Satz von Gauss-Bonnet und damit verwandte Probleme, Jber. Deutsch. Math.-Verein., 69 (1967) 77–88.

5. J. Milnor, Morse Theory, Princeton University Press, 1963.

6. G. Pólya, An elementary analogue of the Gauss-Bonnet theorem, this MONTHLY, 61 (1954) 601–603.

EXISTENCE OF FOUR CONCURRENT NORMALS
TO A SMOOTH CLOSED CURVE*

NARSINGH DEO, California Institute of Technology, and M. S. KLAMKIN, Ford Motor Company†

In a rather elegant lecture on calculus, Guggenheimer [1] establishes, among other related theorems, the existence of a point from which at least four normals can be drawn to an oval. Two such points are the mass centroid and the curvature centroid. Chakerian and Stein [2] have shown that another four normal point is the perimeter centroid. The latter proof is by a simple continuity argument whereas the proofs given by Guggenheimer cannot (at least up to now) be obtained the same way. In this note, we give a simpler intuitive geometric proof for the existence of a four normal point which can be used in elementary courses dealing with continuity methods in convexity or geometry (e.g., a la Yaglom and Boltyanskii [3]).

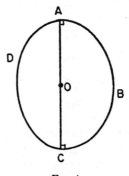

FIG. 1

* From AMERICAN MATHEMATICAL MONTHLY, vol. 77 (1970), pp. 1083–1084.

† Prof. Deo is now at the Indian Institute of Technology, Kanpur, and Prof. Klamkin at Waterloo University.

For an oval (smooth, convex curve), it is well known that there exists at least one largest chord and, furthermore, this chord is perpendicular to the curve at each end.

We now show that the midpoint O of a largest chord AB has the desired property. If ABC in Fig. 1 is circular, our proof is over. If ABC is not circular, then there must exist points whose distance from O is either larger or smaller than AO. The point (or points) P which maximize or minimize the radial distance OP will be such that $\overline{OP}$ is normal to the curve. Geometrically, if we consider the family of concentric circles centered at O, there will exist at least one circle which is tangent to open arc ABC. Since a similar argument applies to the other open arc CDA, our proof is completed.

By a similar argument, one can show that another four normal point is the midpoint of a chord of minimal width.

For the case of C^2 closed nonconvex curves, a largest chord which is perpendicular to the curve at each end still exists. Consequently, the midpoint O will still be a four normal point provided that it falls inside the curve. If O falls on the boundary, then possibly only three distinct normals can be drawn from O. However, there are then at least four points on the curve such that the normals at these points are concurrent (note that if the curve touches itself at O, this point is counted twice). If O falls outside the curve, it is possible that two of the normals are on the same ray. Whether or not these two normals are to be considered distinct is a matter of definition.

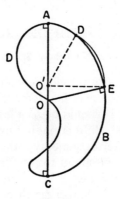

FIG. 2

For the case (Fig. 2) when O falls on the curve and the curve is also simple and contains no circular or straight segments, then for points O' of AC, sufficiently close to O, there exists a normal from O' to the open arc ABC. For the case illustrated, being sufficiently close means that if $\overline{OE}$ is normal to the curve and $OE < OA$; then $O'E$ must be $< O'A$. If $OE > OA$, then $O'E$ must be $> O'C$ (and in this case the extension of circular arc DE will intersect the open arc EBC). Also, there exists a normal from O' to the portion of the arc ADC contained in a circle of center O' and radius $OO' + \epsilon$ but here O' must satisfy $OO' < OA/2$. Hence, we have a segment of four normal points. The same also applies for an oval not containing any circular arcs. It now follows easily that the circle is the only oval with a unique "four-normal" point. Note that if the midpoint O of the maximum length chord AC turns out to be the only point on AC which is a four-normal point (e.g., if ABC was a semi-circle), then we would consider points on the minimum width chord.

Acknowledgment: The authors are grateful to Professor Guggenheimer for his suggestions improving the style of the original note as well as extending its scope.

References

1. H. H. Guggenheimer, Geometrical applications of integral calculus, p. 84 (contained in K. O. May, Lectures on Calculus, Holden-Day, San Francisco, 1967). Guggenheimer notes that since the problems treated in his paper are all of recent origin, it seemed possible to give the names of the first discoverers and a fairly complete bibliography.

2. G. D. Chakerian and S. K. Stein, On the centroid of a homogeneous wire, Mich. Math. J. 11 (1964) 189-192.

3. I. M. Yaglom and V. G. Boltyanskii, Convex Figures, Holt, Rinehart and Winston, New York, 1961.

Editorial Note: The reader may also be interested in these related articles:

H. Guggenheimer, Does there exist a "four normals triangle"?, this MONTHLY, vol. 77 (1970), 177–9.

B. Wegner, Existence of four concurrent normals to a smooth closed hypersurface in E^n, this MONTHLY, vol. 80 (1973), 782–5.

AFTER THE DELUGE*

D. A. MORAN, Michigan State University

The purpose of this note is to provide a somewhat simpler proof than that given by Professor Marston Morse of his elementary theorem about pits, peaks, and passes on the sphere [1]. It should be recalled that Professor Morse views a positive, real-valued, bounded, differentiable function on the sphere as an altitude, measured from the center of some hypothetical spherical planet. Critical points of the function then correspond to pits ($=$minima), peaks ($=$maxima), and passes ($=$saddle points of index -1) on the planet. It is assumed that no critical points more complicated than these three types ever occur, and that no two of these singularities occur at precisely the same altitude. Our viewpoint is essentially the same as this, but we start with the following slightly different

ADDITIONAL HYPOTHESIS: *No pass is as high as a peak, or as low as a pit.*

This hypothesis is easily fulfilled, if we agree to drill deep holes at the bottom of each pit, and raise tall flagpoles atop each peak. Let N_0, N_1, and N_2 be the number of pits, passes, and peaks.

---

* From AMERICAN MATHEMATICAL MONTHLY, vol. 77 (1970), p. 1096.

Now let rain begin to fall on the planet which represents the sphere. Immediately N_0 lakes are created. As the water level rises to the altitude of a pass, a lake can merge with itself, creating an island, or else a lake can merge with another lake, resulting in a net decrease by 1 in the number of lakes. When the water level has risen to inundate every pass, but is not yet as high as any peak, the number of islands will be

$$1 + \text{number of island-increasing passes,}$$

and the number of lakes will be

$$N_0 - \text{number of lake-decreasing passes.}$$

The number of lakes less the number of islands is therefore

$$N_0 - N_1 - 1.$$

On the other hand, at this point in time there is clearly one lake and N_2 islands, so

$$N_0 - N_1 + N_2 = 2,$$

which provides a direct analogue to Euler's theorem that $V - E + F = 2$, where V, E, and F are the numbers of vertices, edges, and faces of any convex polyhedron.

References

1. Marston Morse, *Pits, peaks, and passes* (motion picture film), Modern Learning Aids #3462, New York.

2. George Polya, Induction and Analogy in Mathematics, Princeton Univ. Press, 1954, pp. 163–165.

Editorial Note: Those who have not seen the film may be interested in reading the arguments given in Polya's book [2] and in the article "Networks, ham sandwiches, and putty" by S. S. Cairns which appeared in the Pi Mu Epsilon Journal, Spring 1963.

The novel treatment which appears in Professor Moran's article is generalized by the following fact which is known to students of Morse Theory: given any Morse function f on a compact surface, there is another function g having the same critical points as f (with the same indices), and whose critical points have indices which increase as the critical values increase (i.e., the critical values of index 0 are smaller than those of index 1, etc.). A nice treatment of this result can be found in the book *Topologie des Surfaces* by André Gramain (Presses Universitaires de France, 1971).

DYNAMIC PROOFS OF EUCLIDEAN THEOREMS*

ROSS L. FINNEY, University of Illinois, Urbana

Simple observations about transformations of the plane lead to elegant proofs of unusual Euclidean theorems. The theorems are easily stated, and many of them can be conjectured from hand-drawn pictures. The proofs use neither coordinates nor vectors, and the simplest of them require only a slight familiarity with rotations and translations. We look at some of these first, and then introduce similarity transformations to show that a number of theorems which have appeared in the literature as isolated results are really special cases of two or three general theorems. A pleasant by-product of the proofs given here is that they yield results about quadrilaterals without assuming them to be either convex or simple (compare, for example, Theorems 1 and 3 with their counterparts in [5]).

The first lemma is about Figure 1.

LEMMA 1. *If isosceles triangles ZMX and YMW have right angles at M, then $\overline{YX}$ and $\overline{ZW}$ are perpendicular and congruent.*

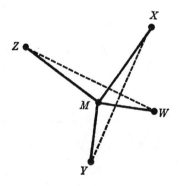

FIG. 1.

Proof. This is because M_{90}, the counterclockwise rotation of 90 degrees about M, moves Y to W and X to Z.

LEMMA 2. *If Z and X are centers of squares that lie on sides of ABC, built towards the exterior of ABC, and if M is the midpoint of the third side, then ZMX is isosceles and has a right angle at M.*

To see why, note that the composite $T = M_{180}X_{90}Z_{90}$ of rotations about Z, X and M is a translation. This is because the degree measures of the rotations add up to an integral multiple of 360. Under this composite, A goes first to B, then to C and finally back to A. But the only translation with a fixed point is the identity transformation. Since $T = I$, the successive images of Z are

$$Z_{90}(Z) = Z,$$

$$X_{90}(Z) = \text{some point which we call } Z',$$

---

* From MATHEMATICS MAGAZINE, vol. 43 (1970), pp. 177–185.

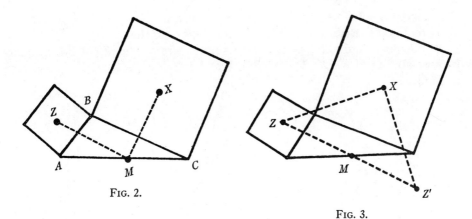

FIG. 2.

FIG. 3.

and

$$M_{180}(Z') = T(Z) = Z.$$

Our picture now looks like Figure 3. Because $X_{90}(Z) = Z'$, we know that $ZX = XZ'$ and that ZXZ' has a right angle at X. We also know that M is the midpoint of $\overline{ZZ'}$ because $M_{180}(Z') = Z$. Thus M is the midpoint of the hypotenuse of a right isosceles triangle, which makes ZMX isosceles with a right angle at M.

The conclusion of Lemma 2 holds if the squares are constructed towards the interior of the triangle instead of towards the exterior. In the proof, one merely replaces the counterclockwise Z_{90} and X_{90} by the clockwise Z_{-90} and X_{-90}.

THEOREM 1. [1, 3] *If X, Y and Z, W are opposite pairs of centers of squares on the sides of a quadrilateral that lie towards the quadrilateral's exterior, then $\overline{YX}$ and $\overline{WZ}$ are perpendicular and congruent.*

Proof. Let M be the midpoint of a diagonal of the quadrilateral, and apply M_{90}.

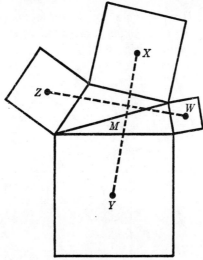

FIG. 4.

Theorem 1 holds also if the squares all lie towards the quadrilateral's interior. The proof is the same: M_{90} throws one segment onto the other.

If the quadrilateral happens to be a parallelogram, then $ZXWY$ is a square, because all four of ZMX, XMW, WMY and YMZ are isosceles and have right angles at M.

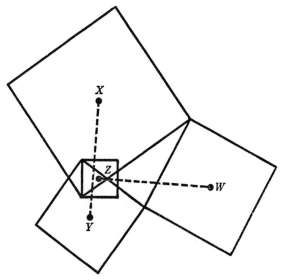

FIG. 5.

It should be emphasized here that the quadrilateral need not be convex, or even simple. To place the squares when the quadrilateral has no obvious interior, one traverses the quadrilateral in one of the two possible directions, laying off squares to the right.

If we shrink an edge of the quadrilateral to a point, we see a theorem about squares on a triangle.

THEOREM 2. [1, 3] *If squares are constructed on the sides of a triangle towards the triangle's exterior, then the segment joining two of the centers is perpendicular and congruent to the segment joining the third center to the vertex opposite it.*

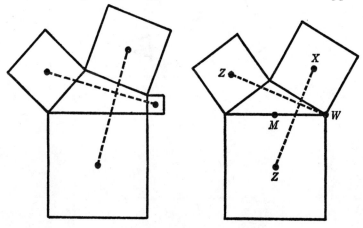

FIG. 6.

Proof. Look at Figure 6, and apply M_{90} as in Lemma 1.

Theorem 2 has the corollary that the lines joining the square centers to the vertices opposite them are concurrent (Figure 7). This is because the lines are altitudes of XYZ.

One of the first theorems of Euclidean geometry is that the midpoints of the sides of a quadrilateral are themselves the vertices of a parallelogram.

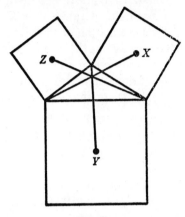

FIG. 7.

THEOREM 3. *The vertices Z, X, W, Y of equilateral triangles built on the sides of a quadrilateral, and lying alternately towards the interior and the exterior of the quadrilateral, are themselves the vertices of a parallelogram.*

Proof. The composite $C_{-60}A_{60}$ is a translation that takes Y to W and Z to X. Thus $\overline{YZ}$ and $\overline{WX}$ are parallel and congruent.

The Euclidean midpoint theorem and Theorem 3 are two of a family of theorems.

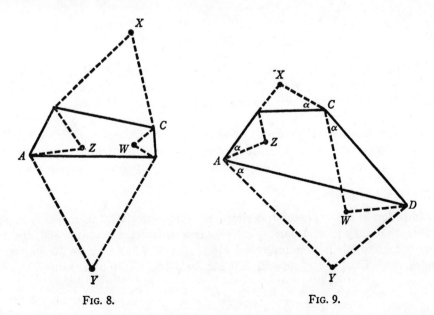

FIG. 8.   FIG. 9.

THEOREM 4. [2] *If Z, X, W, Y are vertices of similar triangles appropriately arranged on the sides of a quadrilateral, then $ZXWY$ is a parallelogram.*

Proof. Let α be the measure of the angles that the four similar triangles have at A and C, and let r denote the ratio of AY to AD. Let $A^{1/r}$ denote the central dilatation with center A that multiplies distance by $1/r$. Let C^r denote the central dilatation with center C that multiplies distance by r. Then the composite $C_{-\alpha}C^rA^{1/r}A_\alpha$ is a translation that takes Y to W and Z to X.

If $\alpha = 0$ we have a slight generalization of the midpoint theorem. If $r = 1$ and $\alpha = 60$, we have Theorem 3. Once again, the quadrilateral need not be convex or simple. Shrinking an edge leads to several nice theorems about triangles.

With similarity transformations, one can prove a theorem (Theorem 5) about triangles that has truly surprising corollaries. Among them are the following three:

COROLLARY 1. *Suppose that 30–60–90 triangles are built on two sides of an arbitrary triangle towards its exterior, as in Figure 10. Let Z and X denote the outer vertices of these triangles, and let M be the midpoint of the remaining side of the given triangle. Then ZMX is equilateral. If, instead, the 30–60–90 triangles lie toward the interior of the given triangle, ZMX is still equilateral.*

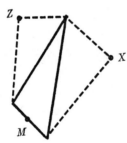 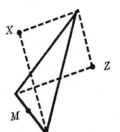

FIG. 10.

COROLLARY 2. (Napoleon's Theorem) *The centers X, Z, M of equilateral triangles constructed on the sides of an arbitrary triangle, and lying towards the triangle's exterior, are themselves the vertices of an equilateral triangle.*

COROLLARY 3. [5, Exercise 23] *Suppose that equilateral triangles are built on the sides of an arbitrary triangle, two towards its exterior and one towards its interior. Let M be the center of the inner one, and Z and X the apexes of the outer ones. Then ZMX is an isosceles triangle, with a 120° angle at M.*

In order to state Theorem 5 simply, we assume that angles are "general," that is, that they have a variety of measures which agree modulo 360, and that they are oriented in a counterclockwise fashion. Thus, Figure 13 shows two angles: angle CBA of measure 60, 420, etc., and angle ABC of measure 300, -60, etc.

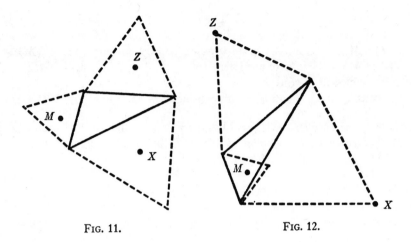

FIG. 11.   FIG. 12.

THEOREM 5. *Let BZC and AXC be nondegenerate similar triangles (with vertices corresponding in the order given) constructed both towards the exterior or both towards the interior of arbitrary triangle ABC. Let the angles BZC and CXA have measure* β. *Let M be the point in the plane that is equidistant from A and B and that is located so that angle BMA has measure* 2β. *Then MZ = MX.*

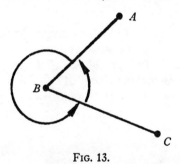

FIG. 13.

Proof. Since BZC and CXA are nondegenerate, $2\beta \neq 360$ and M really does exist. The triangle AMB may be degenerate, however, for if $\beta = 90$ then M is the midpoint of $\overline{AB}$.

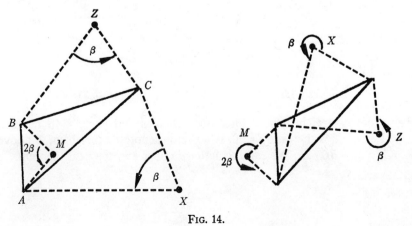

FIG. 14.

Let r denote the ratio ZC/BZ. Then the translation $T = X_\beta X^{1/r} Z' Z_\beta M_{-2\beta}$ is the identity because it leaves A fixed. If we follow M about*, we see that $M_{-2\beta}(M) = M$, that $Z'Z_\beta(M)$ is some point M', say, and that $X_\beta X^{1/r}(M') = T(M) = M$. Thus $MXM'Z$ is a quadrilateral with $XM' = r \cdot MX$, $ZM' = r \cdot MZ$ and with congruent angles at X and Z. This means that triangles MXM' and MZM' are similar. But they have a common side, so they are congruent and $MZ = MX$.

Proofs of the corollaries. If $\beta = 90$ and $r = \sqrt{3}/3$ then MXM' and MZM' are $30 - 60 - 90$ triangles. In particular, $MXM'Z$ has a 60 degree angle at M. This makes ZMX equilateral, which proves Corollary 1.

If $\beta = 120$ and $r = 1$, then Z and X are indistinguishable from M. Consequently $MZ = MX = XZ$, which is Napoleon's theorem.

Corollary 3 is the case $\beta = 60$ and $r = 1$.

If $\beta = -60$ and $r = 1$ we have Corollary 3 with two inner triangles and one outer.

The assignment $\beta = -120$, $r = 1$ produces Napoleon's theorem with three inward triangles.

Lemma 2 is also a corollary of Theorem 5, the squares lying toward the exterior if $\beta = 90$, $r = 1$, and towards the interior if $\beta = -90$, $r = 1$.

We close with a generalization of Lemma 2 that leads to a generalization of Theorem 1.

THEOREM 6. [4] *Suppose that similar isosceles triangles are constructed on the sides of an arbitrary triangle ABC, either both towards the interior of ABC or both towards the exterior. Let X be the apex of one, Z the orthocenter of the other, and let M be the midpoint of the remaining side of ABC. Let β be the measure of the angle at X. Then ZMX has a right angle at M and an angle of measure $\beta/2$ at X.*

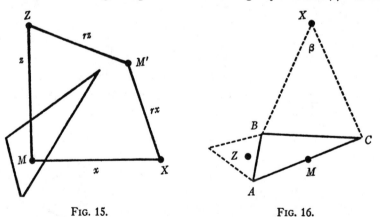

FIG. 15.   FIG. 16.

The proof is straightforward once one observes that angle AZB has measure $180 - \beta$. Lemma 2 is the case $\beta = 90$. If we let h denote the height of X above $\overline{BC}$, then $MZ/XM = BC/2h$. We call this latter number r in the next theorem.

---

* See Figure 15.

THEOREM 7. *Suppose that similar isosceles triangles lie on the sides of a quadrilateral, all towards the interior or all towards the exterior of the quadrilateral. Let X and Y be apexes of one opposite pair of isosceles triangles, and let Z and W be*

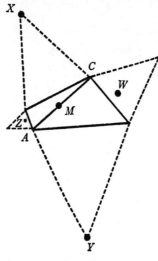

FIG. 17.

the orthocenters of the other two. Then $\overline{YX}$ *and* $\overline{WZ}$ *are perpendicular, and* YX/WZ $= r$.

Proof. Let M be the midpoint of the diagonal $\overline{AC}$. If the triangles lie towards the quadrilateral's exterior, then $M'M_{90}$ takes X to Z and Y to W. If the triangles lie toward the interior, apply $M'M_{-90}$ instead.

Again, the quadrilateral need not be convex or simple, and there is an analogous theorem for triangles.

Acknowledgements: I first saw dynamic proofs of Corollaries 1 and 2 in a course that Nicolaas Kuiper gave at the University of Michigan in 1953. Later that year Kenneth Leisenring showed me Theorem 3, and in 1965 he encouraged me to develop dynamic proofs for the African Mathematics Secondary Program. I am grateful to them.

References

1. M. H. van Aubel, Note concernant les centres des carrés construits sur les côtés d'un polygone quelconque, Nouv. Corresp. Math., 4 (1878) 40–44.

2. ———, Question 56, Mathesis, 1 (1881) 167.

3. C.-A. Laisant, Sur quelques propriétés des polygones, Assoc. Franç. Avanc. Sci. Le Havre, (1877) 142–154.

4. D. St. J. Jesson, Private communication, 1967.

5. M. Yaglom, Geometric Transformations, Random House, New York, 1962.

ON FENCHEL'S THEOREM*

R. A. HORN, Johns Hopkins University

Fenchel's well-known theorem on the total curvature μ of a closed C^2 space curve consists of the inequality $\mu \geqq 2\pi$, together with the statement that equality holds if and only if the curve is a plane convex curve. Our purpose is to deduce this result with a symmetry argument that seems simpler than any proof appearing in the textbooks or recent literature.

Let $G = \{R(s) \mid 0 \leqq s \leqq L\}$ be a closed C^2 space curve with arclength parameter s, unit tangent $T(s) = (T_1, T_2, T_3) \equiv (d/ds)R(s)$, and curvature

$$|\kappa(s)| \equiv \|(d/ds)T(s)\|.$$

The *tangent indicatrix* of G is the curve $\Gamma = \{T(s) \mid 0 \leqq s \leqq L\}$ on the unit sphere S, and the *total curvature* of G is the quantity $\mu \equiv \int_0^L |\kappa(s)| \, ds$. Clearly, the total curvature of G is just the length of its tangent indicatrix. Consider the following two lemmata:

LEMMA 1. *The tangent indicatrix Γ of a closed C^1 space curve G is not contained in any open hemisphere of S. It is contained in a closed hemisphere if and only if G is a plane curve.*

Proof: If Γ were contained in a hemisphere of S, we could perform a rotation to place it in the northern hemisphere. Thus, we may assume $T_3(s) \geqq 0$ for all $s \in [0, L]$ and since G is a closed curve we know that

$$0 = (R(L) - R(0))_3 = \left(\int_0^L T(s) \, ds \right)_3 = \int_0^L T_3(s) \, ds = 0.$$

This shows that $T_3(s)$ cannot be strictly positive, and hence Γ cannot lie in an open hemisphere. Furthermore, since $T_3(s)$ is nonnegative it must vanish identically, i.e., $0 \equiv T_3(s) \equiv (d/ds)R_3(s)$, and hence G must lie in a plane $R_3 = $ constant. Conversely, if G is a plane curve, then Γ lies on a great circle and hence is contained in a closed hemisphere.

LEMMA 2. *Let Γ be a closed rectifiable curve on S. If the length of Γ is less than 2π, then it is contained in an open hemisphere of S; if Γ has length 2π, then it is contained in a closed hemisphere of S.*

Proof. Let P be any point on Γ and let Q be the point of Γ such that the curve segments $\Gamma_1 = PQ$ and $\Gamma_2 = QP$ have equal length, $\Gamma = \Gamma_1 + \Gamma_2$. Rotate S so that P and Q are located symmetrically with respect to the north pole N, i.e., so that either $P = Q = N$ or so that P and Q have the same latitude but have longitudes differing by 180°. If Γ does not now intersect the equator, the conclusions follow. If Γ_1 intersects the equator at some point, construct the unique curve Γ_2' which is symmetric to Γ_1 with respect to N. Then Γ_2' has the same length as Γ_1, the closed curve $\Gamma' \equiv \Gamma_1 + \Gamma_2'$ has the same length as Γ, and there is a pair of antipodal equatorial points on Γ'. But if we join these points with great

* From AMERICAN MATHEMATICAL MONTHLY, vol. 78 (1971), pp. 380–381.

semicircles (which are the geodesics on S), we see that if Γ_1 intersects the equator, then Γ' (and hence Γ) has length at least 2π, and that if Γ_1 crosses the equator into the open southern hemisphere, then Γ' must be strictly longer than 2π.

Thus, if Γ has length less than 2π, then Γ_1 cannot intersect the equator and if the length of Γ is exactly 2π, then Γ_1 cannot cross the equator. Since the same argument applies to Γ_2 we conclude that if Γ has length less than 2π, then it is contained in the open northern hemisphere and that if its length is exactly 2π, then it must lie in the closed northern hemisphere.

Fenchel's inequality now follows immediately from these two results: Since the tangent indicatrix of a closed C^2 space curve cannot lie in an open hemisphere of S its length must be at least 2π. If the length is exactly 2π, then the tangent indicatrix lies in a closed hemisphere of S and hence the original curve must be a plane curve. Using the notion of the *rotation index* as in [1], one can now complete the full proof of Fenchel's Theorem by showing that a plane curve is convex if and only if its tangent indicatrix has length 2π.

REMARKS: This proof grew out of a discussion with Laird E. Taylor about a problem posed by Robert Osserman. It is a pleasure to acknowledge their helpful suggestions as well as those of S. S. Chern, H. Flanders, and the referees.

Contrary to our original belief, this treatment is not the first completely elementary proof of Fenchel's inequality. The referees have remarked that Anthony Morse, A. S. Besicovitch, and H. Flanders have presented elementary proofs in their lectures, and that shortly after Fenchel's original publication [2] in 1929, H. Liebmann [6] published a similar elementary proof which has been almost totally forgotten. In a 1951 article, Fenchel [3] gives references to several different, less elementary, proofs, while recent textbooks such as [4] and [5] have followed a surface theory approach, introduced by K. Voss [7] in 1955. It is interesting to compare these proofs with ours and with each other; Fenchel's elegant pearl has certainly inspired a variety of settings.

This work was partially supported by the National Science Foundation under grant GP-13258.

References*

1 S. S. Chern, ed., Studies in Global Geometry and Analysis, vol. 4, MAA Studies in Mathematics, 1967, 32.

2. W. Fenchel, Über Krümmung und Windung geschlossener Raumkurven, Math. Annalen, 101 (1929) 238–252.

3. ———, The differential geometry of closed space curves, Bull. Amer. Math. Soc., 57 (1951) 44–54.

4. H. W. Guggenheimer, Differential Geometry, McGraw-Hill, New York, 1963, 251–252.

5. D. Laugwitz, Differential and Riemannian Geometry, Academic Press, New York, 1965, 203–204.

6. H. Liebmann, Elementarer Beweis des Fenchelschen Satzes über die Krümmung geschlossener Raumkurven, S.-B. Preuss. Akad. Wiss. (Phys.-Math. Klasse) (1929) 392–393.

7. K. Voss, Eine Bemerkung über die Totalkrümmung geschlossener Raumkurven, Arch. Math., 6 (1955) 259–263.

8. M. Edelstein and B. Schwarz, On the length of linked curves, Israel J. Math., 23 (1976), 94–5.

---

* Prof. Horn provided Reference 8 to update his article for the present volume. It is a nice application of Horn's Lemma 2.—ED.

ON A CHARACTERIZATION OF THE 2-SPHERE*

KRISHNA AMUR, Karnatak University, Mysore State, India

1. Introduction. The purpose of this note is to give a new proof of the following known

Theorem. *If Σ is a smooth, closed, convex, orientable surface in $\mathbf{E}^3$ with the property that at each point the sum of the radii of principal curvatures is constant, that is*

$$2H/K = \text{constant},$$

where H is the mean curvature and K is the Gaussian curvature, then Σ is a sphere.

This is a special case of Liebmann-Süss theorem [1]: *A convex hypersurface in $\mathbf{E}^{n+1}$ is an n-sphere if one of the elementary symmetric functions of the principal radii of curvature is constant.* It also follows as a particular case of a theorem of Chern [2] on a characterization of the n-sphere in $\mathbf{E}^{n+1}$.

In establishing the theorem, we use a parallel surface Σ' whose points are at a constant distance $2H/K$ along the normals to Σ. It turns out that Σ' is diffeomorphic to Σ and that $H' = -H$ and $K' = K$; consequently Σ' also has the properties of Σ. While Σ' may not be imbedded† in $\mathbf{E}^3$, it serves as a useful tool for obtaining information about Σ.

2. Preliminaries. If $\mathbf{n}$ denotes the unit normal to Σ at a point $\mathbf{x}$, the corresponding point of Σ' is given by

$$(2.1) \qquad \mathbf{x}' = \mathbf{x} - a\mathbf{n},$$

where $a = 2H/K = \text{constant}$.‡ We have

$$(2.2) \qquad d\mathbf{x}' = d\mathbf{x} - a\,d\mathbf{n}.$$

Choosing an orthonormal frame $\mathbf{e}_1, \mathbf{e}_2$ in the tangent space to Σ at $\mathbf{x}$ such that $\det(\mathbf{e}_1, \mathbf{e}_2, \mathbf{n}) = +1$, we have

$$(2.3) \qquad d\mathbf{x} = \sigma_1 \mathbf{e}_1 + \sigma_2 \mathbf{e}_2 \quad \text{and} \quad d\mathbf{n} = \omega_1 \mathbf{e}_1 + \omega_2 \mathbf{e}_2,$$

where σ_i and ω_i are 1-forms. From (2.2) we have

$$(2.4) \qquad d\mathbf{x}' = (\sigma_1 - a\omega_1)\mathbf{e}_1 + (\sigma_2 - a\omega_2)\mathbf{e}_2,$$

which shows that we can choose the orthonormal frame $\mathbf{e}_1, \mathbf{e}_2$ in the tangent space to Σ' at $\mathbf{x}'$, and that the normal to Σ' at $\mathbf{x}'$ is $\mathbf{n}$. Denoting the 1-form coefficients of $\mathbf{e}_1$ and $\mathbf{e}_2$ in (2.4) by σ_1' and σ_2' respectively, and the area elements of Σ and Σ' by σ and σ' respectively, we have

$$(2.5) \qquad \sigma' = \sigma_1' \wedge \sigma_2' = (1 - 2aH + a^2K)\sigma_1 \wedge \sigma_2 = \sigma_1 \wedge \sigma_2 = \sigma.$$

---

* From AMERICAN MATHEMATICAL MONTHLY, vol. 78 (1971), pp. 382–383.

† Σ' is an immersed ovaloid, hence imbedded. See below.—ED.

‡ The sign convention for H used here is that which makes H positive for the sphere with outward-pointing normal.—ED.

It follows from (2.5) that the rank of the differential of the map $x \rightarrow x'$ is 2 everywhere. Also from (2.1) it is clear that the map $x \rightarrow x'$ is bijective§. Hence Σ' is diffeomorphic to Σ.

Since Σ and Σ' have the same normal at the corresponding points,

$$(2.6) \qquad K'\sigma' = \omega_1' \wedge \omega_2' = \omega_1 \wedge \omega_2 = K\sigma,$$

which in view of (2.5) implies

$$(2.7) \qquad K = K'.$$

Since from [3],

$$(2.8) \qquad dn \bullet (dx \times n) = 2H\sigma \quad \text{and} \quad dn \bullet (dn \times n) = 2K\sigma,$$

we have

$$
\begin{aligned}
2H'\sigma' &= dn \bullet (dx' \times n) \\
&= dn \bullet [(dx - a\,dn) \times n] \\
&= 2H\sigma - \frac{2H}{K}(2K\sigma) \\
&= -2H\sigma.
\end{aligned}
$$

Hence from (2.5) we get

$$(2.9) \qquad H' = -H.$$

3. Proof of the theorem. Since Σ' is diffeomorphic to Σ, it is smooth, closed, and orientable. We can therefore apply the Minkowski formula

$$(3.1) \qquad \int_\Sigma \sigma = \int_\Sigma Hp\sigma$$

to Σ' to obtain

$$(3.2) \qquad \int_\Sigma \sigma = \int_\Sigma \sigma' = \int_\Sigma H'p'\sigma' = -\int_\Sigma H(p-a)\sigma = -\int_\Sigma Hp\sigma + 2\int_\Sigma \frac{H^2}{K}\sigma,$$

where we have used (2.1), (2.5), and (2.9). Using (3.1) again in (3.2), we get

$$(3.3) \qquad \int_\Sigma \left(\frac{H^2}{K} - 1\right)\sigma = 0.$$

Clearly $H^2 - K \geq 0$ holds, with equality at umbilical points, and since we have assumed that Σ is convex, $K > 0$. The integrand in (3.3) is therefore non-negative, so $H^2 = K$ at all points of Σ. Hence the surface Σ must be a sphere.

---

§ Since it is assumed that Σ is convex, $K > 0$ so $K' > 0$ and Σ' is an immersed ovaloid. By Hadamard's theorem, any such an immersion is an imbedding, so that $x \rightarrow x'$ is injective and Σ' is diffeomorphic to Σ.—ED.

References

1. T. Bonnesen and W. Fenchel, Theorie der konvexen Körper, Springer, Berlin, 1934.

2. S. S. Chern, Integral formulas for hypersurfaces in Euclidean space and their applications to uniqueness theorems, J. Math. Mech., 8 (1959) 947–955.

3. H. Flanders, Differential Forms, Academic Press, New York, 1963, pp. 40–44.

WHAT IS A CONVEX SET?*

VICTOR KLEE, University of Washington

This is a slight expansion (and, of course, a translation) of the author's article [63] on *Convexité* in the new French encyclopedia, *Encyclopedia Universalis*, and appears here with the kind permission of the encyclopedia's publishers. Its purpose is to supply a broad but brief survey, at a rather elementary level, of several aspects of convexity theory. Those aspects are emphasized which appear to the author to be most active at present and to be most accessible to the chosen level of exposition. In the allotted space, the topics covered cannot, of course, be "surveyed" in the usual sense; instead, each is represented by one or more of its highlights. No proofs are included.

For theorems which are often designated in the literature by authors' names, the relevant names are given here, even though in some cases the implied attribution is incomplete. Beyond that, there has been no attempt at attribution. The references are in most cases not to the original or the most definitive sources, but rather to expository treatments or to papers which contain useful collections of additional references. While this policy probably maximizes the utility/size ratio of the bibliography, it has the unfortunate consequence of omitting the names of many prominent workers in the field. Those names can be found in the bibliographies of the references cited. A reference which appears in parentheses at the end of a paragraph contains useful information, or at least useful references, for the entire area of convexity theory with which the paragraph is concerned. In addition to items which are specifically mentioned in the text, the bibliography includes a number of books, monographs, lecture notes, symposium volumes, and survey articles in which the notion of convexity has played an important role. Most of these have appeared in the past fifteen years.

Preparation of this article was partially supported by the Office of Naval Research.

Introduction. The study of **convex sets** is a branch of geometry, analysis, and linear algebra that has numerous connections with other areas of mathematics and serves to unify many apparently diverse mathematical phenomena. It is also relevant to several areas of science and technology.

---

* From AMERICAN MATHEMATICAL MONTHLY, vol. 78 (1971), pp. 616–631.

Though convex sets are defined in various settings (see [27] for a survey), the most useful definitions are based on a notion of betweenness. When E is a space in which such a notion is defined, a subset C of E is called **convex** provided that for each two points x and y of C, C includes all points between x and y. The most important setting, and the only one to be discussed here, is that in which E is a vector space over the real number field R or, in particular, is the n-dimensional Euclidean space E^n, and the points between x and y are those of the line segment xy. Thus, a subset C of a real vector space is convex provided that C contains every segment whose endpoints both belong to C. (For example, a cube in E^3 is convex but its boundary is not, for the boundary does not contain the segment xy unless x and y lie together in some 2-dimensional face of the cube.) The importance of convexity theory stems from the fact that convex sets arise frequently in many areas of mathematics and are often amenable to rather elementary reasoning. Even the infinite-dimensional theory is based to a considerable extent on 2- and 3-dimensional reasoning.

The first systematic study of convexity was made by Minkowski (1864–1909), whose works [71] contain, at least in germinal form, most of the important ideas of the subject. The early developments of convexity theory were finite-dimensional and directed mainly toward the solution of quantitative problems; an excellent survey of them was made by Bonnesen and Fenchel [14] in 1934. Since 1940, however, the combinatorial, qualitative, and dimension-free parts of the theory have tended to predominate, perhaps because of their many applications in other areas of mathematics. After some preliminary material that is relevant to all parts of the theory, the present exposition begins with the quantitative and combinatorial aspects because they are restricted to the finite-dimensional spaces that are most likely to be familiar to the reader. In discussing, later, the qualitative and dimension-free aspects of the theory, some slight familiarity with topological vector spaces is assumed. The reader who lacks this familiarity may restrict his attention to the case of Euclidean n-space E^n.

A fascinating aspect of convexity theory is the large number of easily stated and intuitively appealing unsolved problems that it still contains. A few such problems are included here.

Preliminary Material. Any two distinct points x and y of a real vector space E determine a unique **line**. It consists of all points of the form $(1-\lambda)x+\lambda y$, λ ranging over all real numbers. Those points for which $\lambda \geq 0$ and for which $0 \leq \lambda \leq 1$ form respectively the **ray** from x through y and the **segment** xy. An **affine** set is one that contains all lines determined by pairs of its points; equivalently, it is a translate of a linear subspace. For example, the affine sets in E^3 are the empty set, the onepointed sets, the lines, the planes, and E^3 itself. A **hyperplane** H in E is an affine set of deficiency or codimension 1; that is, H is not properly contained in any affine subset of E other than E itself. In particular, the hyperplanes of E^n are its affine subsets of dimension $n-1$. For any hyperplane H, the complement $E{\sim}H$ is expressible in a unique way as the union of two convex sets. They are called the **open halfspaces** bounded by H and their unions with H are the **closed halfspaces** bounded by H. Two sets X and Y are said to be **separated** by H provided that X lies in one of these closed halfspaces and Y in the other. The set X is **supported** by H at the point x provided that x belongs to X but is separated from X by H. Fig. 1 shows a hyperplane H (in this case, a line) in E^2, separating the convex sets X and Y and supporting X at the point x.

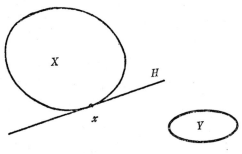

FIG. 1

Intimately related to the notion of a convex set is that of a convex function, which is important in most parts of convexity theory and in several areas of analysis. Let ϕ be a real-valued function whose domain D lies in a real vector space E. Then ϕ is called **convex** provided that D is convex and ϕ satisfies the inequality,

$$\phi((1 - \lambda)x + \lambda y) \leqq (1 - \lambda)\phi(x) + \lambda\phi(y),$$

for all points x and y of D and all numbers λ between 0 and 1. Equivalently, ϕ is a convex function if and only if its **epigraph** G is a convex set, where G is the subset of the product space $E \times R$ consisting of all ordered pairs (x, τ) such

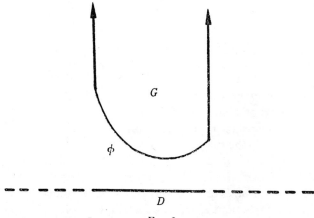

FIG. 2

that $x \in D$ and $\tau \geqq \phi(x)$. Fig. 2 shows a convex function ϕ, its domain D, and its epigraph G.

A function is called **concave** provided that its negative is convex, and **affine** provided that it is both convex and concave. A real-valued function on E is affine if and only if it differs by a constant from a linear function. The hyperplanes H in E are precisely the zero sets of the nonconstant affine functionals f on E. If H is the set of all x for which $f(x) = 0$, then the closed halfspaces bounded by H are determined by the inequalities $f(x) \leqq 0$ and $f(x) \geqq 0$.

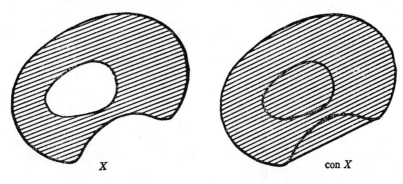

X con X

FIG. 3

The **convex hull** of a set X, denoted here by con X, is the intersection of all convex sets containing X. It is convex, as is any intersection of convex sets, and hence is the *smallest* convex set containing X. (Fig. 3 shows a nonconvex plane set and its convex hull.) Equivalently, con X is the set of all **convex combinations** of X—that is, points of the form $\sum_1^k \lambda_i x_i$, where the x_i's are points of X and the λ_i's are positive numbers whose sum is 1. For any such combination and for any convex function ϕ whose domain contains X,

$$\phi\left(\sum_1^k \lambda_i x_i \right) \leqq \sum_1^k \lambda_i \phi(x_i).$$

The **closed convex hull** of a set is the closure of its convex hull.

A point x of a convex set C is called an **extreme point** of C provided that $C \sim \{x\}$ is convex or, equivalently, that x is not an inner point of any segment in C. More generally, a **face** of C is a convex set $F \subset C$ such that F is not "crossed" by any segment in C—that is, $xy \subset F$ whenever x and y belong to C and F includes an inner point of xy. (For example, a cube in E^3 has six 2-faces, twelve 1-faces (edges), and eight 0-faces (extreme points). It is the convex hull of its set of extreme points.)

The terms *body* and *cone* are used in different ways by various authors. Here **body** always means a bounded closed subset of E^n that has nonempty interior, and **cone** means a set in a real vector space which is a union of rays from the origin O.

Quantitative Aspects. Convexity is a basic notion in the so-called geometry of numbers, and, indeed, it was the latter subject that led Minkowski to many of his investigations. One of his most striking results is that if C is a convex body in E^n which is symmetric about the origin O (that is, $x \in C$ implies $-x \in C$), and if the volume $V(C)$ is at least 2^n, then C includes at least one point other than O whose coordinates are all integers [21].

A **packing** of convex bodies is an arrangement in which no two of the bodies have common interior points. In addition to being of interest for themselves, packing problems are found in number theory, information theory, crystallography, botany, virology, and other areas of science. Often the interest is in packings of maximum density. The densest packing of congruent circular disks in E^2 is one in which the disks are inscribed in nonoverlapping regular hexagons covering E^2; each disk touches six others. (See Fig. 4.) It has long been conjectured that a densest packing of congruent spherical balls in E^3 is the cubic close-packing of Kepler, obtained by imagining the space to be divided into black and white cubes forming a 3-dimensional chessboard, and then placing a ball concentric with each black cube and tangent to each of the twelve edges of the cube; each ball touches twelve others. However, the conjecture has been proved only for packings of congruent balls whose centers form a lattice (if x and y are centers then so is $2x - y$). ([35] [81])

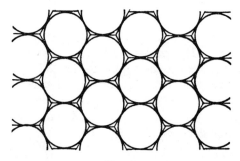

FIG. 4

There is a rich collection of quantitative results involving such measurements of convex bodies as volume, surface area, diameter, etc. Many extremal problems concerning such measurements have spherical balls or simplices as their solutions. (A **simplex** in E^n is the convex hull of $n+1$ points not contained in any hyperplane.) For example, the isoperimetric inequality asserts that if S and V are respectively the surface area and the volume of a convex body C in E^n, and if ω is the volume of an n-dimensional ball of unit radius, then $S^n \geq n\omega V^{n-1}$, with equality if and only if C is a ball. Thus of all bodies with given volume, balls have the least surface area [13] [14] [46] [47]. Any convex body C of E^n lies in a unique ball of minimum radius r. Jung's inequality, of interest in approximation theory, asserts that if C's diameter is d then $r \leq (n/(2n+2))^{1/2}d$, with equality if and only if C is a regular simplex [27] [47]. Loewner's theorem asserts that any body in E^n lies in a unique ellipsoid of minimum volume [28].

Convex sets are prominent in the theory of geometric probability. For example, the following problem was posed (in a different form) by Sylvester. Let C be a convex body of unit volume in E^n and let $n+1$ points be chosen from C, independently and at random. Except in degenerate cases, the convex hull of these points is an n-simplex. What is the expected volume, V_C, of the simplex? For $n=2$, the values of V_C are between $1/12$ and $35/48\pi^2$, attained respectively when C is a triangle and when C is an ellipse. For larger values of n, V_C is known when C is a ball but not when C is a simplex [61].([55] [72][73])

For convex bodies $C_1, \cdots, C_k$ in E^n and positive numbers $\lambda_1, \cdots, \lambda_k$, the set of all points of the form $\lambda_1 x_1 + \cdots + \lambda_k x_k$ with $x_i \in C_i$ is another convex body C, denoted by $\lambda_1 C_1 + \cdots + \lambda_k C_k$. When $C_1, \cdots, C_k$ are fixed, the volume of C is expressible as a homogeneous polynomial of degree n in the parameters $\lambda_1, \cdots, \lambda_k$. Some of the deepest parts of the quantitative theory concern the coefficients of this polynomial, which are called the **mixed volumes** of $C_1, \cdots, C_k$. A basic tool in the study of mixed volumes is the Brunn-Minkowski theorem asserting that for $0 < \lambda < 1$, $(V((1-\lambda)C_1 + \lambda C_2))^{1/n} \geqq (1-\lambda)(V(C_1))^{1/n} + \lambda(V(C_2))^{1/n}$ (that is, the nth root of the volume is a concave function of λ) and characterizing the cases of equality. Inequalities for mixed volumes yield the isoperimetric inequality and other inequalities of immediate geometric interest. ([14] [46] [47])

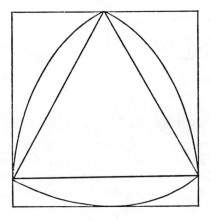

FIG. 5

A convex body C in E^n is said to be of **constant breadth** b provided that b is the distance between any two parallel supporting hyperplanes of C; equivalently, C is of diameter b and the diameter is increased by adding any point of E^n not in C. From the second description it follows that any set of diameter $\leqq b$ in E^n lies in at least one convex body of constant breadth b. Hence the following problem of Borsuk can be reduced to the case in which X is a convex body of constant breadth: Can every set X of diameter 1 in E^n be covered by

$n+1$ sets of diameter <1? The answer is affirmative for $n \leq 3$, unknown for $n > 3$ [42]. Noncircular plane convex bodies of constant breadth have been studied by many mathematicians. Their special properties have led to their use in kinematic linkages and in other mechanisms. Any such body can be placed in a square and then "rotated" while remaining in contact with all four sides of the square. (See Fig. 5.) ([5] [14] [87])

Having mentioned some unsolved problems in E^3 and E^4, we end this section with one in E^2. A **chord** of a convex body is a segment joining two boundary points, and an **equichordal point** is one through which all chords are of equal length. Does any plane convex body have two equichordal points? [60]

Combinatorial Aspects. Much of combinatorial convexity theory deals with intersection properties of convex sets. The intersection C of any family of convex sets is itself convex, though C may be empty. Helly's theorem asserts that C is nonempty if the convex sets are all in E^n, each $n+1$ of them have nonempty intersection, and the family is finite or its members are all compact. There are numerous generalizations and applications of Helly's theorem. From its 1-dimensional form it follows that if C is a cube in E^n then any family of pairwise intersecting translates of C has nonempty intersection. In fact, a convex body C has this intersection property if and only if C is affinely equivalent to a cube. ([27] [48] [87])

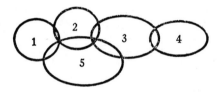 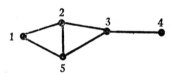

FIG. 6

The problem of determining all intersection properties of convex sets is not trivial even in E^1. It leads to the notion of an interval graph, which has been used in such diverse fields as molecular genetics, psychophysics, archaeology, and ecology. For any family of sets the associated **intersection graph** is an abstract graph having a node for each member of the family, two nodes being joined by an arc of the graph if and only if the corresponding sets intersect. Fig. 6 shows a family of convex sets and its intersection graph. Any finite graph can be realized as the intersection graph of a family of convex bodies in E^3, but not so in E^2 or E^1. An **interval graph** is one that is the intersection graph of a finite family of convex bodies in E^1. Such graphs have been characterized in various ways, but the corresponding problem relative to E^2 is still open. ([84] [86])

Another area of combinatorial research is concerned with the representation of convex hulls. The simplest and most useful result is Carathéodory's theorem, asserting that if $X \subset E^n$ and $u \in \mathrm{con}\, X$ then $u \in \mathrm{con}\, Y$ for some set Y consisting of $n+1$ or fewer points of X. For example, when $X\subset E^2$ any point of con X belongs to X, to a segment determined by two points of X, or to a triangle determined by three points of X. Carathéodory's theorem has many generalizations and applications, including the fact that con X is compact for each compact $X\subset E^n$. ([27] [79])

The most extensive combinatorial developments deal with the facial structure of convex polyhedra. Though terminology has not been standardized, we here use the term **polyhedron** to mean a subset of E^n that is the intersection of a finite number of closed halfspaces. Of special interest are the bounded polyhedra, here called **polytopes,** and the polyhedral cones. By the lemma of Farkas, a set is a polytope if and only if it is the convex hull of a finite set of points, and is a polyhedral cone if and only if it is the convex hull of a finite number of rays from the origin. More generally, the following five conditions on a set P in E^n are equivalent: (i) P is a polyhedron; (ii) P is a closed convex set whose number of faces is finite; (iii) P is the convex hull of a finite system of points and rays; (iv) P is the vector sum $B+C=\{b+c: b\in B,\ c\in C\}$ of a polytope B and a polyhedral cone C; (v) P is the closed convex hull of the union of a polytope B and a translate of a polyhedral cone C. Fig. 7 shows a 2-dimensional polyhedron P and the associated sets B and C. ([56])

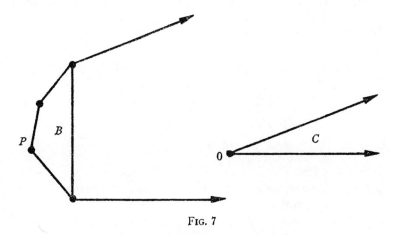

FIG. 7

In the combinatorial study of polyhedra, the first landmark was Euler's 1752 theorem asserting that $v-e+f=2$ for any 3-polytope, where v, e, and f are respectively the numbers of vertices, edges, and 2-faces. (Extreme points of polyhedra are usually called **vertices.**) The generalization of Schläfli and Poincaré asserts that if $f_i(P)$ is the number of i-dimensional faces of an n-dimensional polytope P, then $\sum_{i=0}^{n-1}(-1)^i f_i(P)=1-(-1)^n$. The second landmark was Steinitz's 1934 theorem characterizing the graphs of 3-polytopes, the combinatorial structures formed by vertices and edges, as those that are planar (representable in E^2 without crossings) and 3-connected (between any two vertices there are three independent paths). The first graph of Fig. 8 corresponds to a cube. The second and third graphs of Fig. 8 do not correspond to any 3-polytope, for the first is not 3-connected and the second is not planar. However, the third graph does correspond to a 4-dimensional simplex. Various properties, including n-connectedness, have been established for the graphs of n-polytopes, but no combinatorial characterization is known for $n>3$. ([43])

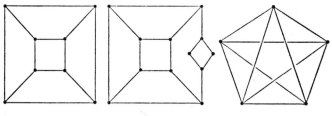

FIG. 8

A third landmark in the combinatorial study of polyhedra was the development, beginning in the late 1940's and still continuing, of computational techniques for minimizing linear functions on polyhedra. These techniques, known as *linear programming*, provide solutions to a wide range of practical optimization problems, and they are also useful for other computations involving polyhedra. Their importance led to a renewed interest in polyhedra and, for example, to rediscovery of the striking fact that for each $k>n$ there is an n-polytope with k vertices such that each $[n/2]$ vertices determine a face. A closely related development is the recent proof [70] that the maximum number of vertices possessed by any n-polyhedron with k $(n-1)$-faces is

$$\binom{k - [(n + 1)/2]}{k - n} + \binom{k - [(n + 2)/2]}{k - n}. \qquad ([26]\ [43]\ [45]\ [59])$$

Qualitative Aspects. The topics to be discussed in this section include normed vector spaces, polarity, separation and support theorems, extreme point theorems, and fixed point theorems.

For a real-valued function ϕ on a real vector space E, any two of the following conditions imply the third: subadditivity ($\phi(x+y) \leqq \phi(x) + \phi(y)$ for all x, $y \in E$); positive homogeneity ($\phi(\lambda x) = \lambda \phi(x)$ for all $x \in E$ and $\lambda \geqq 0$); convexity. A **norm** is a function that satisfies these conditions as well as being symmetric ($\phi(-x) = \phi(x)$) and positive ($x \neq 0$ implies $\phi(x) > 0$). A **normed vector space** consists of a vector space E together with a norm on E. The norm is usually denoted by $\| \ \|$ and leads to a useful notion of distance by defining the distance between x and y as $\|x - y\|$. The **unit ball** of such a space is the set of all points x for which $\|x\| \leqq 1$, while the **unit sphere** is defined by the condition $\|x\| = 1$. When $r = 2$, the function $\|x\|_r = (\sum_1^n |x_i|^r)^{1/r}$ is the usual Euclidean norm for E^n; an inequality of Minkowski asserts that it is a norm for all $r \geqq 1$. In general, the notion of convexity plays a key role in the theory of inequalities, and many useful inequalities assert merely that a certain function is convex. Another important norm for E^n is given by

$$\|x\|_\infty = \max_{1 \le i \le n} |x_i|.$$

The unit ball for $\|\ \|_\infty$ is an n-dimensional cube and for $\|\ \|_1$ is a so-called cross-polytope (a regular octahedron when $n=3$). Fig. 9 shows the unit spheres in E^2 associated with $\|\ \|_r$ for $r=1,\ 3/2,\ 2,\ 3,$ and ∞. ([29])

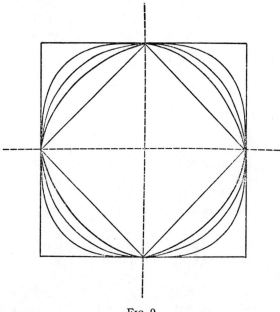

FIG. 9

The study of normed vector spaces is in a sense equivalent to the study of a certain class of convex sets. The equivalence involves the notion of the **gauge function** μ of a set U, where, for each $x \in E$, $\mu(x)$ is defined as the infimum of all numbers $\lambda > 0$ such that $x \in \lambda U$. For any norm $\|\ \|$ on E, the associated unit ball U is a convex set that intersects every line through O in a closed segment having O as its midpoint; further, $\|\ \|$ is the gauge function of U. Conversely, for any convex set U of the sort described, the associated gauge function is a norm for which U is the unit ball. Thus any property of a normed vector space can be described completely in terms of its unit ball or its unit sphere. For example, the **rotundity** of a normed space may be defined by saying that $\|x+y\| < \|x\| + \|y\|$ whenever x and y are not collinear with O, or by saying, equivalently, that the unit sphere does not contain any line segments. **Smoothness** of the space E may be defined by saying that for any two points x and y not equal to 0, the function $\phi(\lambda) = \|x+\lambda y\|$ is differentiable at $\lambda = 0$—or by saying, equivalently, that the unit sphere has at each point a unique supporting hyperplane. Euclidean n-space is both rotund and smooth. The behavior of a normed space can sometimes be improved by **renorming,** which means introducing a new norm that is caught between two positive multiples of the original norm and hence induces the same topology on the space. For example, any separable normed linear space can be renormed so as to be simultaneously smooth and strictly convex. ([24] [25] [29])

All of the quantitative problems mentioned earlier for Euclidean spaces have been studied also for finite-dimensional normed spaces, commonly called **Minkowski spaces.** The analogue of Jung's inequality asserts that if E is an n-dimensional Minkowski space and X is a set of diameter d in E, then there is a point z of E such that $\|x-z\| \leq (n/(n+1))d$ for all $x \in X$ [27]. In contrast to the Euclidean case, the unit ball of a Minkowski space need not be an extreme body so far as the isoperimetric problem for that space is concerned [18]. There are many characterizations of those Minkowski spaces which are equivalent to Euclidean spaces or, in other words, whose unit balls are ellipsoids. Most of these characterizations are related to the fact that, among the Minkowski spaces E of dimension ≥ 3, each of the following conditions characterizes the Euclidean spaces: (a) (Jordan—vonNeumann) $\|x+y\|^2+\|x-y\|^2=2\|x\|^2+2\|y\|^2$; (b) (Blaschke-Kakutani) for any hyperplane H through O in E, there is a linear projection of norm 1 of E onto H–equivalently, there is a line L such that if U is the unit ball then the intersection $U \cap H$ is equal to the intersection of U with the "cylinder" $U+L$. ([29] [58])

So far as the applications of convexity in other parts of mathematics are concerned, separation and support theorems are of special importance. They are widely used in functional analysis and have been used in game theory, in the theory of summability, and even to prove certain coloring theorems of graph theory. Together with Lyapunov's theorem asserting the convexity of the range of a nonatomic vector-valued measure [9] [49] [68], they are among the principal abstract tools of the theory of optimal control [62]. Separation theorems set forth conditions under which two nonempty disjoint convex subsets X and Y of a topological vector space E can be separated by a hyperplane, either in the weak sense defined above or in various stronger senses. It suffices, for example, that E should be finite-dimensional or that one of the sets should have nonempty interior. A consequence is that if C is a convex set whose interior is empty and A is a nonempty affine set disjoint from the interior of C, then A lies in a hyperplane separating A from C. In particular, a closed convex set C with nonempty interior is supported at each of its boundary points (Mazur-Bourgin). Though that conclusion may fail if C's interior is empty, the support points of C are dense in C's boundary if E is a Banach space (Bishop-Phelps) and also if E is locally convex and C weakly compact. If A is an affine subset of a real vector space E, ϕ is a convex function on E, and f is an affine function on A such that $f \leq \phi$ on A, then f can be extended to an affine function g on E with $g \leq \phi$ on E. This result, a slight improvement of the classical Hahn-Banach theorem of functional analysis, follows from the separation theorem applied to the graph of f and the epigraph of ϕ. Because of this and other relationships, separation and support theorems may be regarded as geometric relatives of the Hahn-Banach theorem. ([16] [29] [54] [62] [64] [80] [83])

The notion of polarity is essential in convexity theory and in the theory of topological vector spaces. Let E and F be two spaces that are paired by a bilinear form $\langle \; , \; \rangle$. For example, let $E = F = E^n$ and let $\langle \; , \; \rangle$ be the usual inner product given by

$$\langle (x_1, \cdots, x_n), (y_1, \cdots, y_n) \rangle = x_1 y_1 + \cdots + x_n y_n.$$

For any $X \subset E$ the **polar** X^0 of X is the set of all $y \in F$ such that $\langle x, y \rangle \leq 1$ for all $x \in X$. The polar is always convex, being an intersection of halfspaces. In geometry this notion plays two roles that are dual to each other. When unable to prove directly a theorem about sets $X_1, X_2, \cdots$, one may find it possible to prove an equivalent statement about $X_1^0, X_2^0, \cdots$. On the other hand, having proved an interesting theorem about $X_1, X_2, \cdots$, one may find that the polar form of the theorem is also of interest. For $1 \leq r$ and $1/r + 1/s = 1$, the $\| \; \|_r$ unit ball of E^n is polar to the $\| \; \|_s$ unit ball of E^n. Some polar pairs of this sort appeared in Fig. 9. There are close relationships among the notion of the polar of a convex set, the so-called support function of a convex body, and the notion of the conjugate of a normed linear space. ([16] [29] [43] [54] [64])

Both convex and concave functions occur often in practical optimization problems, and both have properties that are helpful in such problems. Let f be a continuous real-valued function whose domain D lies in a locally convex topological vector space E. If f is convex, then any local minimum x_0 for f is a global minimum [80] [83] [88]. That is to say, if there is a neighborhood U of x_0 such that $f(x_0) \leq f(x)$ for all $x \in U \cap D$, then the same inequality holds for all $x \in D$. This justifies various iterative procedures for finding or approximating x_0. [88] If f is concave and D compact, then f attains a minimum at an extreme point of D. That is one of the reasons for the importance of extreme points in functional analysis. The other reason is contained in theorems of Krein and Milman, asserting that if C is a compact convex subset of a locally convex space, and if $X \subset C$, then C is equal to the closed convex hull of X if and only if the closure of X includes all extreme points of C. Thus C's extreme points form the smallest set by means of which, using convex combinations, all points of C can be approximated. There are extensions of the Krein-Milman theorem to certain noncompact sets, and some sharpening is possible in the finite-dimensional case. For example, if C is a closed convex set in E^n and C contains no line, then C is the convex hull of its extreme points together with its extreme rays. (An **extreme ray** of C is a ray not "crossed" by any segment.) ([3] [6] [22] [29] [54] [64] [76])

When C is a compact convex set in a locally convex space E and X is the set of all extreme points of C, it follows from the Krein-Milman theorem that each point p of C is the barycenter of a probability measure μ carried by the closure $\overline{X}$ of X—that is,

$$f(p) = \int_{x \in \overline{X}} f(x) d\mu(x)$$

for all continuous linear functionals f on E. Several of the integral representation theorems of analysis are consequences of this. However, when X is not closed, it is desirable to have the sharper representation afforded by a measure that is carried by X rather than $\overline{X}$. Choquet's theorem, which has stimulated much research in recent years, asserts that such a representation is always possible when C is metrizable. Uniqueness of the representation, for all $p \in C$, is associated with a useful dimension-free notion of simplex. These ideas have been applied in several fields—for example, in potential theory and in the theory of operator algebras. ([3] [22] [41] [76])

In conclusion, we turn briefly to fixed-point theorems for convex sets, stating two of the simplest but most important ones. Both have been extended in many ways. The Brouwer-Schauder-Tychonov theorem asserts that if C is a compact convex set in a locally convex space and if ϕ is a continuous mapping of C into C, then there is at least one point p of C such that $\phi(p) = p$. The theorem and its relatives are used in many ways, such as proving existence theorems for differential and integral equations, minimax theorems for game theory, and various geometric properties of convex sets. (For example, a compact subset of E^n or of Hilbert space H is convex if and only if each point of the space admits a unique nearest point in the set. It is unknown whether "compact" may be replaced by "closed" in H, though it may be in E^n.) A recent computational development [66] makes it possible to regard this fixed-point theorem as a tool for *constructing* solutions of various sorts of systems, rather than merely establishing their existence. ([15] [30])

The Markov-Kakutani theorem asserts that if C is a compact convex subset of a topological vector space and if Φ is a commuting family of continuous affine transformations of C into C, then there is at least one point $p \in C$ such that $\phi(p) = p$ for all $\phi \in \Phi$. It is used to prove the existence of invariant means on commutative groups, of a finitely additive translation-invariant extension of Lebesgue measure to all bounded subsets of E^n, and in many other ways. Note that local convexity is not required for the Markov-Kakutani theorem. It is unknown whether the assumption can be abandoned in the case of the Krein-Milman extreme point theorem and the Tychonov fixed-point theorem. ([15] [16] [30])

References

1. A. D. Alexandrov, Die innere Geometrie der konvexen Flächen, Akademie-Verlag, Berlin, 1955. (Translated from the 1948 Russian edition.)

2. ———, Konvexe Polyeder, Akademie-Verlag, Berlin, 1958. (Translated from the 1950 Russian edition by H. Abetz, under the scientific editorship of W. Süss.)

3. E. Alfsen, Convex Compact Sets and Boundary Integrals, Springer, Berlin, 1971.

4. R. V. Benson, Euclidean Geometry and Convexity, McGraw-Hill, New York, 1966.

5. L. Beretta and A. Maxia, Insiemi convessi e orbiformi, Univ. Roma e Ist. Naz. Alta Mat. Rend. Mat. (5) 1 (1940) 1–64.

6. C. Berge, Espaces Topologiques: Fonctions Multivoques, Dunod, Paris, 1959.

7. C. Berge and A. Ghouila-Houri, Programming, Games, and Transportation Networks, Methuen, London and Wiley, New York, 1965. (Translated by M. Merrington and C. Ramanujacharyulu from the 1962 French edition.)

8. W. Blaschke, Kreis und Kugel, Teubner, Leipzig, 1916. (Reprinted in 1949 by Chelsea, New York.)

9. E. Bolker, A class of convex bodies, Trans. Amer. Math. Soc., 145 (1969) 323–345.

10. V. G. Boltyanskiĭ and I. Ts. Gohberg, Theorems and Problems of Combinatorial Geometry, (Russian) Nauka, Moscow, 1965.

11. V. G. Boltyanskiĭ and I. M. Yaglom, Convex figures and bodies (Russian), Encyclopedia of Elementary Mathematics, Nauka, Moscow, 1966, pp. 182–269.

12. V. G. Boltyanskiĭ and I. M. Yaglom, Geometric problems on maxima and minima (Russian), Encyclopedia of Elementary Mathematics, Nauka, Moscow, 1966, pp. 270–348.

13. T. Bonnesen, Les Problèmes des Isopérimétriques et des Isépiphanes, Gauthier-Villars, Paris, 1928.

14. —— and W. Fenchel, Theorie der konvexen Körper, Springer, Berlin 1934. (Reprinted by Chelsea, New York, 1948.)

15. F. Bonsall, Lectures on Some Fixed Point Theorems of Functional Analysis, Tata Institute of Fundamental Research, Bombay, 1962.

16. N. Bourbaki, Espaces Vectoriels Topologiques (rev. ed.), Hermann, Paris, 1966, Chaps. 1–2 and 1967, Chaps. 3–5.

17. M. Brückner, Vielecke und Vielfläche, Teubner, Leipzig, 1900.

18. H. Busemann, The foundations of Minkowskian geometry, Comment. Math. Helv., 24 (1950) 156–187.

19. ——, Convex Surfaces, Interscience, Wiley, New York, 1958.

20. ——, The Geometry of Geodesics, Academic Press, New York, 1955.

21. J. W. S. Cassels, An Introduction to the Geometry of Numbers, Springer, Berlin, 1959.

22. G. Choquet, Lectures on Analysis (ed. by J. Marsden, T. Lance, and S. Gelbart), Benjamin, New York, 1969.

23. H. S. M. Coxeter, Regular Polytopes, Pitman, London, 1948. (Second ed. Macmillan, New York and Collier-Macmillan, London, 1963.)

24. D. F. Cudia, Rotundity, in Convexity, Proc. Sympos. Pure Math., Amer. Math. Soc., 7 (1963) 73–97.

25. ——, The geometry of Banach spaces. Smoothness, Trans. Amer. Math. Soc., 110 (1964) 284–314.

26. G. B. Dantzig, Linear Programming and Extensions, Princeton University Press, Princeton, 1963.

27. L. Danzer, B. Grünbaum, and V. Klee, Helly's theorem and its relatives, in Convexity, Proc. Sympos. Pure Math., Amer. Math. Soc., 7 (1963) 101–180. (Also, an expanded version of this, translated into Russian by S. Zalgaller under the editorship of I. M. Yaglom, published as Helly's Theorem and its Applications (Russian), *Mir*, Moscow, 1968.)

28. L. Danzer, D. Laugwitz, and H. Lenz, Über das Löwnersche Ellipsoid und sein Analogon unter den einem Eikörper einbeschriebenen Ellipsoid, Arch. Math. (Basel), 8 (1957) 214–219.

29. M. M. Day, Normed Linear Spaces, Springer, Berlin, 1958.

30. N. Dunford and J. T. Schwartz, Linear Operators, Part I: General Theory, Interscience, New York, 1958.

31. V. Eberhard, Zur Morphologie der Polyeder, Teubner, Leipzig, 1891.

32. H. G. Eggleston, Convexity, Cambridge University Press, New York, 1958.

33. ——, Problems in Euclidean Space: Application of Convexity, Pergamon, New York, 1957.

34. K. Fan, Convex Sets and their Applications, Lecture Notes, Argonne National Laboratory, Illinois, 1959.

35. L. Fejes Tóth, Lagerungen in der Ebene, auf der Kugel und im Raum, Springer, Berlin, 1953.

36. ——, Regular Figures, Pergamon, London, 1964.

37. W. Fenchel, ed., Convex Cones, Sets and Functions, Lecture Notes, Princeton University, 1951.

38. W. Fenchel, Proceedings of the Colloquium on Convexity, Copenhagen, Denmark, 1965, University of Copenhagen, 1967.

39. D. Gale, The Theory of Linear Economic Models, McGraw-Hill, New York, 1960.

40. A. A. Goldstein, Constructive Real Analysis, Harper and Row, New York, 1967.

41. A. Goullet de Rugy, Géométrie des Simplexes, Centre de Documentation Universitaire et S.E.D.E.S. Réunis, Paris, 1968.

42. B. Grünbaum, Borsuk's problem and related questions, in Convexity, Proc. Sympos. Pure Math., vol. 7, Amer. Math. Soc., (1963) 271–284.

43. ———, Convex Polytopes, Interscience-Wiley, London, 1967.

44. ——— and V. Klee, Convexity and Applications (L. Durst, ed.), Proc. CUPM Geometry Conference, Santa Barbara, 1967, MAA Committee on the Undergraduate Program in Mathematics, Berkeley, 1967.

45. ——— and G. C. Shepard, Convex polytopes, Bull. London Math. Soc., 1 (1969) 257–300.

46. H. Hadwiger, Altes und Neues über konvexe Körper, Birkhäuser, Basel, 1955.

47. ———, Vorlesungen über Inhalt, Oberfläche und Isoperimetrie, Springer, Berlin, 1957.

48. ———, H. Debrunner, and V. Klee, Combinatorial Geometry in the Plane, Holt, Rinehart, and Winston, New York, 1964.

49. H. Halkin, On the necessary condition for optimal control of nonlinear systems, J. Analyse Math, 12 (1964) 1–82.

50. A. D. Joffe and V. M. Tikhomirov, Duality of convex functions and extremum problems. Russian Math. Surveys, 23 no. 6 (1968) 53–124. (Translated by T. Garrett from Uspehi Mat. Nauk 23, no. 6 (1968) 51–116.)

51. S. Karlin, Mathematical Methods and Theory in Games, Programming, and Economics, vols. 1–2, McGraw-Hill, New York, 1960.

52. ——— and W. J. Studden, Tchebycheff Systems: with Applications in Analysis and Statistics, Interscience-Wiley, New York, 1966.

53. N. D. Kazarinoff, Geometric Inequalities, Random House, New York, 1961.

54. J. L. Kelley, I. Namioka, et al., Linear Topological Spaces, Van Nostrand, Princeton, N. J., 1963.

55. M. G. Kendall and P. A. P. Moran, Geometrical Probability, Hafner, New York, 1963.

56. V. Klee, Some characterizations of convex polyhedra, Acta Math., 102 (1959) 79–107.

57. ———, ed., Convexity, Proc. Symp. Pure Math., Amer. Math. Soc., 7 (1963).

58. ———, Review of a Paper by W. Rudin and K. T. Smith, Math. Rev., 23A (1962) 376-377.

59. ———, Convex polytopes and linear programming, Proceedings of the IBM Scientific Computing Symposium on Combinatorial Problems, IBM Data Processing Division, White Plains, N. Y., 1966, pp. 123–158.

60. ———, Can a plane convex body have two equichordal points? this MONTHLY, 76 (1969) 54–55.

61. ———, What is the expected volume of a simplex whose vertices are chosen at random from a given convex body? this MONTHLY, 76 (1969) 286–288.

62. ———, Separation and support properties of convex sets—a survey, Control Theory and the Calculus of Variations, Academic Press, New York, 1969, pp. 235–303.

63. ———, Convexité, Encyclopaedia Universalis, Paris, 4 (1970) 982–985.

64. G. Köthe, Topological Vector Spaces I, Springer, New York, 1969. (Translated from the 1965 German edition by D. J. H. Garling.)

65. M. A. Krasnoselskiĭ and Ya B. Rutickiĭ, Convex Functions and Orlicz Spaces, Noordhoff, Groningen, 1961. (Translated by L. F. Boron from the 1958 Russian edition.)

66. H. Kuhn, Approximate Search for Fixed Points, Computing Methods in Optimization Problems—2, Academic Press, New York, 1969, pp. 199–211.

67. ——— and A. W. Tucker, eds., Linear Inequalities and Related Systems, Princeton University Press, 1956.

68. J. Lindenstrauss, A short proof of Liapounoff's convexity theorem, J. Math. Mech., 15 (1966) 971–972.

69. L. A. Lyusternik, Convex Figures and Polyhedra, Dover, New York, 1963. (Translated by T. J. Smith from the 1956 Russian edition.)

70. P. McMullen, The maximum numbers of faces of a convex polytope, Mathematika, 17 (1970) 179–184.

71. H. Minkowski, Gesammelte Abhandlungen, Teubner, Berlin, 1911.

72. P. A. P. Moran, A note on recent research in geometrical probability, J. Appl. Probability, 3 (1966) 453–463.

73. ———, A second note on recent research in geometrical probability, Advances in Appl. Probability, 1(1969) 73–89.

74. J. J. Moreau, Fonctionnelles Convexes, Lecture Notes, Séminaire "Equations aux Dérivées Partielles," College de France, 1966–67.

75. T. S. Motzkin, Beiträge zur Theorie der linearen Ungleichungen, Ph.D. Thesis, Basel, 1933 (Jerusalem, 1936).

76. R. R. Phelps, Lectures on Choquet's Theorem, Van Nostrand, Princeton, N. J., 1966.

77. V. Pogorelov, Die Verbiegung konvexer Flächen, Akademie-Verlag, Berlin, 1958. (Translated from the 1951 Russian edition.)

78. T. Popoviciu, Les Fonctions Convexes, Hermann, Paris, 1945.

79. J. R. Reay, Generalizations of a theorem of Carathéodory, Memoir No. 54, Amer. Math. Soc., 1965.

80. R. T. Rockafellar, Convex Analysis, Princeton University Press, 1970.

81. C. A. Rogers, Packing and Covering, Cambridge University Press, 1964.

82. E. Steinitz and H. Rademacher, Vorlesungen über die Theorie der Polyeder, Springer, Berlin, 1934.

83. J. Stoer and C. Witzgall, Convexity and Optimization in Finite Dimensions I, Springer, Berlin, 1970.

84. A. C. Tucker, Matrix characterizations of circular-arc graphs, Pacific J. Math., 35 (1971) to appear.

85. F. A. Valentine, Convex Sets, McGraw-Hill, New York, 1964.

86. G. Wegner, Eigenschaften der Nerven homologisch-einfacher Familien im R^n, Ph.D. Thesis, Göttingen, 1967.

87. I. M. Yaglom and V. G. Boltyanskiĭ, Convex Figures, Holt, Rinehart, and Winston, New York, 1961.

88. W. Zangwill, Nonlinear Programming: A Unified Approach, Prentice-Hall, Englewood Cliffs, N. J., 1969.

Editorial Note: The reader may be interested in three related articles which appeared in this MONTHLY: E. F. Beckenbach, Convexity properties of surfaces of negative curvature, vol. 55 (1948), 285–301. Truman Botts, Convex sets, vol. 49 (1942), 527–35. L. L. Dines, On convexity, vol. 45 (1938), 199–209.

SOME RECENT RESULTS ON TOPOLOGICAL MANIFOLDS*

REINHARD SCHULTZ, Purdue University

Although topological spaces exist in great variety and can exhibit strikingly unusual properties, the main concern of topology has generally been the study of spaces which are relatively well-behaved. One particularly interesting class of examples is given by those spaces which locally look like Euclidean n-space R^n. Explicitly, a Hausdorff space X is called a **topological n-manifold** (without boundary) if each point of X has an open neighborhood which is homeomorphic to an open subset in R^n. Since open sets in R^m and R^n are homeomorphic if and only if $m = n$, the integer n is a homeomorphism invariant of X and is called the **dimension** of X. In this paper all manifolds under consideration are assumed to be second countable.

Topological manifolds arise naturally in several different ways. For example, they are useful in the qualitative study of differential equations inaugurated by Poincaré (compare [1]). Topological manifolds are also a natural generalization of the mathematical systems studied in non-Euclidean and Riemannian geometry. Many interesting results on topological manifolds are generalizations of older theorems originally proved for these and similar mathematical systems.

During the nineteen sixties important advances in the study of topological manifolds yielded a great deal of information on their basic geometric structure. In particular, two long standing conjectures regarding topological manifolds were shown to be systematically false (see Section 4). One of the most useful results on topological manifolds of dimension $\neq 4, 5$—their description in terms of attaching handles—will be discussed in Section 5. This result allows one to take certain theorems which had previously been proved under additional structural assumptions and generalize them to topological manifolds with only minimal changes in the proofs.

I wish to thank R. Kirby for his detailed comments on an earlier version of this paper.

1. Classification of topological manifolds. Before beginning our discussion, it will be useful to generalize the definition of topological n-manifolds to include the possibility of a boundary. Let R^n_+ be the set of points in R^n whose last coordinate is nonnegative. Then a **topological n-manifold with boundary** is a Hausdorff space X each point of which has an open neighborhood homeomorphic to an open subset of R^n or R^n_+.

Of course, the set of all points having neighborhoods homeomorphic to open subsets of R^n is a topological n-manifold without boundary as previously defined. It is easy to see that the set of such points is open and dense in M; this subset is called the **interior** of M and written Int M. The complement of Int M is called the **boundary** of M and written ∂M; it follows that ∂M is a topological $(n-1)$-

---

* From AMERICAN MATHEMATICAL MONTHLY, vol. 78 (1971), pp. 941–952.

manifold without boundary. The following theorem of M. Brown [9] is extremely important in the study of manifolds with boundary:

THEOREM 1.1. (Collar Neighborhood Theorem) *Let M be a manifold with boundary. Then there is an open neighborhood V of ∂M which is homeomorphic to $\partial M \times [0, 1)$ such that $\partial M \subseteq V$ corresponds to $\partial M \times \{0\}$.*

One of the most immediate problems regarding topological manifolds is their classification up to homeomorphism. The techniques of point set topology suffice for the classification of one-dimensional manifolds; this was completed during the second decade of the twentieth century (see [37] or [41]). There are only four different homeomorphism types of connected one-dimensional manifolds: The open interval, the half-open interval, the closed interval, and the circle.

The study of two-dimensional manifolds is somewhat more difficult and requires a systematic investigation of polyhedra in the Euclidean plane (e.g., see [29], [30], or [41]). One of the earliest results was the Jordan Curve Theorem, first proved correctly by Veblen in 1905 [59]. This theorem was augmented by a result of Schoenflies [48], and we may combine the two theorems into the following single statement:

THEOREM 1.2. (Jordan-Schoenflies Theorem). *Let X be a subset of R^2 which is homeomorphic to a circle. Then $R^2 - X$ has two components, one bounded and one unbounded, and X is the point set-theoretic frontier of each component. The homeomorphism from the unit circle to X extends to a homeomorphism from the unit disk to the closure of the bounded component of $R^2 - X$.*

This theorem is the basic result needed for the following theorem of Radó [45]:

THEOREM 1.3. *Any (unbounded) topological two-dimensional manifold M may be triangulated; i.e., there is a countable locally finite covering $\{T_i\}$ of M by compact subspaces satisfying:*

(i) *There are canonical homeomorphisms h_i from T_i to the solid triangle*

$$\{(x, y) \in R^2 \mid x \geq 0, y \geq 0, \text{ and } x + y \leq 1\}.$$

(ii) *Under these homeomorphisms any nonempty intersection $T_i \cap T_j$ corresponds to either a common side or a common vertex.*

The classification of two-dimensional manifolds up to homeomorphism then follows from a study of triangulated manifolds (e.g., see [18], [30], [46]). Results of Moise imply that the classification of three-dimensional manifolds reduces to a study of triangulated three-dimensional manifolds (e.g., [35], [36]). The classification is fairly well understood, modulo Conjecture 1.4 below.

If X is any arcwise connected space, then X is said to be **simply connected** if any continuous map from the unit circle in R^2 to X (i.e., a *closed curve* in X)

extends to a continuous map of the unit disk. Given this definition, we may state the following conjecture made by Poincaré in 1904 [**44**]:

CONJECTURE 1.4 (Poincaré Conjecture). *Let M be a compact topological 3-manifold without boundary that is simply connected. Then M is homeomorphic to the unit sphere in R^4 (i.e., the 3 dimensional sphere).*

No general classification scheme for compact topological manifolds exists in arbitrary dimension $\geqslant 4$ (compare [**79**] and [**5**, pp. 375–6]), for every finitely presented group arises as the fundamental group of a manifold in each such dimension (see [**14**] or [**30**]), and consequently a classification scheme would yield a solution to the word problem for finitely presented groups (see [**47**, Ch. XII] for a discussion of the latter; also see [**82**], [**83**]). In dimension $\geqslant 5$ the desired manifolds are easy to construct; in this case there is enough space to make construction with circles and disks almost at will. This is not the case in dimension four; both the direct approach for lower dimensions and the general approach for higher dimensions break down, and the study of such manifolds has proved to be very difficult (see [**52**], [**69**], [**75**], [**77**], [**78**], [**85**], [**86**]). Thus the desired 4-manifolds are necessarily given by a more delicate construction [**87**, Aufgaben 2–3, S. 180].

2. Generalized Schoenflies and Poincaré Conjectures. The Jordan curve theorem was soon generalized to higher dimensions by Brouwer ([**7**], also see [**14**, §18] or [**54**]). However, Antoine [**4**] and Alexander [**3**] constructed examples of subspaces X in R^3 that are homeomorphic to the unit sphere in R^3 but are not the frontiers of subspaces homeomorphic to the unit disk*; counterexamples similar to Alexander's exist in all higher dimensions. On the other hand, Alexander also proved that X bounds a disk if it is a polyhedron in R^3 [**2**]. Around 1960 B. Mazur [**31**], M. Morse [**38**], and M. Brown [**8**] proved results implying the following generalization of the Jordan-Schoenflies theorem:

THEOREM 2.1. (Generalized Schoenflies Theorem). *Let X be a subset of R^n that is homeomorphic to the unit sphere, and assume that the closure of the bounded component of $R^n - X$ is a topological n-manifold with boundary. Then the homeomorphism from the sphere to X extends to a homeomorphism from the disk to the closure of the bounded component of $R^n - X$.*

About the same time that the Generalized Schoenflies Theorem was proved, Smale [**53**], Stallings [**55**], and Zeeman [**64**] proved a generalization of the Poincaré Conjecture (1.4 above) in all dimensions greater than four; however, their proofs required additional structure on the manifolds under consideration (i.e., they had to be differential or combinatorial manifolds as defined in Section 3). Several years later Newman gave a proof of this result for topological manifolds using his generalization of Stallings' techniques and arguments of E. H. Connell [**42**]. For completeness, we state the result below:

---

* See [**66**] for a comprehensive survey.

THEOREM 2.2. (Generalized Poincaré Conjecture). *Let M be a compact topological n-manifold ($n \geq 5$) without boundary that is $\frac{1}{2}(n-1)$-connected if n is odd and $\frac{1}{2}n$-connected if n is even. Then M is homeomorphic to the unit sphere in R^{n+1}.*

REMARKS 1. A topological space X is said to be k-**connected** if any continuous map from the unit sphere in R^{k-m+1} (for any $m \geq 0$) extends to a map of the unit disk.

2. We already noted that the three-dimensional case of Theorem 2.2 is unknown; the four-dimensional case is also unknown.

The proof breaks down in dimensions 3 and 4 because in these cases there is not enough room in the manifold to make all the constructions needed in the proof (compare the last paragraph of Section 1).

Smale's proof of the generalized Poincaré Conjecture (most of whose details are independently due to A. H. Wallace [62]) was a central technique in the theory of surgery on manifolds developed by Kervaire, Milnor, S. P. Novikov, W. Browder, and C. T. C. Wall (for a definitive account see Wall's book [61]). Wall's theory in turn was important in studying the following elaboration of the Generalized Schoenflies Conjecture:

CONJECTURE 2.3. (Annulus Conjecture). *Let $A \subseteq R^{n+1}$ be a compact topological $(n+1)$-manifold whose boundary is homeomorphic to a disjoint union of two copies of the unit sphere in R^{n+1}. Then A is homeomorphic to the closed annulus in R^{n+1} bounded by the spheres of radius 1 and 2.*

If this conjecture were false for $n = 1$ or 2, then an argument of Brown and Gluck [10, p. 42] would imply that the compact unbounded topological manifold A' formed from A by gluing together the two components of the boundary of A could not be triangulated. Hence the conjecture is certainly true in these dimensions by *reductio ad absurdum* (more elementary arguments are also possible). In [19] Kirby gave an elegant argument which reduced the proof of the annulus conjecture for $n \geq 4$ to a problem which could be handled by means of Wall's surgery theory. This surgery theoretical problem was solved independently by Wall [60] and W.-C. Hsiang and Shaneson ([15], [16]); thus Conjecture 2.3 is true except possibly in the case $n = 3$.

3. Differentiable and Combinatorial Manifolds. In this section we shall describe the kinds of "additional structure" often associated to topological manifolds and mentioned in the previous sections.

The topological manifolds appearing in analysis and differential geometry usually satisfy the conditions appearing in the following definition:

DEFINITION. A topological n-manifold is **smoothable** if there is a collection of pairs $\{(U_\alpha, h_\alpha)\}_{\alpha \in A}$ satisfying:

(i) U_α is an open subset in R^n.

(ii) The map $h: U_\alpha \to M$ is a homeomorphism onto an open subset.

(iii) The functions $h_\beta^{-1} h_\alpha : h_\alpha^{-1} h_\beta(U_\beta) \to h_\beta^{-1} h_\alpha(U_\alpha)$ are functions of class C^r for some $r \geq 1$.

If U and W are open subsets of Euclidean spaces, recall that a map $f: U \to W$ is a **function of class** C^r if the coordinate functions f^i defined by $f(w) = (f^1(w), \ldots, f^m(w))$ each have all possible partial derivatives of order r and these functions are continuous; a function is C^∞ if it is C^r for every positive integer r. Two collections $\{(U_\alpha, h_\alpha)\}$ and $\{(V_\beta, k_\beta)\}$ satisfying (i)–(iii) are **equivalent** if their union satisfies property (iii); it follows that every collection $\{(U_\alpha, h_\alpha)\}$ is equivalent to a unique maximal collection $\mathcal{C}$ which is called a **smooth atlas** for M of class C^r. A **differentiable** (or **smooth**) n-manifold is a pair $(M, \mathcal{C})$ consisting of a smoothable n-manifold M and a smooth atlas $\mathcal{C}$. We shall always assume that the atlas is smooth of class C^∞, since it is known that any C^r atlas corresponds to a unique C^∞ atlas [40, Sections 4 and 5].

More generally, if Γ is any reasonable family of continuous functions from open sets in R^n to open sets in R^n (technically a **pseudogroup**; see [22]), then it is possible to define a Γ atlas and a Γ n-manifold. In topological investigations Γ is usually taken to be the C^r functions defined above or the piecewise linear (PL) functions defined below. Thus in order to define a piecewise linear n-manifold, it is only necessary to specify which mappings on open subsets of Euclidean space are piecewise linear; this requires a succession of definitions.

DEFINITION. Let $x_0, \cdots, x_n$ be points in R^m such that $x_1 - x_0, \cdots, x_n - x_0$ are linearly independent. Then the n-dimensional **simplex** (or n-simplex) with vertices $x_0, \cdots, x_n$ is the set of all linear combinations $y = \sum t_i x_i$, where each t_i is nonnegative and $\sum t_i = 1$ (the last condition and linear independence imply that the t_i are unique). The x_i are called the **vertices** of the simplex.

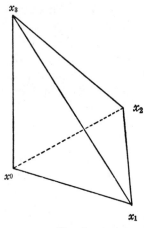

FIG. 1

A simplex is actually a generalized version of a triangle. It is immediate from the definition that a 1-simplex is a line segment and a 2-simplex is a solid triangle. Furthermore, a 3-simplex is a tetrahedron (see Figure 1).

DEFINITION. Let A be a simplex with vertices $a_i (1 \leq i \leq n)$, and let V be any real vector space. A function $f:A \to V$ is **affine linear** provided $y \in A$ and $y = \sum t_i a_i$ with $\sum t_i = 1$ imply $f(y) = \sum t_i f(a_i)$.

DEFINITION. Let U and V be any subsets of R^n. A continuous function $f: U \to V$ is **piecewise linear** (or **PL**) if there is a countable locally finite covering $\mathcal{B}$ of U by simplexes such that f is an affine linear map on each element of $\mathcal{B}$.

REMARK. Any open subset of R_+^n has many countable locally finite coverings by simplexes.

EXAMPLES 1. Let $f:R^n \to R^n$ be an affine transformation; i.e., $f(x) = Lx + y$, where L is a linear transformation. Then f is automatically affine linear on every simplex in R^n (compare [6, p. 272]).

2. Let $f:R^2 \to R^2$ be given by $f(x, y) = (x, y)$ if $y \geqq 0$ and $(x, 2y)$ if $y \leqq 0$. Then f is affine linear on any simplex contained in either the upper or lower half plane.

3. Let f be the map which sends the solid regular pentagon $ABCDE$ to the solid irregular pentagon $A'B'C'D'E'$ in Figure 2 by stretching the triangle OXY into $O'X'Y'$.

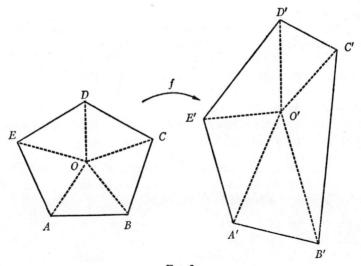

FIG. 2

A fundamental theorem of Cairns and Whitehead states that any smooth manifold determines a basically unique PL manifold ([11], [63], [40, Pt. II]). However, a PL manifold need not be determined by a smooth manifold (a result of Thom [58]), and two distinct smooth manifolds may determine the same PL manifold (a result of Milnor [32]). A comprehensive study of the relationship between smooth and PL manifolds appears in [25].

In the following section we shall discuss the parallel problem regarding the existence and uniqueness of PL manifolds associated to a given topological manifold.

REMARK. For historical reasons the study of PL manifolds and related objects is frequently called **combinatorial topology** and PL manifolds are often called **combinatorial manifolds**.

4. The Triangulation Conjecture and the Hauptvermutung. The following conjectures were formulated (in roughly equivalent form) soon after the establishment of combinatorial topology as a subject in its own right.

TRIANGULATION CONJECTURE. *Any topological n-manifold has a PL atlas.*

HAUPTVERMUTUNG FOR MANIFOLDS. *Any two homeomorphic PL n-manifolds are equivalent as PL manifolds.*

The results quoted in Section 1 imply that the first conjecture is true if $n \leq 3$. Similarly, the second conjecture is true if $n \leq 3$ ($n = 1$, straightforward; $n = 2$, see Papakyriakopoulos [43]; $n = 3$, see Moise [35], [36]). The solution of the generalized Poincaré conjecture in higher dimensions implies that the second conjecture is true for PL manifolds homeomorphic to spheres of dimension at least five. A fairly strong version of the *Hauptvermutung* for simply connected manifolds was proved by Lashof and Rothenberg [26], and Sullivan ([56], [57]); in the next paragraph we shall discuss subsequent results which eliminated the simple connectivity assumption (see Theorem 4.2).

Kirby's reduction of the Annulus Conjecture, other results appearing in [19], and consequences of these results due to Lees [28] led directly to initial results on the Triangulation Conjecture due to Lashof [23]. These theorems and computations of Casson, Wall, Hsiang, and Shaneson ([15], [16], [60]) in turn led to the following strong results on the Triangulation Conjecture and the *Hauptvermutung* due to Lashof and Rothenberg ([27], [76]), and Kirby and Siebenmann [20]:

THEOREM 4.1. *Let M be a topological manifold of dimension at least six (or five in the unbounded case), and assume that the four-dimensional cohomology group $H^4(M; Z_2)$ is zero. Then M has a PL atlas.*

THEOREM 4.2. *Let M be a PL manifold satisfying the above dimensional restriction, and assume that the three-dimensional cohomology group $H^3(M; Z_2)$ is zero. Then any PL manifold homeomorphic to M is equivalent to M as a PL manifold.*

REMARK. For the sake of completeness we shall describe the cohomology groups $H^k(M; Z_2)$ in a geometric manner exploited by Sullivan in his proof of the earlier version of Theorem 4.2; for a more standard description of $H^k(M; Z_2)$ see [14, §23] or any algebraic topology text. If X is any topological space, a **smooth k-manifold in** X is a continuous function $f: V \to X$, where V is a compact smooth k-dimensional manifold. An element in $H^k(X; Z_2)$ is then a function which assigns to each k-manifold in X an element of Z_2 subject to certain consistency conditions which are straightforward but a little too technical to describe here (see [12, §8] or [56] for further discussion; the description does not generalize to odd primes).

The restrictions on cohomology appearing in the above theorems were also shown to be unnecessary if $H^3(\text{Top}/\text{PL}; Z_2) = 0$, where Top/PL is a topological space arising from the geometry of the proof of 4.1 and 4.2. However, Siebenmann (first alone and later jointly with Kirby) constructed examples which implied that $H^3(\text{Top}/\text{PL}; Z_2)$ is nonzero; it followed quickly that both the Triangulation Conjecture and the *Hauptvermutung* were systematically false in every dimension greater than four.

There are very simple manifolds which yield contradictions to the *Hauptvermutung*. For example, consider the cartesian product $S^3 \times T^2$ of the unit sphere in R^4 with the two-dimensional torus T^2. This product is a smooth manifold and consequently determines a unique PL manifold; results of Shaneson combined with $H^3(\text{Top}/\text{PL}; Z_2) \neq 0$ imply the existence of a PL 5-manifold M^5 which is homeomorphic to $S^3 \times T^2$ but inequivalent to $S^3 \times T^2$ as a PL manifold ([49], [50]).

5. Handlebody theory for topological manifolds. In one sense the Kirby-Siebenmann results are disappointing because they disprove two conjectures which would have reduced the study of topological manifolds to combinatorial topology. On the other hand, the results used in the proof of 4.1 and 4.2 yield a convenient method for decomposing topological manifolds of dimension at least six, which will be discussed in this section.

Throughout this section S^p will denote the unit sphere in R^{p+1} and D^{p+1} will denote the unit disk in R^{p+1}. It follows from the definitions that D^{p+1} is a topological $(p+1)$-manifold with boundary, and its boundary is S^p.

DEFINITION. Let V be a topological n-manifold with boundary, and let $f: S^{k-1} \times D^{n-k} \to \partial V$ be a one-to-one continuous mapping. Then the manifold W obtained by **attaching a k-handle to V along** f is the disjoint union of V and $D^k \times D^{n-k}$ modulo the identification of $f(S^{k-1} \times D^{n-k}) \subseteq V$, with $S^{k-1} \times D^{n-k} \subseteq D^k \times D^{n-k}$ (see Figure 3 for an illustration).

This construction dates back to the beginnings of the study of manifolds.

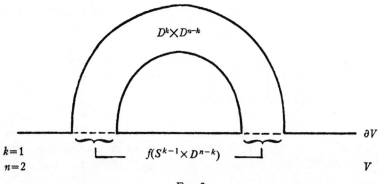

FIG. 3

For example, the classification theorem for compact orientable 2-manifolds may be written as follows (compare [30]):

THEOREM 5.1. *Let M be a compact orientable 2-manifold. Then $M = \partial Q$, where Q is formed by attaching 1-handles to D^3.*

The work of Marston Morse on critical point theory implies that any smooth manifold M may be constructed by successively attaching handles (e.g., see [33], [34], [39]); in terms of the definition below, M has a *handle decomposition*. Standard results of combinatorial topology imply a similar result for PL manifolds [17, p. 226].

In the definition below, $X \cup \partial X \times [0, 1]$ will be interpreted to mean the disjoint union modulo the identification of $y \in \partial X$ with $(y, 0) \in \partial X \times \{0\}$. We shall assume the manifold M discussed below is either unbounded or compact in order to simplify the definition.

DEFINITION. Let M be a topological manifold. A **handle decomposition** of M is a (finite or denumerable) sequence of compact subspaces $\{M_j\}_{j \in J}$ (J a well-ordered subset of the integers) satisfying:

(i) $M = \cup_{j \in J} M_j$ and each M_j is a compact manifold with boundary.

(ii) For all $j \in J$ we have $M_j \subseteq \text{Int } M_{j+1}$; in fact, M_{j+1} is formed by attaching a k-handle to $M_j \cup \partial M_j \times [0, 1]$ (provided j is not maximal in J).

The following result of Kirby and Siebenmann is a straightforward consequence of the arguments used to prove 4.1 and 4.2 [75]:

THEOREM 5.2. *Any topological manifold of dimension greater than five has a handle decomposition.*

This is one case of a general principle implicit in [75]; namely, results which work for smooth and PL manifolds in dimensions greater than five also work for topological manifolds in dimensions greater than five. Some particular examples are the theorems of Siebenmann [51], Farrell [13], Cappell [67], and surgery theory [61] (see also [88]).

Since any topological manifold of dimension $\leqslant 3$ has a PL atlas, and hence a handle decomposition, the only unknown cases occur in dimensions 4 and 5. The nonvanishing of $H^3(\text{Top}/\text{PL}; Z_2)$ implies the following negative result due to Siebenmann [52]:

THEOREM 5.3. *For $n = 4$ or 5 (possibly both) there exists a compact unbounded topological n-manifold that has no handle decomposition.*

Siebenmann also proves in [52] that certain fundamental theorems on smooth and PL manifolds in dimensions greater than five fail somewhere in dimensions three, four, and five. Precise knowledge of where these failures occur would be a useful addition to our relatively meager knowledge of manifolds in these dimensions (compare [78]).

References

1. R. Abraham, Foundations of Mechanics, Benjamin, New York, 1967.

2. J. W. Alexander, On the subdivision of 3-space by a polyhedron, Proc. Nat. Acad. Sci. U.S.A., 10 (1924) 6–8.

3. ———, An example of a simply connected surface bounding a region which is not simply connected, Proc. Nat. Acad. Sci. U.S.A., 10 (1924) 8–10.

4. L. Antoine, Sur l'homéomorphie de deux figures et de leurs voisinages, J. Math. Pures Appl., (8) 4 (1921) 221–325.

5. D. Barden, Simply connected five-manifolds, Ann. of Math., 82 (1965) 365–385.

6. G. Birkhoff and S. MacLane, Survey of Modern Algebra, Third edition, Macmillan, New York, 1965.

7. L. E. J. Brouwer, Beweis des Jordanschen Satzes für den n-dimensionalen Raum, Math. Ann., 71 (1912) 314–319.

8. M. Brown, A proof of the generalized Schoenflies Theorem, Bull. Amer. Math. Soc., 66 (1960) 74–76.

9. ———, Locally flat embeddings of topological manifolds, Ann. of Math. 75 (1962), 331–341.

10. ———, and H. Gluck, Stable structures on manifolds I, II, III, Ann. of Math., 79 (1964) 1–58.

11. S. S. Cairns, Triangulation of the manifold of class one, Bull. Amer. Math. Soc., 41 (1935) 549–552.

12. P. E. Conner and E. E. Floyd, Differentiable Periodic Maps, Ergebnisse der Mathematik und ihrer Grenzgebiete, N. F. Bd. 33. Springer-Verlag, Berlin-Göttingen-Heidelberg, 1964.

13. F. T. Farrell, The obstruction to fibering a manifold over a circle, Bull. Amer. Math. Soc., 73 (1967) 737–740.

14. M. Greenberg, Lectures on Algebraic Topology, Benjamin, New York, 1967.

15. W. C. Hsiang and J. Shaneson, Fake tori, the annulus conjecture, and the conjectures of Kirby, Proc. Nat. Acad. Sci. U.S.A., 63 (1969) 687–691.

16. ———, Fake tori, Topology of Manifolds (Proc. Univ. of Georgia Institute, 1969), 18–51. Markham, Chicago, 1970.

17. J. F. P. Hudson, Piecewise Linear Topology, Benjamin, New York, 1969.

18. B. von Kérékjárto, Vorlesungen über Topologie, Die Grundlehren der mathematischen Wissenschaften, Bd. 1. J. Springer, Berlin, 1923.

19. R. C. Kirby, Stable homeomorphisms and the annulus conjecture, Ann. of Math., 89 (1969) 575–582.

20. ———, and L. C. Siebenmann, On the triangulation of manifolds and the Hauptvermutung, Bull. Amer. Math. Soc., 75 (1969) 742–749.

21. ———, Foundations of Topology, Notices, Amer. Math. Soc., 16 (1969) 848.

22. S. Kobayashi and K. Nomizu, Foundations of Differential Geometry, Vol. I., Interscience Tracts in Pure and Applied Mathematics, vol. 15. Interscience, New York, 1963.

23. R. Lashof, Lees' immersion theorem and the triangulation of manifolds, Bull. Amer. Math. Soc., 75 (1969) 535–538.

24. ———, The immersion approach to triangulation, Topology of Manifolds (Proc. Univ. of Georgia Institute, 1969), 52–56. Markham, Chicago, 1970.

25. ———, and M. Rothenberg, Microbundles and smoothings, Topology, 3 (1964/65) 357–388.

26. ———, Hauptvermutung for manifolds, Conference on the Topology of Manifolds (Michigan State, 1967), 81–105. Prindle, Weber, and Schmidt, Boston, 1968.

27. ———, Triangulation of manifolds I, II, Bull. Amer. Math. Soc., 75 (1965) 750–757.

28. J. Lees, Immersions and surgeries of topological manifolds, Bull. Amer. Math. Soc., 75 (1969) 529–534.

29. S. Lefschetz, Introduction to Topology, Princeton Mathematical Series, vol. 11. Princeton University Press, Princeton, 1949.

30. W. S. Massey, Algebraic Topology; an Introduction, Harcourt, Brace, and World, New York, 1967.

31. B. Mazur, On embeddings of spheres, Acta Math., 105 (1961) 1–17.

32. J. Milnor, On manifolds homeomorphic to the 7-sphere, Ann. of Math., 64 (1956) 399–405.

33. J. Milnor, Morse Theory (notes by M. Spivak and R. Wells), Annals of Mathematics Studies No. 51, Princeton University Press, Princeton, 1963.

34. ———, Lectures on the h-cobordism theorem (notes by L. Siebenmann and J. Sondow), Princeton University Press, Princeton, 1965.

35. E. E. Moise, Affine structures in 3-manifolds, V. The triangulation theorem, Ann. of Math., 56 (1952) 96–114.

36. ———, Affine structures in 3-manifolds, VIII. Invariance of knot types; local tame embedding, Ann. of Math., 59 (1954) 159–170.

37. R. L. Moore, Concerning simple continuous curves, Trans. Amer. Math. Soc., 21 (1920) 333–347.

38. M. Morse, A reduction of the Schoenflies extension problem, Bull. Amer. Math. Soc., 66 (1960) 113–115.

39. ———, and S. S. Cairns, Critical Point Theory in Global Analysis and Differential Topology, Pure and Applied Mathematics Series, Vol. 33, Academic Press, New York, 1969.

40. J. Munkres, Elementary Differential Topology (revised edition), Annals of Mathematics Studies No. 54, Princeton University Press, Princeton, 1966.

41. M. H.A. Newman, Elements of the Topology of Plane Sets of Points, Cambridge University Press, Cambridge, 1951.

42. ———, The engulfing theorem for topological manifolds, Ann. of Math., 84 (1966) 555–571.

43. C. D. Papakyriakopoulos, A new proof for the invariance of the homology groups of a complex (Greek), Bull. Soc. Math. Grèce, 22 (1943) 1–154.

44. H. Poincaré, Cinquième complément à l'Analysis situs, Rend. Circ. Mat. Palermo, 8 (1904) 45–110.

45. T. Radó, Über den Begriff von Riemannsche Fläche, Acta Sci. Math. (Szeged), 2 (1924/26) 101–120.

46. I. Richards, On the classification of noncompact surfaces, Trans. Amer. Math. Soc., 106 (1963) 259–269.

47. J. Rotman, Theory of Groups: an Introduction, Allyn and Bacon, Boston, 1965.

48. A. Schoenflies, Beiträge zur Theorie der Punktmengen III, Math. Ann., 62 (1906) 286–328.

49. J. Shaneson, Wall's surgery obstruction groups for $G \times Z$, Ann. of Math., 90 (1969) 296–334.

50. ———, Non-simply connected surgery and some results in low-dimensional topology, Comment. Math. Helv., 45 (1970) 333–352.

51. L. C. Siebenmann, The obstruction to finding a boundary for an open manifold of dimension greater than five, Ph.D. Thesis, Princeton University, 1965.

52. ———, Disruption of low-dimensional handlebody theory by Rohlin's theorem, Topology of Manifolds (Proc. Univ. of Georgia Institute, 1969), 57–76. Markham, Chicago, 1970.

53. S. Smale, Generalized Poincaré's conjecture in dimensions greater than four, Ann. of Math., 74 (1961) 391–406.

54. E. H. Spanier, Algebraic Topology, McGraw-Hill, New York, 1966.

55. J. R. Stallings, Polyhedral homotopy-spheres, Bull. Amer. Math. Soc., 66 (1960) 485–488.

56. D. Sullivan, On the Hauptvermutung for manifolds, Bull. Amer. Math. Soc., 73 (1967) 598–600.

57. ———, Triangulating and smoothing homotopy equivalences and homeomorphisms, Geometric Topology Seminar Notes, mimeographed, Princeton University, 1967.

58. R. Thom, Les classes caractéristiques de Pontrjagin des variétés triangulées, Symposium internacional de topología algebraica (Mexico, 1956), 54–67. Universidad Nacional Autónoma de México y UNESCO, Mexico, 1958.

59. O. Veblen, Theory of plane curves in nonmetrical analysis situs, Trans. Amer. Math. Soc., 6 (1905) 83–98.

60. C. T. C. Wall, On homotopy tori and the annulus theorem I, Bull. London. Math. Soc., 1 (1969) 95–97.

61. ——, Surgery on Compact Manifolds, London Mathematical Society Monographs No. 1, Academic Press, New York, 1970.

62. A. H. Wallace, A geometric method in differential topology, Bull. Amer. Math. Soc., 68 (1962) 533–542.

63. J. H. C. Whitehead, On C^1 complexes, Ann. of Math., 41 (1940) 809–824.

64. E. C. Zeeman, The Poincaré conjecture for $n \geqslant 5$, Topology of 3-manifolds and Related Topics (Proc. Univ. of Georgia Institute, 1961), 198–204, Prentice-Hall, Englewood Cliffs, N.J., 1962.

Additional References*

65. D. R. Anderson and W.-C. Hsiang, Extending combinatorial PL structures on stratified spaces, Invent. Math., 32 (1976), 179–204.

66. C. E. Burgess and J. W. Cannon, Embeddings of surfaces in E^3, Rocky Mt. Math. J., 1 (1971), 259–344.

67. S. Cappell, A splitting theorem for manifolds, Invent. Math., 33 (1976), 69–170.

68. ——, R. Lashoff, and J. Shaneson, A splitting theorem and the structure of 5-manifolds, Symposia Math., 10 (1972), 47–58.

69. —— and J. Shaneson, Surgery on 4-manifolds and applications, Comment. Math. Helv., 46 (1971), 500–28.

70. —— and J. Shaneson, Some new four-manifolds, Ann. of Math., 104 (1976), 61–72.

71. T. Chapman, Topological invariance of Whitehead torsion, Amer. J. Math., 96 (1974), 488–97.

72. M. Cohen, A Course in Simple Homology Theory, Graduate Texts in Mathematics No. 10, Springer, New York, 1973.

73. R. D. Edwards, The double suspension of a certain homology 3-sphere is S^5, Notices Amer. Math. Soc., 22 (1975), A–334.

74. W.-C. Hsiang, Decomposition formula of Laurent extension in algebraic K-theory and the role of codimensional 1 submanifold in topology, Algebraic K-Theory II: "Classical" Algebraic K-Theory and Connections with Arithmetic (Battelle Institute Conference, Seattle, 1972), Lecture Notes in Mathematics, Vol. 342, 308–27, Springer, New York, 1973.

75. R. C. Kirby and L. C. Siebenmann, Some theorems on topological manifolds, Manifolds-Amsterdam 1970, Lecture Notes in Mathematics, Vol. 197, 1–7, Springer, New York, 1971.

76. R. Lashof, The immersion approach to triangulation and smoothing, Proc. Sympos. Pure Math., Vol. 22, 131–64, American Mathematical Society, Providence, 1971.

77. —— and J. Shaneson, Smoothing 4-manifolds, Invent. Math., 14 (1971), 197–210.

78. T. C. Lawson, Remarks on the four- and five-dimensional s-cobordism conjectures, Duke Math. J., 41 (1974), 639–44.

79. A. A. Markov, Insolubility of the problem of homeomorphy (Russian), Proc. Int. Cong. Math. 1958, 300–6, Cambridge University Press, New York, 1960.

80. R. J. Milgram, Surgery with coefficients, Ann. of Math., 100 (1974), 194–248.

81. C. R. Miller III, On Group-Theoretic Decision Problems and their Classification, Annals of Math. Studies, Vol. 68, Princeton University Press, Princeton, 1971.

82. J. Morgan and D. Sullivan, The transversality characteristic classes and linking cycles in surgery theory, Ann. of Math., 99 (1974), 463–544.

83. M. Rabin, Recursive unsolvability of group-theoretic problems, Ann. of Math., 67 (1958), 172–94.

---

* Prof. Schultz revised his article for the present volume in order to include references 65–91.—Ed.

84. M. Sharlemann, Equivalence of 5-dimensional *s*-cobordisms, Proc. Amer. Math. Soc., 53 (1975), 508–10.

85. ——, Constructing strange manifolds with the dodecahedral space, Duke Math. J., 43 (1976), 33–40.

86. —— and L. Siebenmann, The Hauptvermutung for C^∞ homeomorphisms I. Manifolds-Tokyo 1973, 85–91, University of Tokyo Press, Tokyo, 1975, II. Comp. Math., 29 (1974), 253–64.

87. H. Seifert and W. Threlfall, Lehrbuch der Topologie, Chelsea, New York, 1947.

88. L. C. Siebenmann, Deformations of homeomorphisms of stratified sets, Comment. Math. Helv., 47 (1972), 123–63.

89. ——, L'invariance topologique du type simple d'homotopie (d'après T. Chapman et R. D. Edwards), Sém. Bourbaki 1972/73, Lecture Notes in Mathematics, Vol. 383, 186–209, Springer, New York, 1974.

90. D. Sullivan, Geometric periodicity and invariants of manifolds, Manifolds-Amsterdam 1970, Lecture Notes in Mathematics, Vol. 197, 44–75, Springer, New York, 1971.

91. J. West, Compact ANR's have finite type, Bull. Amer. Math. Soc., 81 (1975), 163–5.

Postscript (1976). Since 1971, work on the central problems of the subject and its underlying machinery has continued. The methods and viewpoints of Kirby-Siebenmann *et al.* have led to major successes on some outstanding classical problems of topology, most notably the topological invariance of Whitehead torsion (T. Chapman [71], [72], and R. D. Edwards [73]), West's proof that compact finite-dimensional ANR's have the homotopy types of finite complexes [91], and R. D. Edwards' proof that some triangulations of topological manifolds (in the sense of [14, p. 106]) do not correspond to PL atlases [73]. An adequate description of these and other advances (e.g., [65], [74]) cannot be given here; the Proceedings of the 1976 A.M.S. Summer Institute on Algebraic and Geometric Topology will undoubtedly present much of this material. On the other hand, despite substantial efforts and some significant discoveries, there have been relatively few overall advances concerning topological manifolds of dimensions three, four, and five. Finally, the definitive version of Kirby and Siebenmann's work is about to appear; the interested reader is encouraged to look there.

ON INVOLUTIONS OF A CIRCLE*

W. F. PFEFFER, University of California, Davis

Many easily expressible geometric theorems about spheres often require proofs which involve rather sophisticated techniques of algebraic topology and thus are completely inaccessible to the undergraduate students. The purpose of this note is to give an elementary proof of the existence of a coincidence point of two free involutions of a circle. Equivalently stated, we shall prove that if two involutions of a circle have no fixed points, their composition always has a fixed point. This is a

---

* From AMERICAN MATHEMATICAL MONTHLY, vol. 79 (1972), pp. 159–160.

fixed point theorem which does not follow from Lefschetz degree considerations.

In a complex plane C we shall consider the unit circle $S^1 = \{z \in C: |z| = 1\}$. A continuous map $\sigma: S^1 \to S^1$ is called an **involution** of S^1 if $\sigma^2 = \sigma \circ \sigma$ is the identity map of S^1. An involution σ of S^1 is called **free** if it has no fixed points, i.e., if $\sigma(x) \neq x$ for all $x \in S^1$. The **antipodal** map $\alpha: S^1 \to S^1$ defined by $\alpha(z) = -z$ is an example of a free involution. If $z_1, z_2 \in C$, we shall denote by (z_1, z_2) the open line segment connecting z_1 and z_2, i.e., $(z_1, z_2) = \{tz_1 + (1-t)z_2: 0 < t < 1\}$.

LEMMA. *Let σ be a free involution of S^1. Then*

$$(x, \sigma(x)) \cap (y, \sigma(y)) \neq \varnothing$$

for every $x, y \in S^1$.

Proof. Choose $x \in S^1$. Since $\sigma(x) \neq x$, $S^1 - \{x, \sigma(x)\} = A \cup B$ where A and B are open connected arcs. Because σ is a continuous map and σ^2 is the identity, either $\sigma(A) = A$ or $\sigma(A) = B$ must hold. If $\sigma(A) = A$ then also

$$\sigma(A \cup \{x, \sigma(x)\}) = A \cup \{x, \sigma(x)\}.$$

But $A \cup \{x, \sigma(x)\}$ is homeomorphic to the closed unit interval $[0, 1]$ and thus there is a $y \in A \cup \{x, \sigma(x)\}$ such that $\sigma(y) = y$; but this is a contradiction. Therefore $\sigma(A) = B$ and the lemma follows.

PROPOSITION. *Let σ_1 and σ_2 be two free involutions of S^1. Then there is a point $z \in S^1$ such that $\sigma_1(z) = \sigma_2(z)$.*

Proof. The angle between two vectors determined by non-zero complex numbers x and y is defined in the usual way as a length of the smaller arc of S^1 determined by points $x/|x|$ and $y/|y|$. If $z \in S^1$ and $i = 1, 2$, we denote by $\theta_i(z)$ the angle between the vectors determined by $\sigma_i(z) - z$ and $z \cdot \sqrt{-1}$. Clearly the function $\theta: S^1 \to (-\pi, \pi)$ defined by $\theta(z) = \theta_1(z) - \theta_2(z)$, $z \in S^1$, is continuous. Choose $z \in S^1$. If $\theta(z) = 0$ then, of course, $\sigma_1(z) = \sigma_2(z)$. Thus without loss of generality we may assume that $\theta(z) > 0$ (see the picture). Since $(z, \sigma_2(z)) \cap (\sigma_1(z), \sigma_2[\sigma_1(z)]) \neq \varnothing$, we have $\theta[\sigma_1(z)] < 0$. From the intermediate value theorem it follows that $\theta(z_0) = 0$ for some $z_0 \in S^1$, and the proof is completed.

COROLLARY. *Let σ be an involution of S^1. Then there is a point $z \in S^1$ such that either $\sigma(z) = z$ or $\sigma(z) = -z$.*

This corollary follows immediately from the previous proposition applied to σ and the antipodal map α.

Note. The above proposition is also valid for higher dimensional spheres. This is deduced, e.g., in [1], 33.6, page 89, as a simple corollary from the generalized Borsuk-Ulam theorem. However, the proof of the generalized Borsuk-Ulam theorem is itself quite intricate and makes extensive use of algebraic topology.

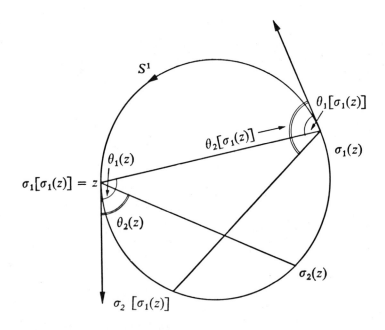

For further elementary properties of involutions of a circle, we refer the reader to [2].

References

1. P. E. Conner and E. E. Floyd, Differentiable Periodic Maps, Springer-Verlag, Berlin, 1964.
2. W. F. Pfeffer, More on involutions of a circle, this MONTHLY, 81 (1974), 613–6.

THE HISTORICAL DEVELOPMENT OF ALGEBRAIC GEOMETRY*

J. DIEUDONNÉ, University of Nice, France and University of Maryland

I. THEMES AND PERIODS

Modern algebraic geometry has deservedly been considered for a long time as an exceedingly complex part of mathematics, drawing practically on every other part to build up its concepts and methods and increasingly becoming an indispensable tool in many seemingly remote theories. It shares with number theory the distinction of having one of the longest and most intricate histories among all branches of our science, of having always attracted the efforts of the best mathematicians in each genera-

* From AMERICAN MATHEMATICAL MONTHLY, vol. 79 (1972), pp. 827–866. An expanded version of this article has appeared as the first volume of a book entitled "Cours de Géométrie algébrique", Presses Universitaires de France, Collection SUP, 1974.

tion, and of still being one of the most active areas of research. Both are perhaps the best candidates for the perfect mathematical theory, according to Hilbert's ideas: if we agree with him that problems are the lifeblood of mathematics, then certainly we may say that algebraic geometry and number theory always have had more open problems than solved ones, and that each progress towards their solution has always brought with it a host of new and exciting methods.

Human minds being unable to grasp complex matters as a whole, I have thought it would be helpful to describe the history of algebraic geometry as a kind of two-dimensional pattern, where many varied trends of thought, belonging to a few big *themes*, weave their way as multicolored threads through the moving succession of years. It should, however, be emphasized from the start that such a presentation inevitably inflicts distortions on reality: these themes constantly react on one another, and any division of time into periods is bound to founder on the fact that periods almost always overlap.

With these reservations, we may first group the main ideas of algebraic geometry as follows:

(A) and (B) The twin themes of *classification* and *transformation*, hardly to be separated, since the general idea behind classification of algebraic varieties is to put together those which can be deduced from each other by some kind of "transformation." Subordinate to these themes are the notion of *invariant*, both of algebraic type and of numerical type (such as dimension, begree, genus, etc.), and the concepts of *correspondence* and of *morphism*, which give precise meanings and extensions to the vague idea of "transformation."

(C) *Infinitely near points*: a thorny problem, which has plagued generations of mathematicians: the definition and classification of singularities, the correct definition of "multiplicity" of intersections, later the concept of "base points" of linear systems, and the recent introduction of rings with nilpotent elements, all belong to that theme.

(D) *Extending the scalars*: a giant step forward in the search for simplicity: the introduction of *complex* points and later of *generic* points were the forerunners of what we now consider as perhaps the most characteristic feature of algebraic geometry, the general idea of *change of basis*.

(E) *Extending the space*: another fruitful method for extracting understandable results from the bewildering chaos of particular cases: *projective geometry* and *n-dimensional geometry* paved the way for the modern concepts of "abstract" varieties and *schemes*.

(F) *Analysis and topology in algebraic geometry*. This theme beautifully exemplifies the cross-fertilization between various branches of mathematics. Out of a problem of integral calculus, the computation of elliptic integrals and of their generalizations, abelian integrals, Riemann developed the concept of Riemann surface (the first non-trivial example of "complex manifold"), invented algebraic topology, and he and his successors showed how these ideas completely renewed the theory of algebraic curves and surfaces. One hundred years later, history repeated itself when A. Weil transferred to algebraic geometry the notion of *fiber bundle*, and Serre the idea of using *sheaves* and their cohomology, which he and H. Cartan had shown to be so effective for complex manifolds.

(G) *Commutative algebra and algebraic geometry.* As we shall see, this has grown into the most important theme for modern algebraic geometry. Since Riemann introduced the field of rational functions on a curve, Kronecker, Dedekind and Weber the concepts of ideals and divisors, commutative algebra has become the workshop where the algebraic geometer goes for his main tools: local rings, valuations, normalization, field theory, and the most recent and most efficient of all, homological algebra.

Needless to say, within the scope of this article, it will be impossible to do more than deal with a few of the highlights of our history, leaving aside a large number of important developments which should be included in a reasonably complete survey.

II. FIRST PERIOD: "PREHISTORY"
(CA. 400 B.C.–1630 A. D.)

If it is true that the Greeks invented geometry as a deductive science, they never (contrary to popular beliefs) made any attempt to divorce it from algebra. On the contrary, one of their main trends was to use geometry to solve algebraic problems, and this is best exemplified in the invention of the conics, the first curves which they thoroughly studied after straight lines and circles. The Greeks knew simple geometric constructions for the root of the equation $x^2 = ab$, a and b being given as *lengths* of segments, and the unknown x being considered as the side of a square; they usually wrote the equation as a "proportion" $a/x = x/b$. The "Delic problem" called for construction of a length x of given cube, $x^3 = a^2b$; this was transformed by Hippocrates of Chio (around 420 B.C.) into a "double proportion" $a/x = x/y = y/b$ for two unknown lengths x, y. Menechmus (ca. 350 B.C.) had the idea of considering the loci given by the two equations $ay = x^2$ and $xy = ab$, whose intersection has as coordinates x, y a solution of the problem. This may seem to involve knowledge of analytic geometry; actually the Greeks made extensive use of coordinates (in particular for the later theory of conics by Apollonius), without however reaching the general point of view of Descartes and Fermat (see below).

This method of solving equations by intersections of curves had in fact already been used in the 5th century B.C., and led to the invention of many curves, both algebraic and transcendental; of course, the distinction between the two kinds of curves could not be perceived during that period, and more generally, there was no attempt at classification, for which no rational basis existed. Besides planes and spheres, the Greeks also studied some surfaces of revolution, such as cones, cylinders, a few types of quadrics and even tori; after having discovered conics "analytically," Menechmus was also the first to recognize that they could be obtained as plane sections of a cone of revolution; and a bold construction of Archytas (late 5th century B.C.) gave a solution of the Delic problem by the intersection of a cone, a cylinder and a torus. Finally, in his astronomical work, Eudoxus was led to describe the intersection of a sphere and a cylinder as the trajectory of a movement conceived as the superposition of two rotations, which may be considered as the first example of a parametric representation of a curve.

III. SECOND PERIOD: "EXPLORATION"
(1630–1795)

For once, this period has a very well-defined starting point, the independent invention by Fermat and Descartes of "analytic geometry," which certainly also marks the true birth of algebraic geometry. The main novelty compared to the way the Greeks used coordinates is that the *same* axes are used for *all* curves (fixed or variable) which are being considered in a problem, and above all the fact that the algebraic notation of Viète and Descartes opens the way to the consideration of arbitrary equations (where the Greeks could not go beyond the third or fourth degree). Within this frame, the distinction between algebraic and transcendental curves immediately emerges; the concept of dimension is already clear to Fermat, who explicitly states that a single equation defines a curve in 2 dimensions, a surface in 3 dimensions, and already hints at the possibility of generalization to higher dimensions. The degree of a plane curve is at once seen to be invariant with respect to a change of coordinates, and Newton knows that it is also invariant under a central projection (an operation which was familiar since the study of conic sections by the Greeks).

Themes A and B

The chief work of that period is one of exploration. Fermat shows that all curves of degree 2 are conics, and Newton classifies all plane cubics with respect to change of coordinates and projections; Euler classifies the quadrics, and the first skew curves, given as intersection of two surfaces, appear in the 18th century. The concept of parametric representation of a curve is fundamental in Newton's approach to calculus, and Euler knows how to get in certain cases a parametric representation from the cartesian equation. A beginning is made in the elucidation of the structure of singular points and inflexion points of algebraic plane curves, although limited to the most elementary cases, so that no general description is yet obtained.

Theme C

The problem of intersection of two algebraic plane curves is already tackled by Newton; he and Leibniz had a clear idea of "elimination" processes expressing the fact that two algebraic equations in one variable have a common root, and using such a process, Newton observed that the abscissas (for instance) of the intersection points of two curves of respective degrees m, n, are given by an equation of degree $\leq mn$. This result was gradually improved during the 18th century, until Bézout, using a refined elimination process, was able to prove that, in general, the equation giving the intersections had exactly the degree mn; however, no general attempt was yet made during that period to attach to each intersection point an integer measuring the "multiplicity" of the intersection, in such a way that the sum of the multiplicities should always be mn. Bézout also generalized his elimination process to 3 dimensions, proving that the points of intersection of three algebraic surfaces of degrees m, n, p are in general given by an equation of degree mnp.

With the beginning of the consideration of algebraic families of algebraic curves a problem in a sense converse to the problem of intersections appeared, namely the determination of a curve of given degree n containing sufficiently many given points. It should be recalled here that this (linear) problem was the starting point for the theory of determinants, and the fact that $n(n + 3)/2$ points in "general position" completely determine a curve of degree n, whereas two curves of degree n have in general n^2 common points, gave the first general example of the concept of rank for a system of linear equations ("Cramer's paradox").

We should finally stress the fact that a number of ideas fully developed during the next period may be traced back (in an embryonic form) to the 17th or 18th century, as we shall see below.

IV. THIRD PERIOD: "THE GOLDEN AGE OF PROJECTIVE GEOMETRY"
(1795–1850)

Here again we have a rather sharp break with the past at the beginning of this period. In the space of a few years, with Monge and his school and especially with Poncelet, a new era begins with the simultaneous introduction of points at infinity and of imaginary points: "geometry" will now, for almost 100 years, exclusively mean geometry in the complex projective plane   Themes $P_2(C)$ or the complex projective 3-dimensional space $P_3(C)$.   D and E In fact, the fundamental idea of (real) projective geometry goes back to Desargues (17th century) who, trying to give mathematical foundations to the methods of "perspective" used by painters and architects, had introduced the concept of "point at infinity," and the use of central projections as a means of getting new theorems from classical results of Euclidean geometry; and although these ideas had inspired Pascal in his work on conics, they had very soon dropped into oblivion, due to the outlandish language of the author and the very limited diffusion of his book (which was for some time believed lost). Other mathematicians in the 18th century, in particular Euler and

Stirling, had hinted at the existence of imaginary points, in order to state general theorems without distinction of various cases. This is precisely what is brilliantly accomplished by the new school: circles now intersect in 4 points as any two conics should, but two of the points are imaginary and at infinity; instead of several kinds of conics and quadrics, all nondegenerate conics (resp. quadrics) are now projectively equivalent; instead of the 72 kinds of cubics enumerated by Newton, only 3 remain projectively distinct; etc.

The chief beneficiaries of these new ideas are at first the theory of conics, quadrics and of linear families of conics and quadrics; but curves and surfaces of degree 3 or 4 are also investigated in this way, revealing beautiful new theorems, such as the configurations of the 9 inflexion points of a plane cubic, the 27 lines on a cubic surface, the 28 bitangents to a plane quartic; the theorem of Salmon, proving the constancy of the cross ratio of the 4 tangents to a cubic issued from a point of the curve, was to gain even more significance later, as the first concrete example of a "module" in Riemann's sense for an algebraic curve.

Although, with Möbius, Plücker and Cayley, projective geometry received a sound algebraic basis by the use of homogeneous coordinates, a general tendency of the projective school was to minimize as much as possible algebraic computations, and to rely instead (beginning with Poncelet) on general heuristic "principles" which they did not bother to justify algebraically. The remarkable success they had in this direction was chiefly due to their skillful use of the idea of geometric *transformation*, which for the first   Theme B time comes to the forefront in geometry, preparing the ground for Klein's famous "Program" linking geometry and the theory of groups. Most of the transformations they consider are linear: for instance, one of their favorite devices in the theory of conics is to consider a conic as the locus of two variable straight lines through two fixed points, one of them being derived from the other by a fixed linear transformation (an idea which, in some particular cases, goes back to Maclaurin). Similarly, in the study of the linear system of conics through 4 fixed points, they investigate the intersections of these conics with a fixed straight line D by considering the (linear) transformation which to a point M of D associates the second point of intersection with D of the conic of the system which contains M. Emboldened by the results obtained in this manner, they inaugurated what was to become the theory of *correspondences*, by considering what they called (α, β)-correspondences, i.e., relations between two points M, M' such that to each point M there exist α points M' related to M, and to each point M' there exist β points M related to M': when M and M' vary on the same projective line, Chasles' "correspondence principle" says that the number of points M (counted with multiplicities) coinciding with one of their transforms is $\alpha + \beta$ unless *every* point of the projective line has that property, a result which it is easy to justify algebraically. A beautiful application is the Poncelet "closure theorem" for polygons inscribed in a conic C and circumscribed to a conic C': for a given integer n, one defines on C

a $(2,2)$-correspondence by assigning to $M \in C$ the nth point M_n in a sequence $M_0 = M, M_1, \cdots, M_n$, where each side $M_i M_{i+1}$ is tangent to C' and the points M_i are on C. It is easily seen that for n even, one has $M = M_n$ if $M_{n/2}$ is a point common to C and C', and for n odd, $M = M_n$ if $M_{(n-1)/2} = M_{(n+1)/2}$, and the tangent to C at that point is also tangent to C'. There are thus at least 4 points M on C such that $M = M_n$, and by the correspondence principle, if there is still *one* more point having that property, then $M = M_n$ for *all* points on C (one uses of course the parametrization of a conic by the projective line).

Later representatives of the projective school (notably Chasles in France, Steiner and von Staudt in Germany), somewhat intoxicated by the elegance of their methods, went so far as to insist that "pure" geometry should be entirely divorced from algebra and even (with von Staudt) from the concept of real number. As could be expected, such efforts did not lead very far, and probably hampered progress in the realization of the importance of linear algebra in classical geometry; it may be, however, that they paved the way for the later "abstract" algebraic geometry over a field different from R or C.

In the general theory of algebraic curves (in $P_2(C)$) and surfaces (in $P_3(C)$), the main problems studied before Riemann are of an enumerative character: to give only one example of such problems, what is the number of conics tangent to 5 given conics in general position? (The correct answer is 3264.) Chasles, and later Schubert and Zeuthen proposed half-empirical   Theme C formulas to solve these problems, based on an intuitive concept of "intersection multiplicity" which could only be justified much later. One of the main ideas of projective geometry, the concept of *duality*, led to the introduction of new "tangential" invariants for algebraic plane curves: the class (number of tangents through a   Theme A point), the number of inflexion points and the number of double tangents, culminating in the famous "Plücker formulas"

$$m' = m(m - 1) - 2d - 3s,$$

$$m = m'(m' - 1) - 2d' - 3s',$$

$$s' - s = 3(m' - m),$$

where m is the degree of the curve, m' its class, d the number of double points, d' the number of double tangents, s the number of cusps, s' the number of inflexion points; no "higher singularities," either punctual or tangential, are supposed to occur.

V. FOURTH PERIOD: "RIEMANN AND BIRATIONAL GEOMETRY"
(1850–1866)

The importance of Riemann in the history of algebraic geometry can hardly be overestimated, but in his two fundamental contributions, the "transcendental" approach *via* abelian integrals and the introduction of the field of rational functions on a curve, he built on basic ideas inherited from the previous period.

The origin of abelian integrals is the study of integrals of type

$$\int \frac{R(t)dt}{\sqrt{P(t)}}$$

where $P(t)$ is a polynomial of degree 3 or 4 and $R(t)$ a rational function; one of these integrals expresses the length of an arc of an ellipse (hence the name "elliptic integrals"). In the first half of the 18th century, Fagnano and Euler, looking for some substitute for the classical formula expressing the sum of two arcs of a circle, when the circle is replaced by an ellipse, found indeed that the sum

$$\int_a^x \frac{dt}{\sqrt{P(t)}} + \int_a^y \frac{dt}{\sqrt{P(t)}}$$

can be written

$$\int_a^z \frac{dt}{\sqrt{P(t)}} + V(x, y),$$

where z is an *algebraic* function of x and y, and V a rational or logarithmic function of x and y, and Euler had similar results for more general integrals.

At the beginning of his famous work of elliptic functions, Abel made a giant step forward by showing that the Fagnano-Euler relations were special cases of a very general theorem: he considers an arbitrary "algebraic function" y of x, defined as a solution of a polynomial equation $F(x, y) = 0$; an "abelian integral" $\int R(x, y)dx$ is an integral in which R is a rational function of x, y, in which y is replaced by the preceding algebraic function (for instance elliptic integrals correspond to $F(x, y) = y^2 - P(x)$). Then, if $G(x, y, a_1, \cdots, a_r) = 0$ is a second polynomial in x, y whose coefficients are rational functions of some parameters $a_1, \cdots, a_r$, and if (x_1, y_1), $(x_2, y_2), \cdots, (x_m, y_m)$ are the points of intersection of the two curves $F=0$, $G=0$, the sum

$$V = \int_{(a,b)}^{(x_1, y_1)} R(x, y)dx + \cdots + \int_{(a,b)}^{(x_m, y_m)} R(x, y)dx$$

is a rational or logarithmic function of the parameters a_j $(1 \leq j \leq r)$*; surprisingly enough, this is little more than an exercise in the theory of symmetric functions of the roots of a polynomial. But Abel does not stop there, and studies in detail the case, in which V is a constant; this leads him to the realization that in that case, *any* sum

$$\int_{(a,b)}^{(x_1,y_1)} R(x,y)dx + \cdots + \int_{(a,b)}^{(x_m,y_m)} R(x,y)dx$$

with *arbitrary* points (x_j, y_j) on the curve $F = 0$, can be expressed as the sum of a *fixed* number δ of values of the same integral, with upper limits algebraic functions of the (x_j, y_j); but, in contrast with the Fagnano-Euler formulas for elliptic integrals, he showed that the number δ may well be > 1, for instance when $F(x, y) = y^2 - P(x)$ with P of degree ≥ 5.

Abel, however, worked exclusively within the framework of analysis, and does not seem to have been acquainted with projective geometry. Furthermore, he obviously had no clear concept of integration in the complex plane (in 1826, Cauchy had hardly begun his work on that subject), and with the exception of a short and inconclusive note, he has no general discussion of the *periods* of his integrals. Thus, although Abel's theorem paved the way for Jacobi's breakthrough in the problem of inversion of hyperelliptic integrals†, Abel himself narrowly missed the concept of integral of the first kind and the definition of the genus of a curve (his failure to take into account the points at infinity has as a consequence the fact that the δ integrals he considers are not necessarily of the first kind).

When Riemann takes up the subject in 1851, the intervening years had seen the great development by Cauchy and his school of the theory of functions of a complex

---

* Of course, the points x_j, y_j usually have complex coordinates; an integral

$$\int_{(a,b)}^{(x_j,y_j)} R(x,y)dx$$

is only properly defined when the path of integration in the complex plane C with extremities a and x_j has been fixed, and y_j is the value taken by y when x varies along the path, y is a continuous function of x and takes the value b at $x = a$. When the path is replaced by another one (with the same extremities), the value of the integral is modified by a "period."

By definition, a logarithmic function of the a_j has the form $\log S (a_1, ..., a_r)$ where S is rational.

† The natural idea of "inverting" the integral $\int_a^x (Q(t)\,dt)/\sqrt{P(t)}=u$ is to study x as a function of u, as Abel and Jacobi had done when P has degree 3 or 4; but Jacobi realized that, due to the existence of 4 periods, no meromorphic function of u could be a solution of the problem. Abel's theorem finally led him to the correct conception of the problem: one considers *two* equations

$$\int_a^x \frac{dt}{\sqrt{P(t)}} + \int_a^y \frac{dt}{\sqrt{P(t)}} = u, \quad \int_a^x \frac{tdt}{\sqrt{P(t)}} + \int_a^y \frac{tdt}{\sqrt{P(t)}} = v,$$

and one "inverts" them by expressing the symmetric functions $x + y$ and xy as functions of u and v; Abel's theorem yields an "addition formula" for these functions, from which one can show that they are meromorphic and quadruply periodic.

variable. Indeed, the starting point of Riemann has nothing to do with algebraic functions, but is the extension of Cauchy's theory to the "surfaces" he introduces in order better to deal with the so-called "multiform" functions of the most general (not necessarily algebraic) type. This was already far beyond the contemporary concepts, and during the 30 years following Riemann, it was the object of long and tedious explanations by the expositors of his theory. But the way Riemann uses this notion in order to attack the problem of abelian integrals is much more original still. Instead of starting   Theme F (as would all his predecessors and most of his immediate succes- sors) from an algebraic equation $F(s, z) = 0$ and the Riemann surface of the algebraic function s of z which it defines, his initial object is an n- sheeted Riemann surface without boundary and with a finite number of ramification points*, given *a priori* without any reference to an algebraic equation (Riemann takes care to complete each sheet with a point at infinity, and thus avoids Abel's difficulties with these points); then he attacks the problem in the most general manner possible: classify the integrals of *all meromorphic functions* on the surface. The work of Cauchy and Puiseux had brought to light the general idea of "periods" of such integrals, generally expressed (as in the example first given by Abel) as an integral taken along an arc joining two ramification points. Here again Riemann breaks entirely new ground: he realizes for the first time that topological concepts are closely related to the problem, and begins by essentially creating the topological study of compact orientable surfaces, attaching to such a surface S an invariantly defined integer $2g$, the minimal number of simple closed curves C_j on S needed to make the complement S' of their union simply connected. Then, instead of studying integrals of meromorphic functions, he *defines directly* integrals of the first and second kinds by their periodicity properties, as functions meromorphic on S', and tending on both sides of each C_j to limits which differ by a quantity k_j constant on C_j (a further reduction of the domain S' is needed to obtain similarly the integrals of the third kind, having logarithmic singularities)†; integrals of the first kind are

---

* The best way to define at least the part of the Riemann surface of a function $s(z)$ (defined by an algebraic relation $F(s, z) = 0$), containing no point at infinity, is to say that it is the subset of C^2 consisting of the pairs (s, z) satisfying the equation $F(s, z) = 0$; there is then no difficulty with the "crossing of sheets." Ramification points are those for which $\partial F/\partial s (s, z) = 0$; Puiseux proved in 1850 that if (s_0, z_0) is such a point, the surface decomposes at that point into a finite number of "branches" such that each branch can be represented by equations of type

$$z - z_0 = t^h, \quad s - s_0 = a_1 t + a_2 t^2 \dots,$$

where t (the "uniformizing parameter") is in a neighborhood of 0 in C and the series converges (the integer h depending on the branch).

This description is only correct, however, when at each ramification point (s_0, z_0) there is only one branch; if not, the point (s_0, z_0) must be replaced by as many points as there are branches; in other words the points of a Riemann surface are the *branches* at the various points of the curve.

† One simply joins the singularity to one of the C_j by an arc, and deletes the arc from S'.

those which have no pole on S. The existence of integrals of the three kinds is proved by Riemann as a consequence of what he calls the "Dirichlet principle," i.e., the existence of a harmonic function in S' taking prescribed values on the boundary (which allows him to prescribe at will the *real parts* of the k_j); and it is also by an ingenious use of the same principle that Riemann obtains the fundamental relation

$$g - 1 = w/2 - n$$

giving the genus in function of the number of sheets n, and the number w of ramification points (supposed to be of a "general" type).

The meromorphic functions on S are then the integrals of the first or second kind whose periods k_j all vanish, and Riemann shows that they may be expressed as rational functions of two of them, linked by an algebraic relation $F(s, z) = 0$, thus recovering the older point of view, but immeasurably enriched with new insights. The choice of these meromorphic functions s, z is in a large measure arbitrary, and this leads Riemann to his   Theme B next big step forward, the general concept of *birational transformation* between two irreducible algebraic curves, corresponding to a biholomorphic mapping of their Riemann surfaces. Here again, Riemann was not without predecessors: already Newton and his followers had introduced quadratic transformations such as

$$x' = 1/x, \ y' = y/x$$

in the plane, and observed that they thus transformed an algebraic curve into a curve of different degree. "Inversion" in the plane and in 3 dimensional space had been intensively studied since the early 1820's, chiefly by "synthetic" geometers; finally, the passage from a plane curve to its transform by duality (exchanging punctual and tangential coordinates) was obviously a birational transformation between two algebraic curves, exchanging degree and class. But the startling novelty of Riemann's approach is of course the fact that to a class of "birationally equivalent" irreducible algebraic curves he was able to attach his topological invariant g, the *genus* of all the curves in the class. But he did not stop there, and by an evaluation (using two different methods)   Theme A of the parameters on which a Riemann surface of genus g depended, he arrived at the conclusion that classes of isomorphic Riemann surfaces of genus $g \geq 2$ were characterized by $3g - 3$ complex parameters varying continuously (for $g = 1$ there is only one parameter, and none for $g = 0$); the precise meaning of this result (the so-called theory of "moduli" of curves) was to remain until very recently among the least clarified concepts of the theory.*

---

* One should emphasize the fact that this only describes the first half of Riemann's paper on abelian integrals; the second part, which solves in a masterly way the inversion problem by the introduction of the general "thêta functions" has been, if anything, even more influential on the development of analysis.

VI. FIFTH PERIOD: "DEVELOPMENT AND CHAOS"
(1866–1920)

The extraordinary wealth of new ideas and methods introduced by Riemann provided inspiration for a steady development of algebraic geometry for over 80 years. But the grandiose synthesis he had envisioned and tried to materialize was almost immediately broken up by his successors. During that period there will be at least two or three schools of algebraic geometry, each using different methods, with little in common even in the fundamental concepts. Riemann's use of analysis, in particular in the "Dirichlet principle," exceeded the possibilities of his time, and he had obviously neglected all the difficulties bound to the existence of singular points on algebraic curves. The first task to which each school of algebraic geometry addressed itself was therefore the systematization of the birational theory of algebraic plane curves, incorporating most of Riemann's results with proofs in conformity with the principles of the school. Then, with varying success, they tried to extend their methods to the theory of algebraic surfaces and higher dimensional algebraic varieties.

VI a: The algebraic approach. Historically, this was the latest one, being initiated by two fundamental papers in 1882, one by Kronecker and one by Dedekind and Weber. But in the light of subsequent history, it is the trend which was to exert the deepest influence on the birth of our modern concepts; in particular, just as Riemann had revealed the close relationship between algebraic varieties and the theory of complex manifolds, Kronecker and Dedekind-Weber brought to light for the first time the deep similarities between algebraic geometry and the burgeoning theory of algebraic numbers, which were to be some of the main driving forces during the next periods. Furthermore, this conception of algebraic geometry is for us the clearest and simplest one, due to our familiarity with abstract algebra; but it was precisely this "abstract" character which made it the least popular and least understood one in its time.

The work of Kronecker and of his immediate followers, Lasker and Macaulay, in the first two decades of the 20th century, was of a very general nature, and its importance only emerged in the later periods: it essentially consisted in setting up and consistently using an elimi-   Theme G
nation method, far more flexible and powerful than the preceding ones, with the help of which it was for the first time possible to give a precise meaning to the concepts of *dimension* and of *irreducible variety** and to show that each variety (defined by an arbitrary system of algebraic equations) in projective *n*-space decomposed in a unique way into a union of irreducible varieties (in general of different dimensions).

---

* An irreducible variety V in $\mathbf{P}_n(\mathbf{C})$ is characterized by the property that if the product PQ of two homogeneous polynomials is 0 in V, then one of the two polynomials P, Q must be 0 in V. The restrictions to V of the rational functions which are defined at one point of V at least then form a field whose transcendence degree over $\mathbf{C}$ is the dimension of V.

The goal of Dedekind and Weber in their fundamental paper was quite different and much more limited; namely, they gave purely algebraic proofs for all the algebraic results of Riemann. They start from the fact that, for Riemann, a class of isomorphic Riemann surfaces corresponds to a *field K* of rational functions, which is a finite extension of the field $C(X)$ of rational fractions in one indeterminate over the complex field; what they set out to do, conversely, if a finite extension K of the field $C(X)$ is given *abstractly*, is to reconstruct a Riemann surface S such that K will be isomorphic to the field of rational functions on S. Their very original and fruitful method may be presented in the following way: *if* the Riemann surface S was already known, at each point $z_0 \in S$, a rational function $f \neq 0$ would have an *order* $v_{z_0}(f)$, namely the integer (positive or negative) which is the degree of the smallest power in the Puiseux development $f(u) = \sum_k a_k u^k$ with respect to a "uniformizing parameter" u (equal to $z - z_0$ if z_0 is not a ramification point, to some power $(z - z_0)^{1/h}$ if z_0 is a ramification point). For a *fixed* $z_0 \in S$, the mapping $f \mapsto v_{z_0}(f)$ of K^* into $\mathbf{Z}$ is what is called a *discrete valuation* on K: we recall that this is by definition a mapping $w: K^* \to \mathbf{Z}$ such that $w(f + g) \geqq \inf(w(f), w(g))$ if $f + g \neq 0$, and $w(fg) = w(f) + w(g)$, which implies $w(1) = 0$ and $w(f^{-1}) = -w(f)$ (w is usually extended to K by taking $w(0) = +\infty$ by convention). What Dedekind and Weber do is to *reverse* this process, and *define* a "point of the Riemann surface S of K" as a *nontrivial discrete valuation* on K (i.e., one which is not identically 0 on K^*: two proportional valuations are then identified).

Now the nontrivial discrete valuations on the field $C(X)$ are easily determined: one of them (the "point at infinity") w_∞ is such that $w_\infty(P) = -\deg(P)$ for any nonzero polynomial $P(X)$; the other ("finite points") correspond bijectively to the points $\zeta \in C$, the corresponding valuation w_ζ being such that $w_\zeta(P)$ is the order of the zero ζ of $P(X)$ (equal to 0 if $P(\zeta) \neq 0$). It can easily be shown that for each discrete valuation w of $C(X)$, there is a finite number of nonproportional valuations v_j on K such that for each j, v_j/e_j reduces to w on $C(X)$, where e_j is an integer $\geqq 1$; one says that the v_j are the points of the Riemann surface S *above* w; the points above w_∞ are again called points at infinity, the other finite points.

The elements $f \in K$ for which $v(f) \geqq 0$ for all *finite* points v of S constitute exactly the elements of K which are *integral** over the ring of polynomials $C[X]$; they form what we now call a *Dedekind ring A*, to which Dedekind's theory of *ideals* may be applied.† The maximal ideals $\mathfrak{P}_v$ of A correspond to the finite points $v \in S$: $\mathfrak{P}_v$ is the set of $f \in A$ for which $v(f) > 0$; the *fractionary ideals* of K are the A-modules $\mathfrak{a}$ contained in K and for which there is an element $c \neq 0$ in A such that $c\mathfrak{a} \subset A$; each of them can be written uniquely as a product $\mathfrak{P}_1^{\alpha_1} \mathfrak{P}_2^{\alpha_2} \cdots \mathfrak{P}_r^{\alpha_r}$, where the $\mathfrak{P}_j$ are maximal ideals of A and the α_j positive or negative integers‡. Another way

---

* Recall that an element x of a ring R is *integral* over a subring S if it satisfies an equation of type $x^m + a_1 x^{m-1} + \cdots + a_m = 0$, with $a_j \in S$.

† Dedekind had developed this theory for algebraic number fields from 1870 on.

‡ Here $\mathfrak{P}_v^{-\alpha} = (\mathfrak{P}_v^{-1})^\alpha$ for all positive integers α where $\mathfrak{P}_v^{-1}$ denotes the inverse of the ideal $\mathfrak{P}_v$, that is, the set of elements f in K such that $v(f) \geqslant -1$.

of stating this result is to say that a fractionary ideal $\mathfrak{a}$ is the set of all $f \in K$ such that $v_j(f) \geq \alpha_j$ for $1 \leq j \leq r$, where the valuations v_j correspond to the maximal ideals $\mathfrak{P}_j$, and $v(f) \geq 0$ for the other finite valuations.

The consideration of the ideals of A, however, leaves the "points at infinity" out of the picture. This led Dedekind and Weber to generalize the concept of ideal and to introduce the notion of *divisor* on K. This is defined as a family $D = (\alpha_v)$ of integers $\alpha_v \in \mathbf{Z}$, where v runs through *all* points of S, and $\alpha_v = 0$ except for a finite number of points: writing $(\alpha_v) + (\beta_v) = (\alpha_v + \beta_v)$ defines the set $\mathscr{D}(K)$ of divisors of K as an *additive group* isomorphic to $\mathbf{Z}^{(S)}$, in which an *order relation* is naturally defined, $(\alpha_v) \leq (\beta_v)$ meaning that $\alpha_v \leq \beta_v$ for all $v \in S$; a divisor $D = (\alpha_v)$ such that $\alpha_v \geq 0$ for all $v \in S$ is called *positive* or *effective*. The *degree* $\deg(D)$ of $D = (\alpha_v)$ is defined as $\sum_{v \in S} \alpha_v$ (positive or negative integer); the *support* of D is the set of the $v \in S$ for which $\alpha_v \neq 0$. One of the problems considered by Riemann was the determination of rational functions on a Riemann surface having poles of orders $\leq \alpha_P$ for prescribed points P (in finite number) on S. Using his bold expression of functions as sums of abelian integrals, he found that there existed rational functions having that property for an *arbitrary* choice of the points P as long as $\sum_P \alpha_P \geq g + 1$, whereas if $\sum_P \alpha_P \leq g$, this was only possible for *special* positions of the points P. This result was completed by his student Roch, and put in its final form by Dedekind and Weber in the following way: the problem is a special case of the study of the set $L(D)$ of rational functions $f \in K$ satisfying the conditions

(1) $$v(f) \geq -\alpha_v \text{ for all } v \in S$$

for a given divisor $D = (\alpha_v)$; it follows from the axioms of valuations that $L(D)$ is a complex vector subspace of K, and it can be shown that this subspace has *finite* dimension $l(D)$.

A fractionary ideal may be described as the union of the increasing family of spaces $L(D_m)$, where $D_m = (\alpha_v)$ is such that the α_v coincide with the $-\alpha_j$ for the v_j, are equal to 0 for the other finite points, and to m for the points at infinity.

The relations (1) can be written in a different way. For each $f \in K^*$, there are only a finite number of valuations $v \in S$ such that $v(f) \neq 0$; let $(f)_0$ (resp. $(f)_\infty$) be the positive divisor $((v(f))^+)$ (resp. $((v(f))^-))$ (in the "transcendental" interpertation, $(f)_0$ is the "divisor of zeroes" and $(f)_\infty$ the "divisor of poles" of the rational function f), and let $(f) = (f)_0 - (f)_\infty$ in the group $\mathscr{D}(F)$; (f) is called the *principal divisor* defined by f. It can be shown that $\deg((f)) = 0$ by purely algebraic arguments (in the transcendental picture, this is merely the *residue theorem*)*; in particular, if $v(f) \geq 0$ for *all* $v \in S$, then $f \in \mathbf{C}$ (only constants are everywhere holomorphic on a Riemann surface) and if in addition $v(f) > 0$ for some v, then $f = 0$. With these

---

* One integrates the differential df/f on the boundary of the simply connected part S' of the Riemann surface, taking into account that each arc of that boundary comes twice in the integral with opposite orientations.

definitions, the relations (1) for $f \neq 0$ are equivalent to the inequality

$$(2) \qquad\qquad\qquad (f) + D \geq 0$$

in the ordered group $\mathscr{D}(K)$.

Principal divisors form a subgroup $\mathscr{P}(K)$ of $\mathscr{D}(K)$ (isomorphic to the group K^*/C^*, two elements of K^* which have the same principal divisor differing by a constant factor by the previous remarks). Divisors belonging to the same *class* in the quotient group $\mathscr{C}(K) = \mathscr{D}(K)/\mathscr{P}(K)$ are called (linearly) *equivalent*: to say that D and D' are equivalent means therefore that there exists $f \neq 0$ such that $D' - D = (f)$; it is clear that $\deg(D') = \deg(D)$ and $l(D') = l(D)$ for equivalent divisors; two elements f, g of $L(D)$ are such that $(f) + D = (g) + D$ if and only if f/g is a constant, in other words, the set $|D|$ of *positive* divisors equivalent to D is identified to the projective space $P(L(D))$ of dimension $l(D) - 1$.

The *Riemann-Roch theorem* is then written in the following way:

$$(3) \qquad\qquad l(D) - l(\Delta - D) = \deg(D) + 1 - g,$$

where g is the genus, and Δ belongs to a well-determined divisor class, called the *canonical class* of K. To define it in the transcendental interpretation, one considers on the Riemann surface S a *meromorphic differential form* ω: at each point P of S, the differential form ω may be written $F(u)du$, where u is the uniformizing parameter in a neighborhood of P and F is meromorphic at P; if δ_P is the order of F at the point P, (δ_P) is a *canonical divisor*, and it does not depend on the choice of the uniformizing parameters. Any other meromorphic differential form may be written $f\omega$ with $f \in K$, hence all canonical divisors belong to the same *class*. There is a purely algebraic definition of Δ (see section VII b), and one proves that $\deg(\Delta) = 2g - 2$ for $g \geq 1$ and $l(\Delta) = g$. Relation (3) implies Riemann's result on the poles of rational functions; more generally, if $\deg(D) \geq g + 1$, (3) implies $l(D) \geq 2$; if $D \geq 0$, $L(D)$ always contains the constant functions, and to say that $l(D) \geq 2$ means that it contains a non constant rational function. From the definition of $L(D)$, it follows that $l(D) = 0$ if $\deg(D) < 0$, hence, by (3), $l(D) = \deg(D) + 1 - g$ if $\deg(D) > 2g - 2$; in particular, for any divisor D such that $\deg(D) > 0$, $l(mD) = m \cdot \deg(D) + 1 - g$ for m large enough (although one may have $l(D) = 0$).

VI b: The Brill-Noether theory of linear systems of points on a curve. An irreducible plane curve Γ without singularity is identified to its Riemann surface, and a positive divisor may therefore be identified with a system of points of Γ, each being counted with a certain "multiplicity" which is a positive integer. Riemann's determination of the "special" systems of at most g points of Γ, which may be the poles of a rational function, had led him (by an extension of some earlier computations of Abel) to define these sets as intersections with Γ of a family of "adjoint" curves of smaller degree, subject to linear conditions on the coefficients of their equations, so that such a family may be considered as given by an equation $\sum_{j=1}^{r} \lambda_j P_j(x, y) = 0$ in nonhomogeneous coordinates, where the P_j are polynomials and the λ_j variable complex parameters. A number of points of intersection of these curves with Γ may be fixed (i.e., independent of the λ_j); as the intersection multiplicity of a common point of Γ and of an arbitrary curve Γ' is immediately defined since Γ has no singular point (it is the same as the intersection multiplicity of Γ' and the tangent to Γ), we may consider for each adjoint curve Γ' of the family the positive divisor $D = \sum_P m_P P - \sum_Q m_Q^0 Q$, where P runs through *all* the intersection points of Γ and Γ', m_P is the corresponding intersection multiplicity, Q runs through the *fixed* intersection points and m_Q^0 is the minimum value of m_Q when the λ_j vary. It is immediate to see that if D_0 is one of these divisors, corresponding to the values λ_j^0 of the parameters, then $D = D_0 + (f)$, where $f = (\sum_j \lambda_j P_j)/(\sum_j \lambda_j^0 P_j)$.

Conversely, given a divisor D_0 (positive or not), if $l(D_0) = r > 0$, the functions $f \in L(D_0)$ may be written $(\sum_{j=1}^{r} \lambda_j P_j(x, y))/Q(x, y)$, where the P_j and Q are polynomials and the λ_j arbitrary complex numbers; the positive divisors $(f) + D_0$ where $f \in L(D)$, are obtained by adding a fixed divisor to the variable divisor obtained as above from the points of intersection of Γ and of the curve $\sum_j \lambda_j P_j(x, y) = 0$.

The study of the vector spaces $L(D)$ attached to divisors is thus essentially equivalent to the study of the systems of points of intersection (with multiplicities) of Γ with the curves Γ' of a system of curves $\sum_j \lambda_j P_j(x, y) = 0$. It is in fact by means of the study of such systems of points, called *"linear series"* or *"linear systems"* on Γ, that the geometric school of Clebsch, Gordan, Brill, and Max Noether described the birational theory of algebraic plane curves after 1866. But they wanted to deal in this way, not only with curves without singularities, but with arbitrary algebraic curves, and linear systems of points are only easy to handle when the curve Γ has no singularities, or at most "nice" singularities such as double points with distinct tangents. One of the first   Theme C efforts of that school was therefore to establish the possibility of finding a birational transformation of an arbitrary irreducible algebraic curve Γ into a plane curve with only double points with distinct tangents; a result proved independently by M. Noether in 1871 and equivalent to a theorem of algebra obtained by Kronecker in 1862. In view of the extension of this result during the later periods, it is worthwhile to note that a slightly weaker theorem may be obtained by a succession of birational transformations of the whole projective plane $P_2(C)$ onto itself of the type

$$x'/yz = y'/zx = z'/xy$$

(for suitable *homogeneous* coordinates), the so-called *quadratic transformations*.

Such a transformation is bijective outside the sides of the triangle having as vertices the points $(1, 0, 0)$, $(0, 1, 0)$, and $(0, 0, 1)$ but sends each point of one side (not a vertex) to the opposite vertex, and is indeterminate at a vertex: however, two points approaching a vertex along distinct lines have transforms which tend to distinct limits on the opposite side, so that the transformation may be said to "blow up" a vertex to the opposite side, and *separates* the branches of a curve having different tangents at a vertex by transforming them to branches through different points of the transformed curve. By repeating conveniently this process, one may show that there is a transformed curve whose singular points are such that each has a number of distinct tangents equal to its multiplicity. To get curves with only double points, one uses birational transformations which are only defined on the given curve (and not in the plane).

   It is during the same period, and in the same school, that n-dimensional algebraic geometry comes into its own for any value of $n \geq 1$ (all algebraic varieties being considered as subvarieties of some $P_n(C)$). As we shall see below, the study of algebraic varieties of dimension ≥ 2 Theme E was to have important repercussions on the theory of algebraic curves, with the concept of algebraic correspondences as subvarieties of a product variety, and the study of abelian varieties. We only mention here another fruitful consequence, the relation between linear series of points and rational mappings of an irreducible curve Γ into a projective space $P_r(C)$: such a mapping can be written

$$\phi : \zeta \to (P_1(\zeta), P_2(\zeta), \cdots, P_{r+1}(\zeta)),$$

where the P_j are homogeneous polynomials in the homogeneous coordinates of ζ, all of the same degree: if Γ' is the image of Γ by ϕ, the points of intersection of Γ by the system of curves $\sum_j \lambda_j P_j = 0$ are the inverse images by ϕ of the points of intersection of Γ' by variable hyperplanes. This observation, in connection with the theory of linear series, enables one to choose the P_j in such a way that ϕ is a birational transformation and Γ' has *no singular points*. Furthermore, the curve Γ' having these properties is uniquely determined up to a birational and *bijective* transformation (one says it is *the* nonsingular *model* of the field of rational functions of Γ).

   VI c: Integrals of differential forms on higher dimensional varieties. As soon as 1870, Cayley, Clebsch and M. Noether inaugurated the study of abelian integrals on irreducible algebraic surfaces, by considering, on a surface S in $P_3(C)$ given by an equation $F(x, y, z) = 0$ in nonhomogeneous coordinates, double integrals of type $\iint R(x, y, z) dx dy$, where   Theme F R is a rational function; after 1885, Picard began a thorough investigation of the properties of these integrals, as well as of simple integrals $\int P(x, y, z) dx + Q(x, y, z) dy$, where P, Q are rational and the differential is exact*. His method, which (conveniently generalized) is still very useful, consists in looking at the sections of the surface by the planes $y = $ const., applying Riemann's theory to abelian integrals on these curves (which in general are irreducible), and studying the way in which they depend on the parameter y; in particular, if p is the genus of the curve for general values of y, the $2p$ periods of the abelian integrals of the first kind satisfy a linear differential equation of order $2p$ (as functions of y), the so called Picard-Fuchs equation, which plays an important part in the theory. The algebraic surfaces considered by these mathematicians were usually supposed to be without singular points, or at most to have only "nice" singularities (double curves with distinct tangent planes except at finitely many points and no singular points except   Theme C finitely many triple points); starting with M. Noether, many attempts were made to prove that any algebraic surface could be transformed into surfaces without singularities (not necessarily immersed in $P_3(C)$, but in higher dimensional projective spaces), but no satisfactory proof was found until much later.

Very early it appeared that the theory of algebraic surfaces exhibited some features which had no counterpart in the theory of algebraic curves. Two irreducible surfaces without singularities may be birationally equivalent without being isomorphic. If p_g denotes the number of linearly independent double integrals of the first kind on an irreducible surface S (i.e., integrals which are finite over any 2-cell of S), the corresponding number   Theme A for a surface S' birationally equivalent to S is not necessarily the same. The number p_g is the obvious counterpart of the genus of a curve; but very soon also, it was realized that the other definition of the genus of a curve, using the "adjoints" of Riemann, also generalized to surfaces, but might give a number p_a different from p_g (see in VIII-a its exact definition in modern terms); p_g was called the *geometric genus* and p_a the *arithmetic genus* of S, and the difference $q = p_g - p_a$ (which is always ≥ 0) the *irregularity* of the surface (for instance,

---

* The exact meaning of a simple integral $\int P(x, y, z)\, dx$ consists in assigning to each piecewise differentiable mapping $t \rightarrow (x(t), y(t), z(t))$ of an interval $[a, b] \subset R$ into S (a "singular 1-simplex") the number $\int_a^b P(x(t), y(t), z(t))\, x'(t)\, dt$. Similarly, the double integral $\iint R(x, y, z)\, dxdy$ assigns to each piecewise differentiable mapping $(u, v) \rightarrow (x(u, v), y(u, v), z(u,v))$ of a triangle $T \subset R^2$ into S (a "singular 2-simplex") the number

$$\iint_T R(x(u,v),\, y(u,v),\, z(u,v))\, \frac{\partial(x, y)}{\partial(u, v)}\, dudv.$$

One can then define in an obvious way the value of simple (resp. double) integrals over 1-chains (resp. 2-chains), i.e., formal linear combinations of 1-simplices (resp. 2-simplices) with coefficients in Z (or in R, or in C). Generalizations to higher dimensions are obvious, once one defines an n-simplex as a piecewise differentiable mapping of the "standard n-simplex" defined by the inequalities $x_j \geq 0$ $(1 \leq j \leq n)$, $x_1 + x_2 + ... + x_n \leq 1$ in R^n.

Cayley found that for *ruled* surfaces $p_g = 0$ and $p_a < 0$ in general).

It soon also became apparent that the properties of abelian integrals on a surface or a higher dimensional variety were to a large extent subordinate to the topological properties of the variety. H. Poincaré had particularly in mind the applications to algebraic geometry when, in 1895, he started to give mathematical substance to Riemann's intuition of higher dimensional "Betti numbers" by inventing the "simplicial" machinery which made rigorous proofs possible*; algebraic varieties (and more generally   Theme F analytic varieties) are amenable to this technique due to the fact that they are *triangulable*, a fact for which Poincaré himself sketched a proof, which was later made entirely rigorous by van der Waerden. Using this machinery and the Picard technique of variable plane sections, Poincaré was able to bring to a satisfactory conclusion previous efforts by Picard and the Italian geometers and to prove that the irregularity q of an algebraic surface without singularity is equal to $R_1/2$, where R_1 is the first Betti number, and also equal to the number of independent simple abelian integrals of the first kind. Around 1920, Lefschetz considerably developed these techniques and generalized them to algebraic varieties of arbitrary dimension, concentrating in particular on the determination of the number of cycles on such a variety V which are homologous to cycles contained in algebraic subvarieties of V: for instance, if V is a projective variety of complex dimension n, and H a hyperplane section of V, the natural mappings $H_i(H,Z) \to H_i(V,Z)$ of homology groups are bijective for $0 \le i \le n-2$ and surjective for $i = n-1$. He also showed that for an algebraic variety V, one had $R_{2p} > 0$, $R_p \ge R_{p-2}$ for $p \le n$ (complex dimension of V) and that the Betti numbers R_{2p+1} of odd dimension were even.

VI d: Linear systems and the Italian school. The definition of divisors, given in VI-a, carries over to any field K finitely generated over C; on a nonsingular model V having K as field of rational functions, the discrete nontrivial valuations of K now correspond to irreducible subvarieties of V of *codimension* 1. It is still true that $\deg((f)) = 0$ for principal divisors, and that $L(D)$ is a finite dimensional subspace of K for all divisors D. The concept of *linear system* of subvarieties of codimension 1 may therefore be associated to the notion of divisor as in VI-b. Around 1890, the Italian school of algebraic geometry, under the leadership of a trio of great geometers: Castelnuovo, Enriques and (slightly later) Severi, embarked upon a program of study of algebraic surfaces (and later higher dimensional varieties) generalizing the Brill-Noether approach *via* linear systems: they chiefly worked

---

* Let us recall that to an n-chain is attached a well determined $(n-1)$ - chain, its boundary; *n-cycles* are the *n*-chains whose boundary is 0, and the *n*-th *homology group* H_n (M, Z) (resp. $H_n(M, R)$, resp. $H_n(M, C)$) of a manifold M, with coefficients in Z (resp. R, C) is the quotient of the group of *n*-cycles with coefficients in Z (resp. R, C) by the subgroup consisting of the boundaries of the $(n + 1)$-chains. The Betti number R_p is the dimension of the real vector space H_p (M, R).

with purely geometric methods, such as projections or intersections of curves and surfaces in projective space, with as little use as possible of methods belonging either to analysis and topology, or to "abstract" algebra.

These limitations implied serious difficulties in the definition of the main concepts and the use of geometric methods. The chief trouble was that whereas on curves one can work almost exclusively with *positive* divisors, this is not the case any more for surfaces: for instance if $p_g = 0$, the canonical divisor (defined as in VI-a, but for meromorphic differential 2-forms) is not equivalent to a positive divisor, hence does not correspond to a linear system of curves. This compelled the Italians to introduce complicated "virtual" notions for linear systems, which obscured the significance of much of their results.

Working under such considerable handicaps, it is amazing to see how many new and deep results were discovered by the Italian geometers. It would be extremely long and intricate to describe these results in their own language (see for instance [16]) and we shall postpone the definition of the most important notions which they introduced until we can use the much simpler modern formulation.

Let us only mention here a few of the beautiful theorems characterizing (up to birational equivalence) simple types of surfaces by   Theme A
the values of the arithmetical genus p_a and new invariants defined by Enriques, the *plurigenera* P_k $(k \geq 2)$: a *rational* surface (i.e., birationally equivalent to a plane) is characterized by the relations $p_a = 0$, $P_2 = 0$, surfaces with $p_a < -1$ are ruled, whereas the surfaces such that $P_4 = P_6 = 0$ are either rational or ruled; finally, a surface for which $p_a = P_3 = 0$ and $P_2 = 1$ is birationally equivalent to the Enriques surface of degree 6 having the 6 edges of a tetrahedron as double lines (it is not a rational surface, although $p_g = 0$).

VII. SIXTH PERIOD: "NEW STRUCTURES IN ALGEBRAIC GEOMETRY"
(1920–1950)

The general trend towards the unification of mathematics by the study of the *structures* underlying each theory, which started to get momentum in the 1920's, was particularly apparent in the development of algebraic geometry; the striking kinships between algebraic varieties and complex manifolds on the one hand, algebraic numbers on the other, which had been discovered in earlier periods, now became organic parts of the fundamental concepts of algebraic geometry. One of the effects of this broadened point of view was to loosen the exclusive grip held until then by projective and birational methods over algebraic geometry, and prepare the way for a far more flexible approach.

VII a: Kählerian varieties and the return to Riemann. Ever since Gauss's fundamental paper of 1826 on the theory of surfaces and Riemann's inaugural lecture of 1854 defining n-dimensional riemannian geometry, the concept of *differential manifold*, defined by "maps" and differentiable "transition functions" between maps*, had gradually become more and more precise as the fundamental topological concepts needed to express them were defined and studied in the last part of the 19th century and the beginning of the 20th. One of the most important developments in that direction was the introduction of the general concept of *exterior differential p-form* on a differential manifold (locally defined by expressions

$$\sum_{i_1 < i_2 < \cdots < i_p} A_{i_1 i_2 \cdots i_p}(x) dx^{i_1} \wedge dx^{i_2} \wedge \cdots \wedge dx^{i_p}$$

in the local coordinates) and of their integrals on *p-chains* (generalizing the earlier notions of "curvilinear" and "surface" integrals), due to H. Poincaré and E. Cartan. At the very beginning of his papers on algebraic topology, Poincaré had pointed out the connection between the homology of a compact differential manifold V and the exterior differential forms on V (of which the classical Stokes' theorem is the simplest example). This was made precise by De Rham's famous theorems in 1931, starting from the duality between chains and forms given by the integral $\langle C, \omega \rangle = \int_C \omega$; due to the generalized Stokes' formula $\langle C, d\omega \rangle = \langle bC, \omega \rangle$ (where b is the boundary and d the exterior derivative), this yields a duality, pairing the real homology groups $H_i(V, R)$ of V and the cohomology groups $H^i(\Lambda)$† , where Λ is the "complex" of exterior differential forms

$$(4) \qquad 0 \xrightarrow{d} \Lambda^1 \xrightarrow{d} \Lambda^2 \xrightarrow{d} \cdots \xrightarrow{d} \Lambda^n \xrightarrow{d} 0 \qquad (n = \dim V),$$

(Λ^j is the R-vector space of the j-forms).

A projective algebraic variety without singularity of (complex) dimension n has a natural underlying structure of differential manifold of dimension $2n$, but in fact it has a much richer structure. In the first place, it is a *complex* manifold, which means that for the "maps" which define the differential structure and which take their values in C^n ($= R^{2n}$), the "transition funct-tions" are *holomorphic*; it follows that the space Λ_C^p of (complex) **Theme F** differential p-forms for $1 \leq p \leq 2n$ decomposes naturally into a direct sum of vector spaces $\Lambda_C^{r,s}$ corresponding to the pairs of integers such that $r + s = p$; for $r \leq n$ and $s \leq n$, the forms in $\Lambda_C^{r,s}$ (called forms of *type* (r, s)) are those which for *complex* local coordinates $z^1, z^2, \cdots, z^n$, are written

$$(5) \qquad \sum A_{j_1 \cdots j_r k_1 \cdots k_s}(x) dz^{j_1} \wedge \cdots \wedge dz^{j_r} \wedge d\bar{z}^{k_1} \wedge \cdots \wedge d\bar{z}^{k_s},$$

where the $A_{j_1 \cdots k_s}$ are differentiable functions with complex values (not holomorphic in general). For $r > n$ or $s > n$, one takes $\Lambda_C^{r,s}$ as reduced to 0 by convention.

But this is not the end of the story. It is possible to define on a projective complex space, and by restriction on any complex compact submanifold of such a space

---

* If M is a differential manifold of dimension $n, \phi: U \rightarrow R^n, \psi: Y \rightarrow R^n$ two maps of open sets U, V, of M onto R^n, the "transition function" from U to V is the mapping (only defined when $U \cap V \neq \emptyset$) $x \mapsto \psi(\phi^{-1}(x))$ of $\phi(U \cap V)$ onto $\psi(U \cap V)$.

† $H^i(\Lambda)$ is the quotient of the kernel $d^{-1}(\Lambda^{i+1})$ by the image $d(\Lambda^{i-1})$.

(which is necessarily an algebraic variety by a theorem of Chow) a riemannian ds^2 which is *kählerian*, i.e., can be written locally as a hermitian form

$$ds^2 = \sum_{j,k} h_{jk}dz^j d\bar{z}^k \text{ with } h_{kj} = \bar{h}_{jk}$$

which has the property that the corresponding exterior 2-form (which is *real* valued)

(6) $$\Omega = (i/2) \sum_{h,j} h_{jk}d\bar{z}^k \wedge dz^j$$

is *exact* (i.e., $d\Omega = 0$).

Beginning around 1930, Hodge, in a series of remarkably original papers, showed how to use these facts to investigate the homology of compact kählerian varieties. On a riemannian manifold, Beltrami had shown that it is possible to define an operator which generalizes the usual laplacian, and therefore enables one to define *harmonic functions* on the manifold. By a very imaginative generalization, Hodge was able to define similarly, on any compact riemannian manifold, the notion of *harmonic exterior differential forms*, and to prove that there existed a unique such form in any cohomology class in any $H^j(\Lambda)$; from that result, he deduced the uniqueness and existence of a harmonic p-form having given periods on homologically independent p-cycles, thus obtaining a complete generalization of Riemann's fundamental result, and showing that Riemann's use of "Dirichlet's principle" was far more than a technical device (fortunately for Hodge, the theory of elliptic partial differential equations had advanced far enough to spare him the difficulties which had plagued Riemann's approach). Turning next to complex kählerian manifolds, the space H^p of harmonic p-forms with complex coefficients splits into a direct sum of $p + 1$ spaces $H^{r,s}$ consisting of (complex) harmonic forms of type (5), for $r + s = p$ (with $H^{r,s} = 0$ if $r > n$ or $s > n$); it can be shown that $H^{p,0}$ consists exactly of the *holomorphic* p-forms (or "differential forms of the first kind"), i.e., those for which in (5), $s = 0$ and the $A_{j_1\cdots j_p}$ are holomorphic. As complex conjugation transforms $H^{r,s}$ into $H^{s,r}$, they have the same dimension, and this shows that the dimension of H^p, i.e., the Betti number R_p, is even when p is odd. On the other hand, one easily verifies that the (real) 2-form Ω defined in (6) is harmonic, as well as all its exterior powers, which proves that $R_{2k} \geq 1$ for every integer k. Finally, $\phi \mapsto \Omega \wedge \phi$ is shown to be an injective mapping of H^p into H^{p+2} for $p \geq n - 2$, from which the inequality $R_{p+2} - R_p \geq 0$ follows; all the Lefschetz's theorems on Betti numbers of algebraic varieties are thus "explained" and shown to belong in fact to the theory of kählerian manifolds (there are compact kählerian manifolds which are not isomorphic to projective algebraic varieties). We shall return to the Hodge's theory when in the next period it merges into sheaf cohomology.

VII b: Abstract algebraic geometry. It is well known that, from 1900 to 1930, the general concepts of algebra (mostly confined until then to real or complex numbers) were developed in a completely abstract setting, the notion of algebraic *structure* (such as group, ring, field, module, etc.) becoming the fundamental one and relegating to second place the nature of the mathematical *objects* on which the structure was defined. It was therefore quite natural to think of an "abstract" extension of algebraic geometry, in which the coefficients of the equations and the coordinates of the points would belong to an arbitrary field. Already Dedekind and Weber, in their 1882 paper, had observed that all their arguments only used the fact that the basic field was algebraically closed (and of characteristic 0, a notion which had not yet been defined then). Even notions which seem linked to analysis, such as derivatives and differentials, had algebraic counterparts: a *derivation* in a commutative ring A is an additive mapping $x \mapsto Dx$ of A into itself such that

$D(xy) = x \cdot Dy + (Dx) \cdot y$, and a *differential* is an A-linear mapping $\omega \colon \mathfrak{D} \to A$ of the A-module of all derivations into A;   Theme G for each $x \in A$, dx is the linear form $D \mapsto Dx$ on $\mathfrak{D}$, and p-forms are defined by the usual methods of exterior algebra.

The motivation for the development of abstract algebraic geometry was therefore a natural outcome of the progress of algebra; after 1930, a more powerful impulse was to come from number theory, as we shall see below.

As it was apparent that a large part of the foundations of classical algebraic geometry came from geometric intuition, more or less justified by appeals to analysis or topology, a thorough examination of the basic concepts, from the exclusive viewpoint of algebra, was necessary in order to carry out an ambitious program of algebraic geometry over an arbitrary field. This groundwork, which at the same time created most of modern commutative algebra, was chiefly due to E. Noether, W. Krull, van der Waerden, and F. K. Schmidt in the period 1920–1940, and to Zariski and A. Weil from 1940 on.

The first two of these mathematicians use the geometric language very sparsely; their results are almost always expressed in the language of rings and ideals, and it was only after 1940 that the importance of their work was properly appreciated: the decomposition into primary ideals in noetherian rings, the properties of integrally closed rings, the extensive use of valuations, the notion of localization and the fundamental properties of local rings are all due to them. (A local ring is a commutative ring A in which there is only one maximal ideal. The typical example consists of the rational functions (elements of the field $C(X)$) for which a given point $\zeta \in C$ is not a pole: they form the local ring of $C(X)$ at the point ζ. A similar remark may be made on the foundational work of Zariski, probably the deepest one in that period; although it is usually expressed in the language of projective geometry, it mostly belongs to local algebra and its central position in algebraic geometry was only recognized in the next period. The contribution of F. K. Schmidt (in connection with his work on number theory which we describe below) essentially consisted in extending the Dedekind-Weber theory to curves defined over an algebraically closed field of *any characteristic*.

The most conspicuous progress realized during that period is the successful definition, in algebraic geometry over an arbitrary field, of the concepts of *generic point* and of *intersection multiplicity*, due to the combined efforts of van der Waerden and A. Weil. The Italians (not to speak of their predecessors) used these notions with a freedom which, to their critics of the orthodox algebraic school, bordered on recklessness. As long as the underlying field was *C*, the notion of "elements in general position" could be easily justified by an appeal to continuity (although the Italians seldom bothered to prove that these elements formed *open sets* in the spaces they considered). On the other hand, Lefschetz had made the elementary but fundamental observation that when two subvarieties *U*, *V* of $P_n(C)$, of complementary dimensions *r* and $n - r$, intersect transversally in simple points, the number of these points is equal (for convenient orientations) to the *intersection number* $(U \cdot V)$ of the *cycles* *U*, *V*, in the sense of algebraic topology; as this number is known to be invariant under homology, it was quite natural to take it as the number of intersections of *U* and *V* (counted with multiplicities) in the most general cases. This justified the extensive use of intersection multiplicity by the Italian geometers, in particular the "self-intersection" number $(C \cdot C)$ of a curve on an algebraic surface. (Unfortunately, the complexity of the Italian definitions was such that it was often impossible to be sure that the same words meant the same things in two different papers; hence the numerous controversies between geometers of that school, such as the one which occurred as late as 1943 between Enriques and Severi, see [4] and [10].)

These foundations of course disappeared in algebraic geometry over an arbitrary field, and this was one of the reasons why no algebraic proofs valid over any field (even of characteristic 0) had been found for the results obtained in the theory of algebraic surfaces by transcendental or geometric methods. In 1926, van der Waerden saw that to gain the freedom which Analysis gave for classical geometry over the complex field, one had only to return to the process which had allowed the passage from real to complex geometry, namely *enlarge* the field *k* to which the coefficients of the equations of a   Theme D variety and the coordinates of its points are supposed to belong: if *K* is any extension of *k*, these equations are still meaningful when the coordinates are taken in *K*. Giving a general form to ideas which went back at least to Gauss, he introduced the idea of *specialization* over *k* of any set of elements $x_1, \cdots, x_m$ in an arbitrary extension *K* of *k*: it is a mapping which to each x_j assigns an element x'_j of an extension *K'* of *k* (which may be equal to *K*), in such a way that for every homogeneous polynomial $P \in k[X_1, \cdots, X_m]$ for which $P(x_1, \cdots, x_m) = 0$, one also has $P(x'_1, \cdots, x'_m) = 0$ (van der Waerden always works in projective spaces, or finite products of such spaces). Suppose then that *V* is an irreducible algebraic variety in $P_n(k)$, and let *K* be the field of rational functions on *V*; one may assume that *V* is not contained in a hyperplane of $P_n(k)$; for $1 \leq j \leq n$, the restriction ξ_j to *V* of the rational function $x \mapsto x^j/x^0$ (where $x^0, x^1, \cdots, x^n$ are homogeneous coordinates of a point, $x \in P_n(k)$) is an element of *K*; if V_K is the variety in $P_n(K)$ defined by the same equations as *V*, the point $(1, \xi_1, \cdots, \xi_n)$

belongs to V_K. Van der Waerden calls this point a *generic point* of V, for it is immediate to check that for *any* extension K' of k, *any* point of $V_{K'}$ is a specialization of $(1, \xi_1, \cdots, \xi_n)$. Such points can then be used in the same way as the "general points" of the Italians, despite their apparently tautological character: any theorem proved for generic points (and of course expressible by algebraic *equations* (not inequalities!) between their coordinates) is valid for *arbitrary* points of corresponding varieties. Van der Waerden then proceeded to apply this new tool with great virtuosity to many problems of algebraic geometry, and in particular to the definition of multiplicity of intersection of two varieties in abstract algebraic geometry, which had not yet been given a meaning except in the case of the intersection of two curves on a surface without singularity. However, Poncelet, as a consequence of his general vague "principle of continuity,"   Theme C had already proposed to define the intersection multiplicity at one point of two subvarieties U, V of complementary dimensions by having V (for instance) *vary* continuously in such a way that for some position V' all the intersection points with U should be *simple*, and counting the number of these points which collapsed to the given point when V' tended to V; in such a way, the total number of intersections (counted with multiplicities) would remain constant ("principle of the conservation of number"), and it is thus that Poncelet proved Bézout's theorem, by observing that a curve C in the plane belonged to the continuous family of all curves of the same degree m, and that in that family there existed curves which degenerated into a system of straight lines, each meeting a fixed curve Γ of degree n in n distinct points. Many mathematicians in the 19th century had extensively used such arguments, and in 1912, Severi had convincingly argued for their essential correctness. The concept introduced by van der Waerden was based on similar ideas: under suitable conditions, the multiplicity of a solution $y = (y_0, \cdots, y_n) \in P_n(k)$ of a system of equations $P_\mu(x, y) = 0$, where x is a point of an irreducible variety V, is the number of the solutions η of the system $P_\mu(\xi, \eta) = 0$, where ξ is the generic point of V, which specialize to y when ξ specializes to x. Using this definition, he was finally able to attach to every irreducible component C of the intersection of two irreducible varieties V, W of an "ambient" nonsingular variety U, an integer $i(C, V \cdot W; U) \geqq 0$, the multiplicity of C in $V \cap W$, provided *all* irreducible components of $V \cap W$ were "proper," i.e., had a dimension equal to $\dim V + \dim W - \dim U$.

Unfortunately, this restriction considerably reduced the usefulness of the notion of multiplicity. Using more powerful algebraic devices, A. Weil could define an intersection multiplicity $i(C, V \cdot W; U)$ when it is *only* supposed that C is proper (the other components of $V \cap W$ can have larger dimensions); furthermore, he showed that this number did not depend on the method used to define it (other, quite different methods, were later given by Chevalley and Samuel), once it possessed the "natural" properties similar to those of the intersection number in algebraic topology; this he showed to be the case for his definition, and it enabled him to develop in abstract algebraic geometry a calculus of "cycles" patterned on the calculus of chains introduced by Poincaré (irreducible subvarieties replacing simplices). In this context, divisors on an irreducible variety of dimension n were the cycles of dimension $n - 1$ (one also says that they have *codimension* 1).

Weil then went on to break away, for the first time, from projective algebraic geometry: for his purposes (see below) he needed constructions of algebraic varieties similar to the "gluing together" constructions of manifolds in algebraic topology or differential geometry, which   Theme E had been familiar since the beginning of the century; he showed that this could be done by using as "transition functions" biregular mappings of complements of subvarieties in affine varieties (the Zariski topology was not yet in use at that time), and he could also define in this context the notion of "complete variety" which is the counterpart of the concept of compact space in "abstract" algebraic geometry (in classical projective geometry, all algebraic subvarieties are complete).

VII c: Zeta functions and correspondences. A. Weil's work was chiefly motivated by problems which had arisen in the early 1920's in number theory. In his thesis of 1923, E. Artin had observed that algebraic congruences modulo a prime p, in 2 variables, i.e., of the form $F(x, y) \equiv 0 \pmod{p}$, where F is a polynomial with integral coefficients, could be interpreted as algebraic equations over the prime field $\boldsymbol{F}_p = \boldsymbol{Z}/p\boldsymbol{Z}$ (and similarly the "higher congruences" in the sense of Dedekind were algebraic equations over an arbitrary *finite* field $\boldsymbol{F}_q$ ($q = p^d$)). He further noticed that the analogy, already exploited by Dedekind and Weber, of finite extensions of the field $\boldsymbol{C}(X)$ with algebraic number fields, was here much closer, since the residual fields of the valuations of a finite extension K of $\boldsymbol{F}_q(X)$ are *finite* fields (extensions of $\boldsymbol{F}_q$) just as for number fields (whereas they are equal to $\boldsymbol{C}$ in classical algebraic geometry). This enabled him to define, in complete analogy with the Riemann-Dedekind zeta function of an algebraic number field, the *zeta function of* K, and to extend to it the classical theory: functional equation and the location of the poles. However, his treatment was entirely algebraic, without any kind of geometric interpretation; a little later, F. K. Schmidt observed that a much simpler and more natural treatment was achieved if one completely modeled the theory after Dedekind and Weber, by introducing *divisors* (or "points of the abstract Riemann surface") instead of ideals; it can then easily be shown that the zeta function can be defined by the equation (for $u = q^s$)

$$\frac{d}{du}(\log Z(u)) = \sum_{m=1}^{\infty} N_m u^{m-1}, \qquad Z(0) = 1,$$

where N_m is the number of points of the curve whose coordinates belong to the extension $\boldsymbol{F}_{q^m}$ of $\boldsymbol{F}_q$ of degree m. It turns out that this function is much simpler than in the classical case; in fact it is a rational function

$$Z(u) = P_{2g}(u)/(1 - u)(1 - qu),$$

where P_{2g} is a polynomial of degree $2g$ (g being the genus of K). F. K. Schmidt further discovered the remarkable fact that the functional equation

$$Z(1/qu) = q^{1-g}u^{2-2g}Z(u)$$

was nothing else but the analytic expression of the Riemann-Roch theorem!

At the same time, arithmeticians had been endeavoring to obtain an evaluation of N_1, the number of points of the nonsingular curve Γ corresponding to K with coordinates in F_q, and had obtained estimates of the form $\left| N_1 - (q + 1) \right| \leq Cq^\alpha$, with C independent of q and $1/2 < \alpha < 1$; they had observed that $\alpha = 1/2$ would be the best possible result. Hasse became interested in the problem and remarked that the result was a consequence of the so-called "Riemann hypothesis for curves over finite fields," namely the fact that all the zeroes of the polynomial P_{2g} lay on the circle $\left| u \right| = q^{1/2}$, this fact implying the inequality

(7) $$\left| N_1 - (q + 1) \right| \leq 2g \cdot q^{1/2}$$

in an elementary way. In 1934, he succeeded in proving this result for $g = 1$, by adapting to the case of finite fields ideas from the theory of complex multiplication of elliptic functions. He and Deuring observed furthermore that an extension to values $g \geq 2$ would have to be based on the theory of correspondences.

This is what A. Weil proceeded to do. An irreducible correspondence between two irreducible curves Γ_1, Γ_2 is an irreducible curve on the surface $\Gamma_1 \times \Gamma_2$, and in general a *correspondence* between Γ_1 and Γ_2 is a *divisor* on $\Gamma_1 \times \Gamma_2$; degenerate correspondences are those of   Theme B
the form $\{x_1\} \times \Gamma_2$ or $\Gamma_1 \times \{x_2\}$ ($x_i \in \Gamma_i$) and linear combinations of such with integral coefficients; correspondences are called *equivalent* if they differ by the sum of a principal divisor and a degenerate correspondence. For $\Gamma_1 = \Gamma_2 = \Gamma$, one defines as in set theory the *composition* $X \circ Y$ of two correspondences; it can be proved that, together with the addition of divisors, this defines on the set of equivalence classes $\mathfrak{A}(\Gamma)$ a structure of *ring* with unit element (the class of the diagonal Δ of $\Gamma \times \Gamma$). The degrees $d(X)$ and $d'(X)$ of a correspondence are defined as the integers, such that the first (resp. second) projection of X is the cycle $d(X) \cdot \Gamma$ (resp. $d'(X) \cdot \Gamma$); on the other hand, for two correspondences X, Y which intersect properly, $I(X \cdot Y)$ is the degree of the cycle $X \cdot Y$. One can then show that the integer

$$S(X) = d(X) + d'(X) - I(X \cdot \Delta)$$

only depends on the equivalence class ξ of X, and has the property of a *trace*, i.e.,

$S(\xi \cdot \eta) = S(\eta \cdot \xi)$ for two elements of $\mathfrak{A}$. Furthermore, to each correspondence X is associated another one X', deduced from X by the symmetry automorphism of $\Gamma \times \Gamma$; if ξ, ξ' are the classes of X and X', one has $S(\xi \cdot \xi') \geqq 0$, equality being only possible for $\xi = 0$ in $\mathfrak{A}$. This theory was first developed in 1885 by Hurwitz, using Riemann's theory of abelian integrals, and the inequality for the trace was obtained by Castelnuovo (of course for the classical case); using his theory of intersection multiplicities, A. Weil was able to extend all these results to curves over arbitrary fields. He then observed that in the Hasse problem, the number N_m was exactly $I(F^m \cdot \Delta)$, where F is the "Frobenius correspondence" which to each point of Γ associates its transform by the automorphism of Γ corresponding to the automorphism $t \mapsto t^q$ of the algebraic closure of F_q; from which it follows by definition that $S(F^m) = 1 + q^m - N_m$, and expressing the inequality $S(\xi \cdot \xi') \geqq 0$ where ξ is the class of $a \cdot \Delta + b \cdot F^m$, for arbitrary integers a, b, one gets $\left| N_m - q^m - 1 \right| \leqq 2g \cdot q^{m/2}$, which generalizes (7) and implies the "Riemann hypothesis."

VII d: Equivalence of divisors and abelian varieties. The introduction of varieties of arbitrary dimension had been particularly useful because it allowed one to view as points in a projective space of sufficiently high dimension geometric objects such as lines, conics, etc. In 1937, Chow and van der Waerden showed quite generally that it is possible to consider the irreducible algebraic subvarieties of given dimension and degree in a given $P_n(k)$ as the points of some algebraic variety in a suitable $P_N(k)$. From this result it follows that it is possible to give a precise meaning (for an arbitrary field k) to the concepts of "specialization of cycles" and of "algebraic family of cycles" which had been used in the classical case by the Italian school. In particular, one can define the concept of *algebraic equivalence* of two divisors D_1, D_2 on a nonsingular variety V as meaning that they belong to a common irreducible algebraic family of divisors. Another concept of equivalence is *numerical equivalence*, meaning that for any curve C on V, the intersection numbers $(D_1 \cdot C)$ and $(D_2 \cdot C)$ are equal. If one denotes by G, G_n, G_a, G_l the group of divisors on V and its subgroups formed of divisors equivalent to 0 for numerical, algebraic and linear equivalence, one has $G \supset G_n \supset G_a \supset G_l$. Severi for the classical case, and Matsusaka for arbitrary characteristic proved that the group G_n/G_a is finite. A deeper result, proved by Severi for complex algebraic surfaces, following earlier results of Picard, is that the group G/G_n is a free finitely generated commutative group Z^ρ; this result was extended by Néron for arbitrary fields and in any dimension. Finally, it was known since Riemann that for an irreducible algebraic curve over C, the group G_a/G_l was naturally endowed with a structure of g-dimensional algebraic nonsingular variety (g being the genus of the curve) which, as a topological group, is isomorphic to a *complex torus* C^g/Γ, where Γ is a lattice in C^g (discrete group isomorphic to Z^{2g}); this variety is called the *Jacobian* of the curve, and it had been used since Clebsch to study the geometry on an algebraic curve. In general, a complex torus C^n/Γ, where Γ is a lattice in C^n (isomorphic to Z^{2n}) can only be given the structure of an algebraic variety if the lattice Γ satisfies certain bilinear relations which had been already found by Riemann; it is then called an *abelian variety*. The work of Picard and his successors proved that for an arbitrary nonsingular algebraic variety V over C, the group G_a/G_l was again equipped with a structure of abelian variety, called the *Picard variety* of V. Following his work on the Riemann hypothesis, A. Weil developed the general theory of abelian varieties over an arbitrary field (as "abstract" varieties), and was able to define the Jacobian of a curve. Later work of Chow and Matsusaka proved that abelian varieties can still be imbedded in projective space in the general case, and extended to any field the definition of the Picard variety.

VIII. SEVENTH PERIOD: "SHEAVES AND SCHEMES"
(1950–   )

After 1945, the considerable progress brought in algebraic topology, differential topology and the theory of complex manifolds by the introduction of sheaves and spectral sequences (both due to J. Leray) completely renewed the concepts and methods of algebraic geometry, both "classical" and "abstract," simplifying old definitions and results and opening new ways leading to the solution of old problems.

VIII a: The Riemann-Roch theorem for higher dimensional varieties and sheaf cohomology. The Riemann-Roch problem for an irreducible algebraic variety V is the computation of the dimension $l(D)$ of the vector space $L(D)$ for an arbitrary divisor D on V by some formula similar to the Riemann-Roch theorem for curves (3). The Italian geometers had attacked the problem for surfaces, but succeeded only in getting a *lower bound* for $l(D)$, expressed in terms of $\deg(D)$ and birational invariants of the surface S, of D and of $\Delta - D$ (where Δ is a canonical divisor).

In the 1930's, study of differential geometry and in particular of E. Cartan's method of moving frames had finally led to the definition of *vector bundles* over a differential manifold M: such   Theme F
a bundle is a differential manifold **E** with a projection $p: \mathbf{E} \to M$
such that the fibers $p^{-1}(x)$ for any $x \in M$ are real (resp. complex)
vector spaces of fixed dimension r (the *rank* of **E**), and locally on M, **E** looks like the product of M and $\mathbf{R}^r$ (resp. $\mathbf{C}^r$); in other words each point of M has an open neighborhood U for which there is a diffeomorphism ϕ transforming $p^{-1}(U)$ onto $U \times \mathbf{R}^r$ (resp. $U \times \mathbf{C}^r$) in such a way that ϕ transforms *linearly* each fiber $p^{-1}(x)$ into $\{x\} \times \mathbf{R}^r$ (resp. $\{x\} \times \mathbf{C}^r$). A *section* of **E** is a differentiable mapping $s: x \mapsto s(x)$ of M into **E** such that $s(x) \in p^{-1}(x)$ for every $x \in M$. Over a complex manifold M, one can similarly define holomorphic vector bundles by taking **E** as a complex manifold, the projection p being holomorphic, the fibers $p^{-1}(x)$ complex vector spaces, and ϕ (in the above definition) being also holomorphic. Important examples of vector bundles are the *tangent bundle* $\mathbf{T}(M)$, where the fiber $p^{-1}(x)$ consists of the tangent vectors to M at x (so that the rank is $\dim(M)$), and the *bundle of p-covectors* on M, whose sections are the exterior differential p-forms on M (see VII a).

The concept of *divisor* can be generalized to arbitrary complex manifolds M: if (U_α) is an open covering of M, one considers in each U_α a meromorphic function

h_α, such that in $U_\alpha \cap U_\beta$, h_β/h_α is holomorphic and $\neq 0$ everywhere; two such systems (h_α), (h'_λ) corresponding to coverings (U_α), (U'_λ) are identified if h_α/h'_λ is holomorphic and $\neq 0$ in $U_\alpha \cap U'_\lambda$ for any pair (α, λ) of indices, and these classes of systems (h_α) are called divisors on M. One sees that for projective algebraic varieties over C, this notion coincides with the old one: for instance, if $M = P_n(C)$, and $D = \sum_k m_k S_k$ is a divisor on M, where each S_k is an irreducible hypersurface defined by an equation $F_k(x_0, x_1, \cdots, x_n) = 0$, F_k being an irreducible homogeneous polynomial of degree d_k, one covers $P_n(C)$ with the $n + 1$ open sets U_j $(0 \leq j \leq n)$, U_j being defined by the relation $x_j \neq 0$; one can then take as meromorphic function h_j in U_j the function

$$x \mapsto x_j^{-d} \prod_k (F_k(x_0, \cdots, x_n))^{m_k}$$

with $d = \sum_k m_k d_k$. In 1950, A. Weil observed that to a divisor D on a complex manifold M was naturally attached a complex vector bundle of rank 1 (what one calls a *line bundle*) $\mathbf{B}(D)$: with the previous notations, one "glues together" the complex manifolds $U_\alpha \times C$ by taking as "transition function" from U_α to U_β the function $(x, z) \mapsto (x, (h_\beta(x)/h_\alpha(x))z)$, holomorphic in $(U_\alpha \cap U_\beta) \times C$. Furthermore, if s is a holomorphic section of $\mathbf{B}(D)$, the restrictions s_α of s to U_α are such that in $U_\alpha \cap U_\beta$ one has $s_\beta = (h_\beta/h_\alpha)s_\alpha$, hence there is a meromorphic function f on M such that the restriction of f to U_α is s_α/h_α for each α; for an algebraic variety M this is equivalent to $(f) + D \geq 0$, and therefore $L(D)$ can be interpreted as the vector space $\Gamma(\mathbf{B}(D))$ of all *holomorphic sections* of the line bundle $\mathbf{B}(D)$. For instance, if $M = P_n(C)$, and $D = H$, a hyperplane in $P_n(C)$, the transition functions for $\mathbf{B}(H)$ are

$$(x, z) \mapsto \left(x, \frac{x_k}{x_j} z \right)$$

in $U_j \cap U_k$ (with the notations introduced above), and $\Gamma(\mathbf{B}(H))$ is the vector space of all linear forms $(x_0, \cdots, x_n) \mapsto \lambda_0 x_0 + \cdots + \lambda_n x_n$ in C^{n+1}.

Now to each complex vector bundle $\mathbf{E}$ over a differential manifold M of dimension n are attached, for each even integer $2j \leq n$, well determined elements $c_j(\mathbf{E})$ of the cohomology group $H^{2j}(M, Z)$ called the *Chern classes* of $\mathbf{E}$*; when M is a complex manifold of real dimension $2n$, the Chern classes of $\mathbf{T}(M)$ are simply written c_j

---

* One can define the concept of *direct sum* of vector bundles over M by defining it locally in an obvious way; for any differentiable map $f: M' \rightarrow M$, one defines the "pullback" $f^*(\mathbf{E})$ of a vector bundle $\mathbf{E}$ over M as the submanifold of the product $M' \times \mathbf{E}$ consisting of the pairs (x', z) such that $f(x') = p(z)$. The Chern classes of $\mathbf{E}$ can then be characterized by the following conditions, where one writes $c(\mathbf{E})$ for the sum $\sum_{j=0}^\infty c_j(\mathbf{E})$ (the sum is finite since the groups $H^{2j}(M, Z)$ are 0 for $2j > \dim M$; one writes by convention $c_0(\mathbf{E}) = 1$): (i) $c(f^*(\mathbf{E})) = f^*(c(\mathbf{E}))$, where on the right hand side $f^*: H^*(M, Z) \rightarrow H^*(M', Z)$ is the natural mapping deduced from $f: M' \rightarrow M$.

(ii) $c(\mathbf{E}_1 \oplus \mathbf{E}_2 \oplus \cdots \oplus \mathbf{E}_m) = c(\mathbf{E}_1)c(\mathbf{E}_2)\ldots c(\mathbf{E}_m)$ for any direct sum of vector bundles $\mathbf{E}_j$ over M^1 (product taken in the cohomology ring $H^*(M, Z)$).

(iii) $c(\mathbf{B}(H)) = 1 + h_n$ for a hyperplane $H \subset P_n(C)$, $h_n \in H^2(P_n(C), Z)$ being the cohomology class corresponding to the homology class of the $(2n - 2)$–cycle H by Poincaré duality.

$(1 \leqq j \leqq n)$ and called the *Chern classes of M*; the number $\langle c_n, M \rangle$ (where M is considered as $2n$-cycle) is the Euler-Poincaré characteristic

$$\chi(M) = \sum_{j=0}^{2n} (-1)^j R_j .$$

Using the interpretation of divisors by line bundles and Hodge's theory of harmonic forms, Kodaira was able in 1951 to obtain, for compact kählerian manifolds of complex dimension 2, a "Riemann-Roch formula" in which the missing terms from the formula found by the Italian geometers were expressed by means of Chern classes; in 1952 he found a similar formula for kählerian manifolds of dimension 3.

Meanwhile, H. Cartan and Serre had discovered that Leray's concept of *sheaf* led to a remarkably simple and suggestive expression of the main results of the theory of complex manifolds. The holomorphic functions in open sets of such a manifold M satisfy Leray's axioms: if $\mathcal{O}(U)$ is the set of the complex functions holomorphic in the open set $U \subset M$, then, for every open covering (V_α) of U, a function $f \in \mathcal{O}(U)$ is entirely determined by its restrictions $f \mid V_\alpha \in \mathcal{O}(V_\alpha)$, and conversely, given for each α an $f_\alpha \in \mathcal{O}(V_\alpha)$ such that f_α and f_β have the same restriction to $V_\alpha \cap V_\beta$ for all pairs (α, β), there exists an $f \in \mathcal{O}(U)$ such that $f \mid V_\alpha = f_\alpha$ for all α. The sheaf thus defined is called the structural sheaf of M and written $\mathcal{O}_M$; one writes $H^0(U, \mathcal{O}_M)$ instead of $\mathcal{O}(U)$. More generally, for any complex vector bundle **E** over M, one defines the sheaf $\mathcal{O}(\mathbf{E})$ by replacing $\mathcal{O}(U)$ by the set of sections $\Gamma(U, \mathbf{E})$ of **E** above U, written $H^0(U, \mathcal{O}(\mathbf{E}))$); in particular one writes Ω_X^p the sheaf corresponding to the complex bundle of p-covectors on M, so that $H^0(U, \Omega_X^p)$ is the set of holomorphic exterior differential p-forms on U; for a divisor D on M, one writes $\mathcal{O}_X(D)$ instead of $\mathcal{O}(\mathbf{B}(D))$.

There are many types of sheaves other than those derived from vector bundles, and the usefulness of sheaves derives from this versatility and from the many operations one can do with sheaves. In the first place, to a sheaf of groups $\mathcal{F}$ over M and to each point $x \in M$ is associated a group, the *stalk* $\mathcal{F}_x$ of $\mathcal{F}$ at x: for $\mathcal{O}(\mathbf{E})$, $\mathcal{O}(\mathbf{E})_x$ consists of the equivalence classes of sections of **E** over neighborhoods of x for the following relation: two sections are equivalent if they coincide on a neighborhood of x ("germs of sections"); the general definition of $\mathcal{F}_x$ is similar. For a sheaf of abelian groups $\mathcal{G}$ and a sheaf $\mathcal{N} \subset \mathcal{G}$ such that $\mathcal{N}_x$ is a subgroup of $\mathcal{G}_x$ for each x, one can then define a quotient sheaf $\mathcal{G}/\mathcal{N}$ such that $(\mathcal{G}/\mathcal{N})_x = \mathcal{G}_x/\mathcal{N}_x$. Each stalk $(\mathcal{O}_X)_x$ (written $\mathcal{O}_x$) is a local ring, and if $\mathcal{F}, \mathcal{G}$ are two sheaves such that $\mathcal{F}_x$ and $\mathcal{G}_x$ are $\mathcal{O}_x$-modules, then one can define a sheaf $\mathcal{F} \otimes \mathcal{G}$ such that $(\mathcal{F} \otimes \mathcal{G})_x = \mathcal{F}_x \oplus_{\mathcal{O}_x} \mathcal{G}_x$; one has $\mathcal{O}_x(D + D') =$ Theme G $\mathcal{O}_x(D) \otimes \mathcal{O}_x(D')$ for divisors D, D'. The chief interest of sheaf theory is that sheaves of groups may be used to replace the *coefficients* in cohomology groups by "local coefficients" varying with $x \in M$. The cohomology groups $H^j(M, \mathcal{F})$ which one thus defines for each integer $j \geqq 1$ (one also writes $H^j(\mathcal{F})$) have the fundamental property that for any exact sequence of sheaves of abelian groups $0 \to \mathcal{N} \to \mathcal{G} \to \mathcal{G}/\mathcal{N} \to 0$, one has a "long exact sequence"

(8) $0 \to H^0(\mathcal{N}) \to H^0(\mathcal{G}) \to H^0(\mathcal{G}/\mathcal{N}) \to H^1(\mathcal{N}) \to H^1(\mathcal{G}) \to H^1(\mathcal{G}/\mathcal{N}) \to H^2(\mathcal{N}) \to \cdots$

Once these new tools were introduced in analysis it was soon recognized that the invariants introduced by the Italian school and by Hodge were easily expressed by sheaf cohomology. In the first place, if M is a compact connected kählerian variety of dimension n, Dolbeault and Serre proved that the corresponding space $H^{r,s}$ of harmonic forms of type (r,s) (see VII-a) is isomorphic to $H^s(\Omega_M^r)$; furthermore, for any divisor D on M, Serre discovered that there is a natural duality pairing the spaces

$$H^j(\mathcal{O}_M(D)) \text{ and } H^{n-j}(\Omega_M^n \otimes \mathcal{O}_M(-D)) = H^{n-j}(\mathcal{O}_M(\Delta - D))$$

"explaining" the intervention of the canonical divisor Δ in Riemann-Roch's theorem (3) (one has written $\Omega_M^n = \mathcal{O}_M(\Delta)$). By definition, the *geometric genus* of M can be written

(9) $$p_g = \dim(H^0(\Omega_M^n)) \text{ and also } p_g = \dim(H^n(\mathcal{O}_M))$$

by the isomorphism of $H^{r,s}$ and $H^{s,r}$; one has similar invariants for holomorphic exterior forms of all degrees $< n$. The *arithmetic genus* turns out to be the number

(10) $$p_a = \dim H^n(\mathcal{O}_M) - \dim H^{n-1}(\mathcal{O}_M) + \cdots + (-1)^{n-1}\dim H^1(\mathcal{O}_M)$$

and the plurigenera are given by

(11) $$p_k = \dim H^0(\mathcal{O}_M(k\Delta)).$$

In 1937, Eger and Todd introduced, on an algebraic nonsingular projective variety M of complex dimension n, "canonical" equivalence classes of algebraic cycles of dimension $n - j$, which later were recognized to correspond exactly *via* Poincaré duality, to the Chern classes c_j of M; furthermore, Todd discovered that the arithmetic genus of M could be computed by the formula

(12) $$(-1)^n p_a + 1 = \langle T_n(c_1, \cdots, c_n), M \rangle,$$

where T_n is a polynomial with rational coefficients in the Chern classes, defined by the following device: in the power series

$$\prod_{j=1}^{n} \frac{\gamma_j z}{1 - \exp(\gamma_j z)}$$

one considers the coefficient of z^n, which is a symmetric polynomial in the variables γ_j, and one expresses it in terms of the elementary symmetric functions of the γ_j; then one replaces each elementary symmetric function σ_j by c_j. For instance, the first three Todd polynomials are

$$T_1(c_1) = c_1/2, \quad T_2(c_1,c_2) = (c_2 + c_1^2)/12,$$

$$T_3(c_1,c_2,c_3) = c_2 c_1/24.$$

In 1954, Hirzebruch generalized both Todd's result and the Riemann-Roch formulas of Kodaira by proving that for any divisor D on M, the expression

$$\dim H^0(\mathcal{O}_M(D)) - \dim H^1(\mathcal{O}_M(D)) + \cdots + (-1)^n \dim H^n(\mathcal{O}_M(D))$$

could be expressed as $\langle P(f,c_1,\cdots,c_n), M \rangle$, where f is the first Chern class of the bundle $\mathbf{B}(D)$, and P a polynomial which is obtained by the same device as above, starting from the power series

$$e^{fz} \prod_j \frac{\gamma_j z}{1 - \exp(\gamma_j z)}.$$

It was later recognized that in fact, Hirzebruch's formula was a particular case of a much more general theorem valid for all compact differential manifolds, the Atiyah-Singer index formula.

The Hirzebruch formula enables one to solve the Riemann-Roch problem when all cohomology groups $H^j(\mathcal{O}_M(D))$ are reduced to 0 for $j \geq 1$. Kodaira found sufficient conditions for this fact to hold; for instance, it is true when one replaces D by $D + mH$ where H is the intersection of M and a hyperplane (in the projective space where M is imbedded) and $m > 0$ is large enough. He has also obtained a fundamental criterion for a compact kählerian manifold M to be isomorphic to a projective algebraic variety: there must exist on M a kählerian metric such that the cohomology class of the form Ω (equation (6)) in $H^2(M, \mathbf{R})$ belongs to $H^2(M, \mathbf{Q})$.

VIII b: The Serre varieties. In 1942, Zariski began a deep study of singularities of projective algebraic varieties over any field, in view of proving a desingularization theorem (which he succeeded to do for dimension ≤ 3 and over a field of characteristic 0); for that purpose, he used for the first time the general theory of valuations*, developed 10 years earlier by Krull. In the course of this work, he introduced the generalization of the "abstract Riemann surface" of Dedekind-Weber for an arbitrary field K of algebraic functions over a field k, defining it to be the set V of all valuations of K which vanish on k^*; but in addition, using ideas introduced a few years earlier by M. Stone, he defined on V (by purely

Theme C

---

* The only difference between the definition of a general valuation and the definition of a discrete valuation (see VI–a) is that the valuation may take its value in an *arbitrary* totally ordered group. For instance, the group $\mathbf{Z} \times \mathbf{Z}$ may be totally ordered by writing $(m, n) < (m', n')$ if either $m < m'$, or $m = m'$, and $n < n'$ ("lexicographic ordering"); one may then define on $\mathbf{C}(X, Y)$ a valuation with value in that totally ordered group by taking for $w(P)$, where P is a polynomial $\neq 0$, the smallest (m, n) in $\mathbf{Z} \times \mathbf{Z}$ for which the term in $X^m Y^n$ in P has a nonzero coefficient.

algebraic considerations) a *topology* for which V became quasi-compact, although that topology is not Hausdorff in general: for instance, in the case of dimension 1, considered by Dedekind-Weber, the closed sets are V and all the finite subsets of V.

By 1950 A. Weil observed that this "Zariski topology" could be defined on his "abstract varieties" (see VII-b); not only did it appreciably improve the exposition of the theory by allowing one to use a "geometric" language, but it also made possible a definition of *vector bundles* modeled on the classical one, and to extend to abstract varieties the relations between divisors and line bundles (see VIII-a). Going one step further, Serre, in 1955, had the idea to transfer in the same way the theory of sheaves to abstract varieties, using the Zariski topology instead of the usual one in Leray's definition. At the same time, he observed that the concept of sheaf made possible a much simpler definition of "abstract varieties," using the general idea of "ringed space" of H. Cartan, i.e., a topological space X on which is given a sheaf of rings $\mathcal{O}_X$; the advantage of this kind of structure is that it lends itself very easily to "gluing" ringed spaces along open subsets, the verification of the conditions of compatibility being usually trivial. In Serre's case the "pieces" which are glued together are *affine varieties* over an algebraically closed field k of arbitrary characteristic: such a variety X is a (Zariski) closed set of some k^n (i.e., defined by polynomial equations), and $\mathcal{O}_X$ is the sheaf of rings such that for each open set $U \subset X$, $\mathcal{O}(U) = H^0(U, \mathcal{O}_X)$ consists of the restrictions to U of the rational functions $P(X)/Q(X)$ on k^n which are defined (i.e., $Q(x) \neq 0$) at every $x \in U$. Of course cohomology groups $H^j(\mathscr{F})$ can still be defined when $\mathscr{F}$ is a sheaf of modules over the rings $\mathcal{O}_x$; they are vector spaces over k and Serre computed the groups $H^j(\mathcal{O}_M(mH))$ for $M = P_n(k)$ and H a hyperplane ($m \in \mathbf{Z}$); he also extended to arbitrary fields and to projective varieties his duality theorem; but when k has characteristic $p > 0$, most of the results obtained in the classical case by the methods of Lefschetz and Hodge fail to generalize: for instance, the dimension of $H^r(\Omega_X^s)$ and of $H^s(\Omega_X^r)$ for a projective variety X are not necessarily equal. Nevertheless, Grothendieck and Washnitzer were able independently to extend Hirzebruch's formula to fields k of arbitrary characteristic, and Grothendieck, by the introduction of his "K-theory," gave a far reaching generalization of that formula. Finally, when k is the complex field, Serre showed that the cohomology groups obtained by using the Zariski topology coincided with the classical ones.

Being chiefly interested in cohomology, Serre did not dwell at length on the general properties of his varieties; these were investigated in detail by Chevalley almost simultaneously (in a different language, which we do not reproduce here). One of the points which should be emphasized is that with Serre and still more with Chevalley, birational geometry fades out of the picture and the concept of *morphism* comes to the fore. Until then, the center of interest was the theory of *complete* varieties, and it is only seldom that a correspondence between two such

Theme G

Theme B

varieties X, Y, even if it assigns only one point of Y to a point of X (a $(1, n)$-correspondence in classical language), is defined at *every* point of X. A morphism $f: X \to Y$, where X and Y are Serre varieties, is on the contrary a *mapping of X* into Y, which is continuous for the Zariski topologies and such that for every point $x \in X$ and every affine neighborhood V of $y = f(x)$, there is an affine neighborhood U of x such that $f(U) \subset V$ and, for every function $s \in H^0(V, \mathcal{O}_Y)$, the function $x \mapsto s(f(x))$ defined in U, belongs to $H^0(U, \mathcal{O}_x)$. The main results of Chevalley are general theorems on morphisms and studies of special types of morphisms using results of commutative algebra going back to E. Noether and Krull. It had been known for a long time that the image $f(X)$ of X by a morphism $f: X \to Y$ was not even locally closed in Y in general; Chevalley showed however that when X is irreducible, $f(X)$ always contains a set which is open and dense in the subspace $\overline{f(X)}$ of Y. Another of Chevalley's results is that if X and Y are irreducible, and for each $x \in X$ one writes $e(x)$ the maximum of the dimensions of the irreducible components of $f^{-1}(f(x))$ which contain x, then the mapping $x \mapsto e(x)$ is upper semi-continuous in X (in other words, when x' is close enough to x, $e(x')$ is never $< e(x)$).

Chevalley also showed how important concepts introduced by Zariski in the 1940's, and which A. Weil had already used in his theory of abstract varieties, led to very suggestive theorems on morphisms. For projective varieties, Zariski had observed that the "regularity" properties of a point $x \in X$ were linked very closely to the structure of the *local ring* $\mathcal{O}_x$ of the variety X at that point: x only belongs to one irreducible component if $\mathcal{O}_x$ has no zero divisors, and x is *simple* if $\mathcal{O}_x$ is a *regular* local ring (i.e., $\mathcal{O}_x$ is an integral domain whose field of fractions has a transcendence degree over the base field k (always assumed to be algebraically closed) equal to the dimension over k of the vector space $\mathfrak{m}_x/\mathfrak{m}_x^2$, where $\mathfrak{m}_x$ is the maximal ideal of $\mathcal{O}_x$). A property, of which Zariski was the first to grasp the geometric significance, is the fact for $\mathcal{O}_x$ to be *integrally closed* in its field of fractions, in which case x is said to be *normal*. Zariski showed that simple (or normal) points of an irreducible variety formed an open dense set, and that the complement of the set of normal points has codimension *at least* 2. Furthermore, Zariski defined for each projective irreducible variety X its "normalization;" this can easily be extended to Serre varieties: for any finite extension L of the field of rational functions K of X, there is a variety X' and a morphism $p: X' \to X$ such that for each affine open set U of X, $p^{-1}(U)$ is an affine open set of X' and the ring $H^0(p^{-1}(U), \mathcal{O}_{X'})$ is the integral closure in L of the ring $H^0(U, \mathcal{O}_X)$; X' is called the normalization of X *in* L, and simply the normalization of X if $L = K$. The normalization of X is of course birationally equivalent to X, and its singular points form a subvariety of codimension ≥ 2; in particular, if X is a curve, X' has no singular points, and this is the simplest "desingularization" of a curve (valid in every characteristic).

The climax of Zariski's investigations on normal varieties had been his "main theorem" expressed in the language of birational correspondences; Chevalley showed that it implies a far more intuitive result about morphisms: suppose X

and Y are irreducible and *normal* varieties, $f: X \to Y$ is a morphism such that $f(X)$ is dense in Y and each set $f^{-1}(y)$ is *finite* for $y \in Y$. Then f factorizes in $X \xrightarrow{g} Y' \xrightarrow{p} Y$ where Y' is the normalization of Y in the field of rational functions of X, and g is an *isomorphism* of X onto an *open* subvariety of Y'.

Finally, Chevalley defined the notion of *complete* variety in a much simpler way than before: X is complete if, for every variety Y, the second projection $X \times Y \to Y$ is a *closed* mapping.

The interest of Chevalley in such theorems was spurred by the theory of *algebraic groups*, which he and A. Borel brought to a high level of development during the 1950's; in that theory, both affine and complete varieties play an important part and the preceding theorems are powerful tools.

VIII c: Schemes and topologies. Until the 1950's, no one seems to have tried to give an *intrinsic* definition of an affine variety over an *algebraically closed field k*, independent of any imbedding of the variety in some "affine space" k^n, although the tools to do so were available since the 1890's. In his work on invariant theory, Hilbert had proved his famous "*Nullstellensatz,*" one of the forms of which is that the maximal ideals of the algebra of polynomials $k[X_1, \cdots, X_n]$ are in one-to-one correspondence with the elements $z = (\zeta_1, \cdots, \zeta_n) \in k^n$, such an element corresponding to the ideal generated by the polynomials $X_1 - \zeta_1, \cdots, X_n - \zeta_n$. Just as Riemann attached to a projective curve the field of rational functions on that curve, so one may attach to an affine variety $V \subset k^n$ the ring $R(V)$ of the restrictions to V of all *polynomial* functions on k^n; this ring is a finitely generated algebra over k, which has no nilpotent elements (one says it is *reduced*); and by Hilbert's *Nullstellensatz*, the points of V are in one-to-one correspondence with the maximal ideals of $R(V)$. Conversely, it is readily seen that *any* reduced and finitely generated k-algebra has the form $R(V)$ for an affine variety determined up to isomorphism. Furthermore, when V is irreducible, it is even possible to define the sheaf $\mathcal{O}_V$ directly from the ring $R(V)$: for any open (Zariski) subset U of V which is defined as the set of points x such that $f(x) \neq 0$ for some $f \in R(V)$, one defines $\mathcal{O}(U)$ as the ring of rational functions of type g/f^m for $g \in R(V)$ and m a positive integer, and it is easy to see that this defines completely $\mathcal{O}_X$. Finally, if V, W are two affine varieties over k, we have seen above that to a morphism $f: V \to W$ corresponds a k-algebra homomorphism $R(f): R(W) \to R(V)$; but the converse is also true, for Hilbert's *Nullstellensatz* implies that for any such homomorphism $\phi: R(W) \to R(V)$, the inverse image $\phi^{-1}(\mathfrak{m})$ of a maximal ideal of $R(V)$ is again a maximal ideal in $R(W)$, and $\mathfrak{m} \mapsto \phi^{-1}(\mathfrak{m})$ is the morphism corresponding to ϕ. In the language of categories, which was beginning to be used in the late 1950's, the category of affine varieties over k was *equivalent* to the *dual* of the category of reduced finitely generated (commutative) k-algebras.

Following a suggestion of Cartier, A. Grothendieck undertook around 1957 a gigantic program aiming at a vast generalization of algebraic geometry, absorbing

all previous developments and starting from the category of *all* commutative rings
(with unit) instead of reduced finitely generated algebras over an algebraically
closed field. If one wanted to define a category which would be equivalent to the dual
of the category of all commutative rings, a nontrivial modification was needed from
the start, since if $\phi: A \to B$ is a homomorphism of rings (sending unit element on
unit element), the inverse image $\phi^{-1}(\mathfrak{m})$ of a maximal ideal of B is not in general
a maximal ideal of A, whereas the inverse image $\phi^{-1}(\mathfrak{P})$ of a *prime* ideal of B is
always a prime ideal of A. It was thus necessary to take as the set replacing the
affine variety the *spectrum* of A, i.e., the set $\mathrm{Spec}(A)$ of all *prime* ideals of A; closed
sets in $\mathrm{Spec}(A)$ are defined as sets of prime ideals containing a given (arbitrary)
ideal of A, hence a "Zariski topology" for which, however, finite sets are no longer
closed in general; finally, using work of Chevalley and Uzkov on localization dating
from the 1940's, it is possible to give a meaning to g/f^m even when f is a zero-
divisor of A, hence to define the sheaf $\mathcal{O}_X$ on $X = \mathrm{Spec}(A)$ in the same way as for
affine varieties. The ringed spaces thus obtained are called *affine schemes* and they
form a category equivalent to the dual of the category of all commutative rings;
finally, the usual "gluing process" for ringed spaces yields the category of *schemes*
by replacing affine varieties by affine schemes.

The experience of the last 10 years has convinced the specialists that, in spite
of the much greater amount of commutative algebra techniques which it requires,
the theory of schemes is the context in which the problems of algebraic geometry
are best understood and attacked. Among the features which distinguish it from
previous conceptual frames for algebraic geometry, let us mention only the few
following ones:

(1) The notion of *generic point*, which had disappeared from the Serre-Chevalley
theory, is now reintroduced in a natural way: for instance, if A is an integral domain,
its (unique) generic point is the prime ideal (0) in $\mathrm{Spec}(A)$; its "generic" property is
expressed by the fact that its *closure* is the *whole* space $\mathrm{Spec}(A)$, and thus continuity arguments in the Italian style (but in the Zariski topology!) are now again
available.

(2) The predominance of "relative" versus "absolute" notions, or, put in a
different way, the fact that most of the times what is studied is not a scheme but a
morphism of schemes $f: X \to S$, where S is often quite arbitrary (one also says
that the study of such morphisms, for fixed S, is the study of "S-schemes"). This is
particularly apparent when it comes to imposing *finiteness conditions* (without *any*
such condition, there is very little likelihood of ever getting any deep result):
Grothendieck has shown that, except for cohomological notions, one may usually
allow the "base scheme" S to be free from finiteness assumptions (such as being
noetherian, or of finite dimension, etc.), and the results only depend on finiteness
conditions for the morphism f; this allows considerable freedom in the "change
of bases" (see below).

(3) Given two "S-schemes" $f: X \to S$, $g: Y \to S$, there is an essentially unique
triplet consisting in an S-scheme $X \times_S Y$ and two morphisms $p_1: X \times_S Y \to X$,

p_2: $X \times {}_SY \to Y$ such that $f \circ p_1 = g \circ p_2$, which is the "categorical" *product* of X and Y over S: this means that, given two morphisms $u: Z \to X$, $v: Z \to Y$ such that $f \circ u = g \circ v$, there is a unique morphism $w: Z \to X \times {}_SY$ such that $u = p_1 \circ w$ and $v = p_2 \circ w$ (there is no similar result for Serre varieties; it easily follows from the existence of the tensor product $B \otimes_A C$ of arbitrary A-algebras, where A is any ring).

Most of the time this fundamental process is applied to study the morphism $f: X \to S$ by replacing the "base" S by another one Y, in such a way that the new morphism p_2, which is now written $f_{(Y)}: X_{(Y)} \to Y$ (the notation $X_{(Y)}$ replacing $X \times {}_SY$) can be more easily handled. This "change of base" is probably the most powerful tool in the theory of schemes, generalizing in a bewildering variety of ways the old idea of "extending the scalars." To give only one example, consider at any point $s \in S$ the residual field $k(s) = \mathcal{O}_s/\mathfrak{m}_s$ of the local ring $\mathcal{O}_s$ at that point; then $X_s = X \times {}_S \operatorname{Spec}(k(s))$ has as underlying space the "fiber" $f^{-1}(s)$ in X and (provided f satisfies finiteness conditions) it can be considered as a "variety" over the field $k(s)$ (in a slightly more general sense than with Serre). In this way, an S-scheme X may be considered as a "family of varieties" X_s parametrized by S (generalizing the old Picard method (see VI-c)) and many properties of S-schemes may be obtained by a study of the fibers X_s.

(4) It may seem strange at first that one should consider affine schemes $\operatorname{Spec}(A)$ even when A has *nilpotent elements* other than 0; but in fact, this also corresponds to geometric facts which were not taken into account by older theories. For instance, consider the parabola $y^2 - x = 0$ in C^2 and the mapping which projects it on the x-axis; in the language of schemes, we consider the affine schemes $U = \operatorname{Spec}(C[X, Y]/(Y^2 - X))$, $V = \operatorname{Spec}(C[X])$ and the morphism $p: U \to V$ which corresponds to the natural injection $C[X] \to C[X, Y]/(Y^2 - X)$ which sends X onto the class of X. A maximal ideal $(X - \zeta)$ in $C[X]$ is identified with the point $\zeta \in C$, and the fiber $V_\zeta = p^{-1}(\zeta)$ is the affine scheme $\operatorname{Spec}(C[Y]/(Y^2 - \zeta))$; now, if $\zeta \neq 0$, the ring $C[Y]/(Y^2 - \zeta)$ is isomorphic to the direct sum of two fields isomorphic to C, corresponding to the fact that the fiber has two distinct points; but if $\zeta = 0$, $C[Y]/(Y^2)$ has nilpotent elements: the two points have become "infinitely near" one another. It turns out that this is a general phenomenon: nilpotent elements in the local rings of a scheme are the algebraic counterpart of "infinitesimal" properties, and their presence allows a much more natural and flexible treatment of these properties than in classical algebraic geometry (see e.g. [8]).

(5) If we return to the concept of affine Serre variety, corresponding to a reduced finitely generated algebra A over an algebraically closed field k, the points of the variety are not *all* points of $\operatorname{Spec}(A)$, but only the *closed* ones, corresponding to all homomorphisms $A \to k$ which are k-homomorphisms, i.e., such that the composition with the natural mapping $k \to A$ gives the identity on k; similarly, if one wants to consider the points of variety "with coordinates in a field K extension of k" (see VII-b), one has to consider homomorphisms $A \to K$ which by composition $k \to A \to K$ give the homomorphism defining the extension K of k. This idea has been greatly

generalized by Grothendieck: for an S-scheme $X \to S$ the "points of X in an arbitrary S-scheme T" (or more briefly the "T-points" of X) are by definition the morphisms $T \to X$ which, composed with $X \to S$, give the structural morphism $T \to S$; if we denote by $\mathrm{Mor}_S(T, X)$ the set of these "S-morphisms," it can easily be shown that $T \mapsto \mathrm{Mor}_S(T, X)$ is a *functor* from the category of S-schemes to the category of sets, and that the knowledge of that functor entirely determines the S-scheme X, which is said to "represent" the functor. This idea has become a very fruitful principle allowing the definition of schemes by the functor which they "represent," which is generally much easier (provided one has general theorems establishing the "representability" of functors); in particular, one transfers in that way to the theory of schemes many classical constructions such as projective spaces, Grassmannians, Chow varieties, Picard varieties, and one is able to give a general meaning to the concept of "moduli" introduced by Riemann for curves.

(6) It was early recognized that the Zariski topology on schemes had some unpleasant features regarding "principal fiber bundles": natural definitions of S-schemes $X \to S$, which in classical geometry gave principal fiber bundles X over S, did not have in general the property of being "locally" products of a (Zariski) neighborhood and a "typical fiber" (one says that they are not "locally trivial" for the Zariski topology).

However, Serre observed that in important cases, a mild "extension of the base" $T \to S$, where T is an "etale covering" of S (which corresponds in classical geometry to an unramified covering with finitely many sheets) was enough to restore "local triviality." Starting from this remark, Grothendieck conceived the idea of replacing the Zariski topology on S by a new structure, called "*etale topology*," which is not any more a topology in the usual sense; essentially it consists in replacing the usual open subsets of S (or rather their natural injections $U \to S$) by etale coverings of S (one may say that the open sets are now "out of the space" instead of being parts of it). The important fact is that he was able to transfer to this new concept the definition of sheaves and of sheaf cohomology, and to show that this "etale cohomology" can partly remedy to the defects of the usual (Zariski) sheaf cohomology for varieties over a field of characteristic $p > 0$.

IX. OPEN PROBLEMS

To have some idea of the dozens of problems on which algebraic geometers are now working, one may consult for instance the various reports in [18], [19], or [20]. We will conclude by mentioning very briefly some of the most conspicuous ones.

(1) The famous problem of "desingularization" of algebraic varieties over a field k has been solved by Hironaka in all dimensions, when k has characteristic 0, and this result has become a very powerful tool in many problems of algebraic geometry, both classical and "abstract." For fields of characteristic $p > 0$, the problem is still open in dimensions ≥ 3; for dimension 2, the desingularization theorem has been proved by Abhyankar in all characteristics.

(2) The problem of Riemann's "moduli" has attracted much attention during the last 20 years, both in classical and in abstract geometry: the general idea is to

prove the existence of a variety (or scheme) whose points would correspond to isomorphism classes of curves of a given genus over a given field; the most comprehensive results to date are those of Mumford, who has proved the existence of such a scheme; but much remains to be done regarding the properties of that scheme. One has similar results when curves of given genus are replaced by abelian varieties of given dimension; but already for algebraic surfaces, very little progress has been made on similar problems. Even when one considers "local" problems, i.e., how algebraic structures depending on parameters may "deform" in the neighborhood of a point in the parameter space, the results are far from final.

(3) In spite of the progresses brought by "etale cohomology" (and other similar theories based on other types of "Grothendieck topologies"), the cohomological properties of varieties over a field of characteristic $p > 0$ are not yet well understood, and nothing has yet satisfactorily replaced the abelian integrals in that case. Central in these problems are the "Weil conjectures" which he formulated as extensions to algebraic varieties of arbitrary dimension of his work on the zeta function of algebraic curves over finite fields; some of them were proved by Grothendieck and M. Artin, using etale cohomology, and a complete verification was recently given by Deligne [21] and [23].

(4) In classical algebraic geometry, the theory of integrals of "second" of "third" kinds on projective algebraic varieties of arbitrary dimension is still incomplete, although much advanced recently by the work of Leray, Hodge-Atiyah and Griffiths on the concept of "residue." Generalizations of the Hodge theory to non compact algebraic varieties (over C) with singularities have recently been started by Deligne and others.

(5) One would expect that the precise knowledge of divisors under various "equivalence" concepts (see VII-d) should extend to "cycles" of arbitrary condimension, but even in the classical case that theory is still in an embryonic stage.

(6) Finally, the beautiful results of Castelnuovo and Enriques on the characterization of classes of surfaces by properties of their invariants have been greatly extended by Kodaira and Shafarevich [11], Bombieri [22] and [24], and generalized by Mumford to surfaces over an algebraically closed field of characteristic $p > 0$ [19], but much remains to be done, and practically no comparable results have been obtained in higher dimensions.

References*

1. M. Baldassari, Algebraic varieties, Ergeb. der Math., Heft 12, Springer, Berlin-Göttingen-Heidelberg, 1956.

2. J. Dieudonné, Algebraic geometry, Advances in Math., 3 (1969) 233–321.

3. ———, Fondements de la géométrie algébrique moderne, Advances in Math., 3 (1969) 322–413.

4. F. Enriques, Sui sistemi continui di curve appartenenti ad una superficie algèbrica, Comm. Math. Helv., 15 (1943) 227–237.

5. F. Hirzebruch, Topological methods in algebraic geometry, Springer, Berlin-Heidelberg-New York, 3rd ed., 1966.

* References 21–24 were provided by the author for the present volume.—ED.

6. S. Lefschetz, L'Analysis Situs et la Géométrie algébrique, Gauthier-Villars, Paris, 1924.

7. D. Mumford, Geometric invariant theory, Erg. der Math. Heft 34, Springer, Berlin-Heidelberg-New York, 1965.

8. ——, Lectures on curves on an algebraic surface, Princeton Univ. Press, Princeton, 1966.

9. ——, Abelian varieties, Oxford Univ. Press, Oxford, 1970.

10. F. Severi, Intorno ai sistemi continui di curve sopra una superficie algèbrica, Comm. Math. Helv., 15 (1943) 238–248.

11. I. Shafarevich et al. Algebraic surfaces, Proc. Steklov Inst. of Math., Amer. Math. Soc., 1967.

12. B. L. van der Waerden, Einführung in die algebraische Geometrie, Springer, Berlin, 1939.

13. A. Weil, Foundations of algebraic geometry, Amer. Math. Soc. Coll. Publ., 29 (1946).

14. ——, Sur les courbes algébriques et les variétés qui s'en déduisent, Hermann, Paris, 1948.

15. ——, Introduction à l'étude des variétés kählériennes, Hermann, Paris, 1958.

16. O. Zariski, Algebraic surfaces, Erg. der Math. 2nd ed., Springer, Berlin-Heidelberg-New York, 1971.

17. ——, An introduction to the theory of algebraic surfaces, Lecture Notes in Math., 83, Springer, Berlin-Heidelberg-New York, 1969.

18. Dix exposés sur la cohomologie des schémas, North-Holland, Amsterdam-London, 1968.

19. Global Analysis (Papers in honor of K. Kodaira), Princeton Univ. Press, 1969.

20. Actes du Congrès international des mathématiciens, Nice, 1970, vol. I et II, Gauthier-Villars, Paris, 1971.

21. P. Deligne, La conjecture de Weil I, Publications Mathématiques IHES No. 43 (1974), 273–307, extensively reviewed in Math. Rev. 49 #5013.

22. H. Popp, editor, Classification of Algebraic Varieties and Compact Complex Manifolds, Lecture Notes in Math. 412, Springer-Verlag, Berlin-Heidelberg-New York, 1974.

23. J. P. Serre, Valeurs propres des endomorphismes de Frobenius [d'après P. Deligne], Exposé 446, Séminaire Bourbaki 1973/74, Lecture Notes in Math. 431, 190–204, Springer Verlag, Berlin-Heidelberg-New York, 1975.

24. K. Veno, Classification Theory of Algebraic Varieties and Compact Complex Spaces, Lecture Notes in Math. 439, Springer-Verlag, Berlin-Heidelberg-New York, 1975.

Editorial Note: See also Solomon Lefschetz' article, "The early development of algebraic geometry", reprinted on pp. 168–177 of this volume, and S. S. Abhyankar's article, "Historical ramblings in algebraic geometry and related algebra", this MONTHLY, vol. 83 (1976), 409–48.

AN ISOPERIMETRIC INEQUALITY FOR POLYHEDRA*

H. H. JOHNSON and J. OSAKA

There are many formulas and inequalities from differential geometry which can be extended to polygons and polyhedra when the basic notions of curvature, mean curvature, etc., are suitably defined. Moreover, these concepts are usually very intuitive and geometrically evident when defined on polyhedral surfaces, since these are discrete. Proofs which are quite technical for surface theory often have elementary analogues on polyhedra. Such material could well be introduced to students at a much earlier level than is possible in differential geometry.

---

* From AMERICAN MATHEMATICAL MONTHLY, vol. 81 (1974), pp. 58–61.

In this note we define total and mean curvature for polyhedra in such a way as to extend an isoperimetric inequality of W. T. Reid [3] to these surfaces. Reid's proof used delicate analytic inequalities. Our proof is entirely elementary, using only vector algebra and the classical isoperimetric inequality for plane curves.

Reid's theorem concerns a surface S of area A bounded by a simple closed curve of length L in 3-space R^3. If the origin is assumed to be any point on the boundary and if X is the position vector to a point on S having mean curvature vector H, let $B = \int_S X \cdot H dA$. Then Reid's inequality is $L^2 \geqq 4\pi(A + B)$. When S is in a plane, $B = 0$ and Reid's inequality reduces to the classical isoperimetric inequality for plane areas.

Definitions. A **face** is a plane region bounded by a simple closed polygon $P = A_1 A_2 \cdots A_n$ where the A_i are the vertices, the line segments $A_i A_{i+1}$ and $A_n A_1$ are the edges. **Simple** means that edges meet only if they share a common vertex.

A **polyhedron** Π is a finite collection of faces satisfying: (1) two faces meet in at most a common edge or vertex, (2) no edge belongs to more than two faces, (3) if Q and Q' are faces which meet at a vertex A, it is possible to connect Q and Q' by a sequence $Q = Q_1, Q_2, \cdots, Q_r = Q'$ of faces having A as a vertex, in which Q_i and Q_{i+1} share a common edge, and (4) any two faces Q and Q' of Π can be connected by a sequence of faces where adjacent faces in the sequence share a common edge. The **boundary** of Π is the collection of all edges which belong to only one face. We also assume that Π is **orientable**. That is, it is possible to choose a "top side" to each of its polygons by means of a unit normal vector, so that along each common edge these top sides agree. We also assume that the origin is a vertex on the boundary. The right-hand rule determines the orientation of the boundary.

Let $X_1 X_2$ be an edge common to two faces. Choose unit normals to the faces, agreeing with the given orientation of Π, and label them Y_1, Y_2 so that in the orientation given to its face by Y_1, the edge is oriented from X_1 to X_2 (and in the orientation given to its face by Y_2, the edge is oriented from X_2 to X_1).* We define $\frac{1}{2}(X_2 - X_1) \times (Y_2 - Y_1)$ to be the mean curvature vector $H(X_1, X_2)$ along the edge $X_1 X_2$. This measures the bending of Π along $X_1 X_2$. When Y_1 and Y_2 are parallel, so that no bending occurs, $H(X_1, X_2) = 0$. It is easy to check that $H(X_1, X_2)$ does not depend on the accident of choosing X_1 first.

In order to interpret Reid's inequality we must examine $H(H_1, X_2) \cdot X$, where X is a position vector to some point of the line segment $X_1 X_2$.

LEMMA 1. *If X is any vector on the line segment $X_1 X_2$, then*

$$H(X_1, X_2) \cdot X = \tfrac{1}{2}[X_2, Y_2 - Y_1, X_1] = \tfrac{1}{2}\{X_2 \times (Y_2 - Y_1)\} \cdot X_1.$$

Proof. Let $X = tX_1 + (1 - t)X_2, 0 \leqq t \leqq 1$. Then

$$H(X_1, X_2) \cdot X = \tfrac{1}{2}(X_2 - X_1) \times (Y_2 - Y_1) \cdot (tX_1 + (1 - t)X_2),$$

and the result follows by elementary vector algebra.

---

* The authors wish to thank the editors for pointing out this correction of the original paper.

In particular, $H(X_1, X_2) \cdot X$ is independent of X.

Let now B be the sum of these numbers $H(X_1, X_2) \cdot X$, where X is any vector on $X_1 X_2$, summed over all edges $X_1 X_2$ of the polyhedron Π. We shall prove that if Π has a simple boundary of length L, and if the area of Π is A, then $L^2 \geqq 4\pi(A + B)$.

First, we can subdivide the polygons of Π by adding extra edges so that each new polygon is a triangle. See Benson [1, p. 9]. This certainly does not change L or A. Also, B is unchanged, since $H(X_1, X_2) = 0$ if $X_1 X_2$ is one of the new edges, for the unit vectors Y_1 and Y_2 to the polygons on each side of the edge are equal. Hence, we may assume that all polygons in Π are triangles.

PROPOSITION. *The area of triangle $X_1 X_2 X_3$ equals $\frac{1}{2}[(X_1 - X_3), (X_2 - X_3), Y]$ where Y is the unit normal to the triangle.*

This follows from the definition of cross product and its geometric interpretation as the area of a parallelogram formed by $X_1 - X_3$ and $X_2 - X_3$.

THEOREM 1. *$A + B = C$, where C is the sum, over all edges $X'X''$ along the boundary of Π, of the triple product $\frac{1}{2}[X', X'', Y]$, where Y is the orienting unit vector on the face having $X'X''$ as boundary, and the boundary is oriented from X' to X''.* (This theorem is true even if the boundary is not simple. A simple boundary is one consisting of a single polygon without self-intersections.)

Proof. Proof by induction on the number n of triangles in Π. For $n = 1$, we have only one triangle $X_1 X_2 X_3$, and

$$A + B = A = \frac{1}{2}[X_1 - X_3, X_2 - X_3, Y]$$

$$= \frac{1}{2}([X_1, X_2, Y] + [X_2, X_3, Y] + [X_3, X_1, Y]) = C.$$

Now assume the theorem true for all polyhedra which are decomposable into k triangles, and suppose Π has $k + 1$ triangles. Let Π' be obtained by removing from Π one triangle, say $X_1 X_2 X_3$. Then if A', B', C' are the corresponding quantities for Π', $A' + B' = C'$ by induction.

Case 1. Suppose triangle $X_1 X_2 X_3$ has only one edge in common with Π', say $X_1 X_3$. Suppose $X_1 X_3$ is oriented in Π' from X_1 to X_3. Then Y, the unit normal to triangle $X_1 X_2 X_3$, orients this triangle in the order X_1 to X_2 to X_3. Hence,

$$A = A' + \frac{1}{2}[X_2 - X_1, X_3 - X_1, Y]$$

$$= A' + \frac{1}{2}([X_2, X_3, Y] + [X_3, X_1, Y] + [X_1, X_2, Y]).$$

Since Π has one more interior edge, $X_1 X_3$, than Π', let Y' be the unit normal to the triangle of Π' which adjoins $X_1 X_3$. Then $B = B' + H(X_1, X_3) \cdot X_1 = B' + \frac{1}{2}[X_3, Y - Y', X_1]$.

Finally, $C = C' + \frac{1}{2}[X_1, X_2, Y] + \frac{1}{2}[X_2, X_3, Y] - \frac{1}{2}[X_1, X_3, Y']$. Hence,

$$A + B = A' + \frac{1}{2}([X_2, X_3, Y] + [X_3, X_1, Y] + [X_1, X_2, Y]) + B' + \frac{1}{2}[X_3, Y, X_1]$$

$$- \frac{1}{2}[X_3, Y', X_1]$$

$$= C' + \frac{1}{2}[X_2, X_3, Y] + \frac{1}{2}[X_1, X_2, Y] - \frac{1}{2}[X_1, X_3, Y'] = C.$$

The other two cases, where triangle $X_1 X_2 X_3$ has two and then three edges in common with Π', are resolved similarly.

THEOREM 2. *If the boundary of Π is simple, and if 0 is a vertex on the boundary, then $L^2 \geqq 4\pi(A + B)$.*

Proof. Let the boundary be the polygon in space, $O, X_1, X_2, \cdots, X_n$, oriented in this order. Then

$$A + B = C = \frac{1}{2}([X_1, X_2, Y_1] + [X_2, X_3, Y_2] + \cdots + [X_n, O, Y_n]).$$

Now, $[X_i, X_{i+1}, Y_i] \leqq |X_i \times X_{i+1}|$, and $|X_i \times X_{i+1}|$ is twice the area of triangle $OX_i X_{i+1}$. Hence, $A + B \leqq S$, where S is the sum of the areas of the triangles $OX_i X_{i+1}$.

On the other hand, L is the sum of the lengths of sides $|X_1| + |X_2 - X_1| + |X_3 - X_2| + \cdots + |X_n|$. We can compare these quantities by supposing the triangles $OX_i X_{i-1}$ all lie in the same plane, as a series of adjacent triangles sharing adjacent edges OX_i and the common vertex O. When the sum of the angles $X_i O X_{i-1}$ is less than 2π, S is the area and L is the length of the boundary of this region. But then $L^2 \geqq 4\pi S$ by the classical isoperimetric inequality. All that remains is to consider what occurs if the triangles overlap, which happens if the sum of the angles at 0 is greater than 2π.

But then one obtains a plane figure consisting of an outer closed polygon with its area S_1 and length L_1 together with one or more inner polygons and their areas. For each of these the isoperimetric inequality holds: $L_i^2 \geqq 4\pi S_i$. Hence,

$$(L_1 + \cdots + L_r)^2 \geqq L_1^2 + \cdots + L_r^2 \geqq 4\pi(S_1 + \cdots + S_r).$$

Hence the theorem is proved.

It may be noted that results of this kind are technically new. They are not included in Reid's theorem where differentiability of the surface is assumed.

Finally, Reid's theorem has been generalized to hypersurfaces in R^n by K. Hanes [2]. It would be illuminating and instructive to extend this result to polyhedra in R^n.

References

1. R. V. Benson, Euclidean Geometry and Convexity, McGraw-Hill, New York, 1966.
2. K. Hanes, Ph.D. Thesis, University of Washington, Seattle, Wash., 1970.
3. W. T. Reid, J. Math. Mech., 8 (1959) 897–906.

A CHARACTERIZATION OF CURVES OF CONSTANT WIDTH*

G. D. CHAKERIAN, University of California at Davis

Let C be a simple closed rectifiable plane curve, and let f be a continuous involution of C without fixed points. That is, f is a continuous mapping assigning to each point $x \in C$ another point $f(x) \in C$, with $f(f(x)) = x$ and $f(x) \neq x$ for all x. Let $s(x) = |f(x) - x| = $ the length of the chord joining x to $f(x)$, and $s = \min s(x)$, $x \in C$. By the continuity of $s(x)$ and the compactness of C, the minimum value is achieved. The main result of this note is the following theorem.

THEOREM. *If C, f, and s are as above, and $p(C)$ is the perimeter of C, then*

$$(1) \qquad\qquad p(C) \geqq \pi s.$$

Moreover, equality holds in (1) if and only if C is a convex curve of constant width and f is the involution such that each chord joining x to $f(x)$ is a diametral chord (defined below) of C.

One of the conjectures in [3] was that if f is the particular involution such that x and $f(x)$ bisect the perimeter of C, then (1) holds, with equality possible only for the circle. Our theorem confirms this conjecture since, as proved in [2, Theorem 2.5], the only curve of constant width each of whose diametral chords bisect the perimeter is the circle.

The proof of the theorem depends on some simple results about convex curves which we now discuss. By a *supporting line* of a plane convex curve K we mean a line meeting K but such that K is contained in one of the closed half-planes determined by that line. If G is any line making angle θ with the horizontal, K will have two supporting lines parallel to G, and the distance between these supporting lines, denoted by $w(\theta)$, is the *width* of K in the direction θ. The function $w(\theta)$ is continuous and, by a theorem of Cauchy (see [1, p. 65]), the perimeter of K is given by the formula

$$(2) \qquad\qquad p(K) = \tfrac{1}{2} \int_0^{2\pi} w(\theta)d\theta.$$

In other words, the perimeter of a plane convex curve is equal to π times its mean width. If $\Delta(K)$ denotes the minimum width of K, $\Delta(K) = \min w(\theta)$, $0 \leqq \theta \leqq 2\pi$, it follows immediately from (2) that

$$(3) \qquad\qquad p(K) \geqq \pi\Delta(K),$$

and equality can hold in (3) if and only if $w(\theta)$ is constant, that is, K is a curve of constant width. By a *diametral chord* of K we mean a chord of K whose endpoints

---

* From AMERICAN MATHEMATICAL MONTHLY, vol. 81 (1974), pp. 153–155.

lie on two parallel supporting lines. It is easily shown (see [1, p. 51]) that $\Delta(K)$ is also the minimum length of all diametral chords of K. For the proof of the main theorem it will be important to observe that of all chords of K parallel to a fixed direction, a diametral chord in that direction has maximum length. The following two lemmas will complete our preparation for the proof of the theorem.

LEMMA 1. *If K is the boundary of the smallest convex set containing C, then $p(K) \leqq p(C)$, with equality if and only if $C = K$.*

Proof. The plausibility of the result is seen when one thinks of K as an elastic band looped tightly around C. Since a rigorous proof is difficult to find in the literature, the following argument, based on ideas from integral geometry, might be of interest. If to each line G in the plane we assign coordinates (p, θ), where p is the distance from G to the origin and θ the angle with the horizontal, then a motion invariant measure on the set of lines in the plane is obtained by defining the density $dG = dp\, d\theta$. A formula of Crofton asserts (see [4], or [1, p. 69])

$$(4) \qquad p(C) = \tfrac{1}{2} \int n(G \cap C) dG,$$

where $n(G \cap C)$ is the number of intersections of G with C, and the integration is over all lines. Now observe that any line G meeting K in exactly two points must also meet C in at least two points. Otherwise C would be contained in a closed half-plane H determined by G and hence in a proper closed convex subset of the set bounded by K, contradicting the definition of K. Thus, except possibly for supporting lines of K (which constitute a set of lines of measure zero), $n(G \cap K) \leqq n(G \cap C)$. Applying (4) we obtain $p(K) \leqq p(C)$. If $C \neq K$, it is not difficult to show that there exists a set of lines G of positive measure for which $n(G \cap K) < n(G \cap C)$, so $p(K) < p(C)$.

LEMMA 2. *Let C be a simple closed plane curve with an involution f as described above. Then given any line G, there is some $x \in C$ such that the chord from x to $f(x)$ is parallel to G.*

Proof. Let $v(x) = f(x) - x$. The lemma follows immediately from observing that the vectors $v(x)$ vary continuously, never vanish, and $v(f(x)) = -v(x)$.

The proof of the main theorem now follows quickly. Let K be the boundary of the smallest convex set containing C, and let $\Delta(K)$ be the minimum width of K. By Lemma 2, there exists $x_0 \in C$ such that the chord joining x_0 to $f(x_0)$ is parallel to a diametral chord of K of length $\Delta(K)$. Thus $s(x_0) = |f(x_0) - x_0|$ is less than or equal to the length of a chord of K parallel to a diametral chord of length $\Delta(K)$, and

$$(5) \qquad \Delta(K) \geqq s(x_0) \geqq \min s(x) = s.$$

By Lemma 1, we have $p(C) \geqq p(K)$. Applying (3) and (5) we have,

$$(6) \qquad p(C) \geqq p(K) \geqq \pi\Delta(K) \geqq \pi s,$$

establishing (1). If equality holds in (1), then it holds throughout (6). Then $p(C)$ $= p(K)$, so $C = K$. Further, equality then holds in (3), so C is a curve of constant width Δ. Then *every* diametral chord of C has length Δ and $s(x) \leq \Delta$ for all x (for Δ is then the maximum chord length in every direction). But equality throughout (6) also implies $\Delta = s = \min s(x)$, so $s(x) \geq \Delta$ for all x. Hence $s(x) \equiv \Delta$, so the chord joining x to $f(x)$ is a diametral chord for all x, completing the proof.

The author would like to acknowledge his indebtedness to G. T. Sallee for stimulating discussions about the main theorem.

References

1. T. Bonnesen and W. Fenchel, Theorie der konvexen Körper, Springer, Berlin, 1934; Chelsea, New York, 1948 and 1971.

2. P. C. Hammer and T. J. Smith, Conditions equivalent to central symmetry of convex curves, Proc. Cambridge Phil. Soc., 60 (1964) 779-785.

3. H. Herda, A conjectured characterization of circles, this MONTHLY, 78 (1971) 888-889.

4. H. Steinhaus, Length, shape and area, Colloq. Math., 3 (1954) 1-13.

Editorial Note: An extensive bibliography of related papers may be found in the survey article of H. Herda, A characterization of circles and other closed curves, this MONTHLY, vol. 81 (1974), 146-9. Prof. Chakerian has also called our attention to Roger Fenn's article, Sliding a chord and the width and breadth of a closed curve, J. London Math. Soc. (2), vol. 5 (1972), pp. 39-47.

SUPPLEMENTARY BIBLIOGRAPHY

The reader may be also interested in these articles from the AMERICAN MATHEMATICAL MONTHLY and MATHEMATICS MAGAZINE. The journals are abbreviated AMM and MM respectively.

1. J. C. Alexander, On the connected sum of projective planes, tori, and Klein bottles, AMM, 78(1971), 185–7.

2. G. H. Atneosen, An application of simplicial homology theory, AMM, 77(1970), 381–4.

3. F. Bagemihl, On indecomposable polyhedra, AMM, 55(1948), 411–3.

4. N. H. Ball, On ovals, AMM, 37(1930), 348–53.

5. J. P. Ballantine, A mathematical treatment of perspective, AMM, 31(1924), 90–1.

6. A. A. Bennett, The imaginary points of geometry, AMM, 29(1922), 145–9.

7. ———, Review of Granville's "The Fourth Dimension and the Bible," AMM, 30(1923), 35.

8. C. Berge, Graph theory, AMM, 71(1964), 471–81.

9. G. A. Bliss, Mathematical interpretations of geometrical and physical phenomena, AMM, 40(1933), 472–82.

10. R. F. Brown, Elementary consequences of the noncontractibility of the circle, AMM, 81(1974), 247–52.

11. R. L. W. Brown, The Klein bottle as an eggbeater, MM, 46(1973), 244–50.

12. R. H. Bruck, Recent advances in foundations of Euclidean plane geometry, AMM, 62(1955), Aug.–Sept. Supplement, 2–18.

13. R. C. Buck and E. F. Buck, Equipartition of convex sets, MM, 22(1948–9), 195–8.

14. J. Callahan, Singularities and plane maps, AMM, 81(1974), 211–40.

15. J. W. Cell, Solid angles, AMM, 48(1941), 136–8.

16. E. Cerutti and P. J. Davis, Formac meets Pappus, AMM, 76(1969), 895–905.

17. G. D. Chakerian, Sylvester's proof on collinear points and a relative, AMM, 77(1970), 164–7.

18. G. D. Chakerian and M. S. Klamkin, Inequalities for sums of distances, AMM, 80(1973), 1009–17.

19. ———, Minimal covers for closed curves, MM, 46(1973), 55–61.

20. M. Cohen, Foliations of 3-manifolds, AMM, 81(1974), 462–73.

21. A. J. Coleman, Curves on a surface, AMM, 47(1940), 212–20.

22. S. R. Conrad, Another simple solution to the butterfly problem, MM, 46(1973), 278–80.

23. J. L. Coolidge, The rise and fall of projective geometry, AMM, 41(1934), 217–28.

24. H. S. M. Coxeter, A problem of collinear points, AMM, 55(1948), 26–8.

·25. ——, A problem of Apollonius, AMM, 75(1968), 5–15.

26. P. H. Daus, Collineations and central projections, AMM, 45(1938), 294–8.

27. N. G. deBruijn, Filling boxes with bricks, AMM, 76(1969), 37–40.

28. L. E. Dickson, The simplest model for illustrating conic sections, AMM, 1(1894), 261.

29. ——, Rational triangles and quadrilaterals, AMM, 28(1921), 244–50.

30. L. W. Dowling, Projective geometry—fields of research, AMM, 32(1925), 486–92.

31. R. A. Duke, How is a graph's Betti number related to its genus? AMM, 78(1971), 386–8.

32. C. M. Fulton, The centroid in absolute geometry, AMM, 67(1960), 885.

33. B. A. Fusaro, The area of a hypersphere in Riemannian space, AMM, 80(1973), 179–84.

34. D. Gans, A circular model for the Euclidean plane, AMM, 61(1954), 26–30.

35. ——, An introduction to elliptic geometry, AMM, 62(1955), Aug.–Sept. Supplement, 66–75.

36. ——, Models of projective and Euclidean space, AMM, 65(1958), 749–56.

37. ——, A new model of the hyperbolic plane, AMM, 73(1966), 291–5.

38. J. Gerriets and G. Poole, Convex regions which cover arcs of constant length, AMM, 81(1974), 36–41.

39. W. B. Givens, The division of angles into equal parts and polygon construction, AMM, 45(1938), 653–6.

40. M. Goldberg, The converse Malfatti problem, MM, 41(1968), 262–6.

41. L. M. Graves, A finite Bolyai-Lobachevsky plane, AMM, 69(1962), 130–2.

42. G. B. Halsted, Proving the false, AMM, 9(1902), 129–31.

43. S. B. Jackson, A development of the Jordan curve theorem and the Schoenflies theorem for polygons, AMM, 75(1968), 989–98.

44. R. A. Johnson, Directed angles in elementary geometry, AMM, 24(1917), 101–5.

45. L. S. Johnston, Remarks on the geometry of the triangle, AMM, 42(1935), 235–7.

46. ——, Further remarks on the geometry of the triangle, AMM, 42(1935), 612.

47. E. Kasner, The group generated by central symmetries with applications to polygons, AMM, 10(1903), 57–63.

48. —— and A. Kalish, The geometry of the circular horn triangle, MM, 18(1943–4), 299–304.

49. L. M. Kelly, The neglected synthetic approach, AMM, 55(1948), 24–6.

50. P. J. Kelly, Von Aubel's quadrilateral theorem, MM, 39(1966), 35–7.

51. J. W. Kenelly and Andrew Sobczyk, Canonical placement of simplices, MM, 41(1968), 150–2.

52. R. B. Kershner, On paving the plane, AMM, 75(1968), 839–44.

53. C. Keyser, Review of Manning's "Geometry of Four Dimensions," AMM, 22(1915), 127–8.

54. M. S. Klamkin, On ruled and developable surfaces of revolution, MM, 27(1953–4), 207–8.

55. ———, Vector proofs in solid geometry, AMM, 77(1970), 1051–65.

56. S. Lefschetz, On remarkable points of curves, AMM, 19(1912), 27.

57. V. F. Lenzen, Physical geometry, AMM, 46(1939), 324–44.

58. E. O. Lovett, Sophus Lie's transformation groups, AMM, 4(1897), 237–42, 270–5, 308–13, and 5(1898), 2–9, 75–82.

59. C. C. MacDuffee, Euclidean invariants of second degree curves, AMM, 33(1926), 243–52.

60. G. A. Miller, On the group of the figures of elementary geometry, AMM, 10(1903), 215–8.

61. R. S. Millman and A. K. Stehney, The geometry of connections, AMM, 80(1973), 475–500.

62. F. V. Morley, A curve of pursuit, AMM, 28(1921), 54–61.

63. J. C. C. Nitsche, The smallest sphere containing a rectifiable curve, AMM, 78(1971), 881–2.

64. R. Osborn, Some historical and philosophic aspects of geometry, MM, 24(1950–1), 77–82.

65. T. G. Ostrom, Ovals and finite Bolyai-Lobachevski planes, AMM, 69(1962), 899–901.

66. E. R. Ott, Finite projective geometries, AMM, 44(1937), 86–92.

67. B. C. Patterson, The inversive plane, AMM, 48(1941), 589–99.

68. F. Pavlick, If n lines in the Euclidean plane meet in 2 points then they meet in at least $n-1$ points, MM, 46(1973), 221–3.

69. D. Pedoe, Thinking geometrically, AMM, 77(1970), 711–21.

70. J. F. Ramaley, Buffon's noodle problem, AMM, 76(1969), 916–8.

71. N. E. Rutt, The sources of Euclid, MM, 11(1936–7), 374–81.

72. W. L. Schaaf, Art and mathematics: a brief guide to source materials, AMM, 58(1951), 167–77.

73. P. Scherk, Some concepts of conformal geometry, AMM, 67(1960), 1–30.

74. K. Tan, Various proofs of Newton's theorem, MM, 39(1966), 45–58.

75. F. A. Valentine, A characteristic property of the circle in the Minkowski plane, AMM, 58(1951), 484–7.

76. ———, Visible shorelines, AMM, 77(1970), 146–52.

77. J. H. Weaver, The duplication problem, AMM, 23(1914), 106–13.

78. R. M. Winger, Infinite and imaginary elements in algebra and geometry, AMM, 29(1922), 290–7.

79. R. E. M. Wong, Geometry in preparation for high school mathematics teachers, AMM, 77(1970), 70–8.

80. R. C. Yates, The trisection problem, MM, 15(1940–1), 129–42, 191–202, 278–93, and 16(1941–2), 20–8, 171–82.

81. ———, The catenary and the tractrix, AMM, 66(1959), 500–5.

82. A. Zirakzadeh, A mapping of the projective n-space on the projective plane, AMM, 70(1963), 399–401.

PROBLEMS AND SOLUTIONS

These problems were chosen from the AMERICAN MATHEMATICAL MONTHLY and MATHEMATICS MAGAZINE to challenge the reader with interesting questions from past eras. The problems from each journal are printed in chronological order of their appearance, with those from the MONTHLY since 1933 separated into "elementary" and "advanced" categories as in the MONTHLY itself.

Volume and page references are given in brackets following the problem number. Solutions and information on related problems begin on page 305. An asterisk indicates that only one of the solutions which originally appeared is reprinted here.

Problems from THE AMERICAN MATHEMATICAL MONTHLY, 1894–1932

Geom. 1 [*1*, 22]. *Proposed by B. F. Finkel*

Show that the bisectors of the angles formed by producing the sides of an inscribed quadrilateral intersect each other at right angles.

Geom. 14 [*1*, 24]. *Proposed by Henry Heaton*

Through two given points to pass four circles tangent to two given circles.

Geom. 20 [*1*, 51]. *Proposed by G. B. Halsted*

Demonstrate by pure spherical geometry that spherical tangents from any point in the produced spherical chord common to two intersecting circles on a sphere are equal.

Geom. 113 [*5*, 305]. *Proposed by T. W. Palmer*

Given three concentric circles. Draw a straight line from the inner to the outer circumference that shall be bisected by the middle circumference.

Geom. 236 [*11*, 196]. *Proposed by J. R. Hitt*

If two sides of a triangle pass through fixed points, the third side touches a fixed circle.

Geom. 357 [*16*, 216]. *Proposed by E. R. Hoyt*

A room is 30 feet long, 12 feet wide, and 12 feet high. At one end of the room, 3 feet from the floor, and midway from the sides, is a spider. At the other end, 9 feet from the floor, and midway from the sides, is a fly. Determine the shortest path by way of the floor, ends, sides, and ceiling, the spider can take to capture the fly.

Geom. 375 [*17*, 178]. *Proposed by C. N. Schmall*

From a point P on a circle there are drawn three chords PA, PB, PC. Show that the circles described on these chords as diameters intersect again in three collinear points.

2899 [*28*, 228]. *Proposed by Norman Anning*

A, B, C, and P are any four coplanar points. P describes a sextant about A when the line AP turns about A through $+60°$. Show that P moves in a closed curve when it describes sextants in succession *either* about $A, B, A, B, A, \ldots$ *or* about $A, B, C, A, B, C, \ldots$.

2945 [*29*, 29]. *Proposed by T. M. Blakslee*

A point P in the plane of the triangle ABC rotates in a given direction around the vertices taken in either cyclical order, in each case through an angle equal to the corresponding angle of the triangle. That is, for example, AP rotates around A through an angle equal to the angle A of the triangle; then BP around B through an angle equal to the angle B, and so on. Prove that P coincides with its original position at the end of six of these rotations. (See problem 2899, *1921*, 228.)

3223 [*33*, 481]. *Proposed by Paul Capron*

A circle of radius b and a straight line at a distance a from the center of the circle, lie in the same plane; the circle is revolved about the line generating a torus. A plane Π is passed through the axis l, intersecting the torus in two circles S_1 and S_2; a plane Σ is passed perpendicular to Π and containing the common interior tangent of S_1 and S_2. Show that Σ intersects the torus in two circles.

3325 [*35*, 261]. *Proposed by Paul Capron*

Two circles, S_1 and S_2 with centers O_1 and O_2 intersect at A; O_1A meets S_2 at K_2, O_2A meets S_1 at K_1. Show that the triangle K_1AK_2 is similar to the triangle O_1AO_2.

3482 [*38*, 170]. *Proposed by Nathan Altshiller-Court*

If the circumscribed and the inscribed spheres of a tetrahedron are concentric, the sum of the face angles of each trihedral angle of the tetrahedron is equal to two right angles.

3565 [*39*, 430]. *Proposed by Orrin Frink, Jr.*

Find the ellipse of least area circumscribing a given triangle.

Elementary problems from
THE AMERICAN MATHEMATICAL MONTHLY, since 1933

E16 [*40*, 51]. *Proposed by G. A. Yanosik*

Prove that the envelope of the circles whose diameters run from points on a parabola to its focus, is the straight line tangent to the parabola at its vertex.

E113 [*41*, 517]. *Proposed by E. T. Krach*

Prove that if three circles are so arranged that their six external tangents are real (each tangent touching two circles), then the three points of intersection of the three pairs of corresponding tangents are collinear.

E150 [*42*, 246]. *Proposed by Maud Willey*

Points M and N trisect side BC of triangle ABC, so that $BM = MN = NC$. A line parallel to AC meets lines AB, AM and AN in points D, E and F, respectively. Show that $EF = 3DE$.

E337 [*45*, 319]. *Proposed by V. Thébault*

(S) and (T) are two fixed intersecting spheres, and (X) and (Y) are wo variable spheres, tangent to each other at P, and each tangent to both (S) and (T). Show that the locus of P is the surfaces of two more spheres which cut each other orthogonally. Examine the cases in which (S) and (T) are internally or externally tangent, or have no points in common.

E591 [*50*, 560]. **Intersections of Orthogonal Circles**, *proposed by C. J. Coe*

Two given coplanar circles, (A_1) and (A_2), are cut orthogonally by a third circle, (B). Prove that a line joining either intersection point on (A_1) to either intersection point on (A_2) will pass through one of two points on the line of centers $A_1 A_2$, these two points being the same for all choices of the orthogonal circle (B).

E630 [*51*, 348 and 405]. **Parallel Curves,** *proposed by R. A. Rosenbaum*

For any point P of a given closed convex curve C, let P' be that point on the exterior normal to C at P for which $PP' = k$, a constant. The locus of P' is a curve C', parallel to C. Let s, s' be the respective lengths of C, C', and A, A' the areas within these curves. Show that

$$s' = s + 2\pi k,$$
$$A' = A + sk + \pi k^2.$$

E661 [*52*, 159]. **The Conformal Points of a Projection,** *proposed by Howard Eves*

A plane p is projected from a point L onto a plane p'. Find those points on p for which all angles on p having such a point for vertex are invariant under the projection.

E665 [*52*, 159]. **Circles Covering a Given Curve,** *proposed by L. A. Santaló*

Let C be a closed convex plane curve with continuous radius of curvature R. Let R_M be the greatest value of R. Given $\lambda \geqq R_M$, show that the area F_λ covered by the centers of circles of radius λ which contain C in their interior is given by

$$F_\lambda = F - L\lambda + \pi\lambda^2,$$

where L and F are the length and area of C.

E741 [*53*, 532]. **A Pythagorean Inequality,** *proposed by William Scott*

Prove that in a (non-degenerate) right spherical triangle with hypotenuse c and legs a, b we have $a^2 + b^2 > c^2$.

E774 [*54*, 281]. **Lines Cutting Off One-third the Area of a Triangle,** *proposed by Norman Anning*

Consider points on the median of a triangle. Through the centroid no straight line can be drawn which will cut off one-third of the area. Through a point four-fifths of the distance from vertex to base, four such lines can be drawn. Find points on the median at which the number of possible lines changes.

E788 [*54*, 471]. **An Impossible Journey,** *proposed by Leo Moser*

Consider a map on a spherical surface where the countries are determined by n great circles of which no three are concurrent. Show that if n is a multiple of four it is impossible to make a trip visiting each country once and only once, if travelling along a boundary or crossing at a boundary point of more than two countries is forbidden.

E888 [*56*, 632]. **Passing a Cube through a Cube of Same Size,** *proposed by H. D. Grossman*

Show how to cut a hole in a cube through which another cube of equal size can pass.

E929 [*57*, 483]. A Product of Reflections, *proposed by H. S. Shapiro*

Given three non-concurrent straight lines l_1, l_2, l_3 in the plane. Let T_i denote reflection in l_i and set $T = T_1 T_2 T_3$. Show that T^2 is a translation.

E1224 [*63*, 421]. A Fixed Point Theorem, *proposed by G. K. Wenceslas*

Let G be a finite group of rigid motions (or more generally affine transformations) of a Euclidean space. Then there is a point of the space which is left fixed by all the transformations of G.

E1228 [*63*, 491]. An Affine Invariant, *proposed by Viktors Linis*

Let $n(P)$ be the number of distinct lines L through a point P such that L divides the area of a given triangle in two equal parts. Show that the locus of all points P with $n(P) \geqq 2$ is a region the ratio of whose area to the area of the given triangle is an absolute constant.

E1423 [*67*, 593]. Closed Self-intersecting Curves, *proposed by A. Zirakzadeh*

Consider a plane closed curve possessing only ordinary points and a finite number of double points. Assign a positive direction to the curve and, starting from an ordinary point A, trace the curve in the given positive direction. Assign number 1 to the first double point met, 2 to the second double point met, and so on until you come back to A. Prove that of the two integers assigned to any double point, one is always even and the other always odd.

E1515 [*69*, 312]. A Decomposition of the Euclidean Plane, *proposed by Joachim Lambek and Leo Moser*

Show that for every positive integer n, the Euclidean plane, considered as a point set, can be decomposed into n congruent connected subsets.

Advanced problems from
THE AMERICAN MATHEMATICAL MONTHLY, since 1933

3829 [*44*, 332]. Midpoints of Chords, *proposed by J. D. Hill*

Let C be a simple closed rectifiable plane curve and P an arbitrary point inside of C. (a) Show that there exist two points A and B on C such that P bisects the chord AB. (b) Does this property remain true if the curve is nonrectifiable?

4036 [*49*, 340]. **An Oval and its Normal Expansion,** *proposed by L. A. Santaló*

Let C_1 be an oval with a continuously varying radius of curvature R; at each point of C_1 a normal of length R is drawn exteriorly giving points of a second curve C_2 (which may not be convex); and let A be the area enclosed between the two curves. From a chosen fixed point a vector is drawn parallel to the normal at a point of C_1 and of length R for that point, thus giving as the point varies on C_1 a curve C_3 having the area A_3 and length L_3. If L_2 is the length of C_2 and A_1 is the area of C_1, show that

$$\text{(a)} \quad A = 3A_3; \qquad \text{(b)} \quad L_2 L_3 \geqq 8\pi A_1;$$

the equality in (b) is true only when C_1 is a circle.

4087 [*50*, 391]. **Affine Geometry,** *proposed by Betty Dick and B. M. Stewart*

Let P be a plane polygon with vertices $A_1, A_2, \ldots, A_n$, and consider $A_{n+k} = A_k$. Determine points $B_1, B_2, \ldots, B_n$ such that B_i is on the side $A_i A_{i+1}$ with $A_i A_{i+1} = k \cdot A_i B_i$, where k is a fixed real number. Let the lines $A_i B_{i+1}$ and $A_{i+1} B_{i+2}$ intersect in the point C_i, thus determining the polygon Q with the vertices $C_1, C_2, \ldots, C_n$. Let $R(k)$ be the ratio of the area of Q to the area of P, wherein not to restrict the type of polygon we use Klein's definition of area, *Elementary Mathematics from an Advanced Standpoint, Geometry*, p. 10. (1) Show that k and $R(k)$ are invariants under affine transformations. (2) As a corollary to (1) show that for any triangle we have $R(k) = (k-2)^2/(k^2 - k + 1)$. (3) For any parallelogram we have $R(k) = (k-1)^2/(k^2 + 1)$. (4) Show that $R(k)$ does not have the same value for all quadrilaterals.

This problem is a generalization of the so-called Problem of Steinhaus which asserts for any triangle $R(3) = 1/7$. An early reference, suggestive of this problem, is problem 276 in *Mathesis*.

4262 [*54*, 418]. **Rectifiable Plane Curves,** *proposed by L. A. Santaló*

Let C be a rectifiable plane curve of length L, contained within a given circle of radius R. Prove that there is a circle of radius $\rho \geqq R$ which cuts C in n points, where

$$\text{(1)} \qquad n \geqq L/\pi R.$$

In particular there is a line which cuts C in n points, where n satisfies (1). If $\rho < R$, the inequality (1) must be replaced by

$$\text{(2)} \qquad n \geqq \frac{4L\rho}{\pi(R+\rho)^2}.$$

See L. A. Santaló, A theorem and an inequality referring to rectifiable curves, *American Journal of Mathematics*, 1941, p. 635.

4739 [*64*, 369]. **Intersecting Octahedra,** *proposed by V. L. Klee, Jr.*

Suppose C is a closed convex subset of the Euclidean space E^3 whose boundary is a regular octahedron, and that C_1, C_2, and C_3 are translates of

$C(i.e., C_i = C + x_i$ for some $x_i \in E^3$). Then, if each of the intersections $C_1 \cap C_2, C_2 \cap C_3$, and $C_3 \cap C_1$ is nonempty, must $C_1 \cap C_2 \cap C_3$ be nonempty?

4927 [67, 809]. Transformation of *n*-space, *proposed by David Gale*

Suppose f is a continuous transformation of n-space into itself, and suppose there is a positive number λ such that $|f(x) - f(y)| \geqq \lambda|x - y|$ for all points x and y, where $|x - y|$ is the Euclidean distance. Prove that f maps onto all of n-space.

5527 [74, 1014]. Projections of a Simple Closed Curve in 3-Space, *proposed by Frederic Cunningham, Jr.*

Prove that for every simple closed curve in 3-space there is a plane projection of it having at least two double points (or a multiple point of order greater than two). Give an example of such a curve for which no plane projection has more than two double points (or one triple point).

Problems from MATHEMATICS MAGAZINE

159 [12, 53]. *Proposed by T. Dantzig*

Two planes, α, β, make an angle θ with each other. A triangle in α is in perspective from a point P with another triangle in β. Show that as θ varies P describes a circle.

214 [12, 253]. *Proposed by David Amidon*

On the altitude AD of triangle ABC, select an arbitrary point P. Let BP meet AC in E and CP meet AB in F. Show that (a) angle EDF is bisected by AD and (b) EF and BC meet in a fixed point however P may be chosen.

391 [15, 261]. *Proposed by D. L. MacKay*

According to Louis C. Karpinski, the following theorem, ascribed to Archimedes by Albiruni (c. 1000 A. D.) was recently brought to light in connection with an Arabic work on Trigonometry and is now believed to have been the basis of Greek Trigonometry before Ptolemy (150 A. D.).

Given cords AB and $BC, AB > BC$, prove that if M is the midpoint of the arc ABC, then MN, perpendicular to AB at N, bisects the broken line ABC.

38 [23, 207]. Five Intersecting Great Circles, *proposed by Leo Moser*

Show that 5 or more great circles on a sphere, no 3 of which are concurrent, determine at least one spherical polygon having 5 or more sides.

146 [*26*, 222]. **A Spherical Locus,** *proposed by D. L. Silverman*

What is the locus of the center of a circle, radius r, which touches each of three mutually perpendicular planes?

Solutions to problems from
THE AMERICAN MATHEMATICAL MONTHLY, 1894–1932

Geom. 1 [*1*, 22]. *Solution by the Proposer*

Let $ABCD$ be the inscribed quadrilateral and FO and EO the bisectors of the angles F and E, respectively, formed by producing the sides of the quadrilateral. Denote the angle EAF by A; AFB, by F; BFE, by F'; AED, by E; DEF, by E'; FCE, $=DCB$, by C; and FOE, by O. Then $A+C=2$ rt. angles., (1); being opposite angles of an inscribed quadrilateral. Also, in the triangle $AFE, A+F+F'+E+E'=2$ rt. angles..(2); in the triangle $FOE, \frac{1}{2}F+F'+\frac{1}{2}E+E'+O=2$ rt. angles....(3); and, in the triangle $FCE, F'+E'+C=2$ rt. angles....(4). Multiplying (3) by two and subtracting (4) from the resulting equation, we have $F+F'+E+E'+2O-C=2$ rt. angles..(5). Subtracting (5) from (2), we have $A+C-2O=0$, whence $2O=A+C=2$ rt. angles. $\therefore O=$ a rt. angle.   Q.E.D.

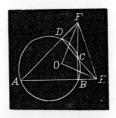

Geom 14 [*1*, 232]. *Solution by the Proposer*

In the figure, C and C' the intersection of the common tangents to the two circles are known as the external and internal centers of similitude.

It is not necessary to demonstrate here the following well known properties:

I. If $CKLMN$ be any secant line from C, then $CL \times CM = CK \times CN = CH \times CI$.

II. If $POC'QR$ be any line through C', secant to both circles, $C'O \times C'R = C'Q \times C'P = C'L \times C'S$.

III. If any circle be drawn to which both circles are either internally or externally tangent, the line through the points of tangency will pass through the external center of similitude, C.

IV. If any circle be drawn to which one of the given circles is internally tangent, and the other externally tangent, the line joining the points of tangency will pass through the internal center of similitude, C'.

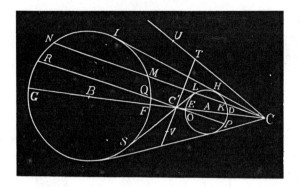

Let T be the given point. On the line through CT take a point U such that $CT \times CU = CH \times CI$. This may be done by passing a circle through I, T, and H. It will cut CT in U.

The circle to which both circles are either internally or externally tangent and which passes through T, will also pass through U.

This follows from *I.* and *III.*

If through U and T we pass two circles tangent to either of the given circles, (Prob. 17.) they will be tangent to the other.

On the line through $C'T$ take a point V such that $C'T \times C'V = C'S \times C'L$. This may be done by passing a circle through T, L, and S. It will cut $C'T$ in V.

The circle to which one of the given circles is tangent internally and the other externally and which passes through T will also pass through V. This follows from II. and IV.

If through V and T we pass a circle to which one of the given circles is internally tangent, the other will touch it externally and *vice versa*.

Geom. 20 [*15*, 10]. *Solution by G. B. M. Zerr*

No solution of this problem has yet appeared in the MONTHLY. A simple geometrical solution such as is possible for the corresponding problem in planes is not possible for this problem. The following solution is quite simple.

Let P be a point on the common chord DE; PB, PC the tangents, O the pole of one circle. Let $PE = R, PD = r, PC = \rho, PB = \rho', PO = \delta, OD = OE = OC = \beta, \angle EPO = \phi$.

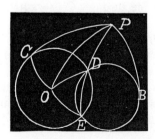

Then

$$\cos\beta = \cos R \cos\delta + \sin R \sin\delta \cos\phi \ldots \tag{1},$$

$$\cos\beta = \cos r \cos\delta + \sin r \sin\delta \cos\phi \ldots \tag{2},$$

$$\cos\delta = \cos\beta \cos\rho \ldots \tag{3},$$

cos ϕ from (1) in (2) gives $\cos\beta(\sin R - \sin r) = \cos\delta\sin(R-r)\ldots(4)$.
$\cos\delta$ from (3) in (4) gives $\cos\rho = (\sin R - \sin r)/\sin(R-r)$.
Similarly, $\cos\rho' = (\sin R - \sin r)/\sin(R-r)$. $\quad \therefore \rho = \rho'$.
These equations reduce to $\tan^2\frac{1}{2}\rho = \tan^2\frac{1}{2}\rho' = \tan\frac{1}{2}R\tan\frac{1}{2}r$.

Geom. 113 [6, 176]*. *Solution by C. Hornung*

Let C be the common center and CP the radius of the inner circle, and CQ that of the outer circle.

Bisect CP at M and with one-half of CQ as radius and M as center describe a circle cutting the middle circumference at O or O'. Draw PO and produce to the outer circumference at Q. Then POQ is the required line. For, $\therefore PM = MC$ and $MO = \frac{1}{2}CQ, \therefore PO = OQ$.

According as $\frac{1}{2}$radius of outer circle is greater than, equal to, or less than the radius of the middle circle increased by $\frac{1}{2}$radius of inner circle, there are two solutions, one solution, or no solution.

Geom. 236 [12, 48]*. *Solution by the Proposer*

Let the sides AB, BC of triangle ABC pass through the points D, E, respectively. Draw BH perpendicular to AC, from F, intersection of circumcircle of triangle DBE with BH, draw diameter FK. Draw BK, KL perpendicular to AC. Arc DBE is locus of B. Since $\angle ABH$ is constant, F is a fixed point. $\therefore K$ is fixed. FBK is a right angle. $\therefore BHLK$ is a rectangle. $\therefore KL = BH$, a constant. Hence AC touches the circle with center K and radius KL.

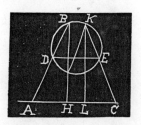

Geom. 357 [*17*, 67]. *Solution by G. B. M. Zerr*

Suppose the six sides of the room spread out in one plane as in the figure, the floor being the second rectangle from the bottom, and let x=distance of spider from floor, $12-x$ the distance of the fly, $x<6$. There are three courses for the spider to take.

First, the route $SC=30+x+12-x=42$ feet...(1).

Second, the route $SB=\sqrt{[(Sb)^2+(Bb)^2]}=\sqrt{[(36+x)^2+(18-x)^2]}$...(2).

Third, the route $SA=\sqrt{[(Sa)^2+(Aa)^2]}=\sqrt{[(30+2x)^2+(24)^2]}$...(3).

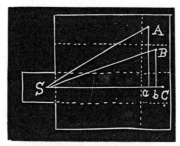

Let $(30+2x)^2+576=(36+x)^2+(18-x)^2$.

Then $x^2+42x=72$, and $x=1.6495$ feet.

Let $(36+x)^2+(18-x)^2=(42)^2$.

Then $x^2+18x=72$, and $x=3.3693$ feet.

Then if $x<1.6495$ feet, the third route is shortest.

If $x>1.6495$ and <3.3693, the second route is shortest.

If $x>3.3693$, the first route is the shortest.

Since $x=3$, the spider takes the second route, and travels $\sqrt{[(39)^2+(15)^2]}=3\sqrt{[13^2+5^2]}=3\sqrt{(194)}<[3\sqrt{(196)}=3\times14=42]$ feet.

This problem is generalized in E1056, this MONTHLY [*60*, 627].

Geom. 375 [*18*, 91]*. *Solution by S. Lefschetz*

If we transform by inversion, the pole of inversion being in P, the transformed of the three circles of diameters PA, PB, and PC are perpendiculars at PA, PB, and PC in A', B', and C', points where these three lines meet the line obtained by transformation of the given circle. These three perpendiculars envelop a parabola of focus P; therefore, the circle circumscribed to the triangle they form passes through P,—a well known property of the parabola. By transforming back, we obtain a straight line and the proposition is thus proved.

Notice the similarity to Geom. 266, this MONTHLY, [*12*, 205].

2899 [*29*, 358]*. *Solution by C. F. Gummer*

These are particular cases of a more general theorem. Consider n coplanar points $A_1, A_2, \ldots, A_n$; and let a point in the plane, starting at P, revolve successively about $A_1, A_2, \ldots, A_n, A_1, \ldots,$ through angles $\phi_1, \phi_2, \ldots, \phi_n, \phi_1, \ldots,$ until all the points $A_1, \ldots, A_n$ have been used m times each. The resultant displacement from P after the first set of rotations about $A_1, A_2, \ldots, A_n$ is the sum of a number of vectors, namely PA_1, A_1A_n and those obtained by turning A_nA_{n-1} through $\phi_n, A_{n-1}A_{n-2}$ through $\phi_{n-1} + \phi_n, \ldots A_2A_1$ through $\phi_2 + \phi_3 + \cdots \phi_n$ and finally A_1P through $\phi_1 + \phi_2 + \cdots \phi_n$. If Φ denotes the sum of the n angles, the vector sum after m sets of rotations contains two vectors of length PA_1 inclined at an angle $\pi + m\Phi$, and it contains m vectors of length A_iA_{i-1} (including the case of A_1A_n) inclined successively at the angle Φ so that they may be regarded as a series of equal chords placed end to end in a circle of suitable size. This is true for each value of i from 2 to n and for A_1A_n.

Three cases occur:

(1) $\Phi/(2\pi)$ *is rational but not integral.* A value of m other than 1 may then be found so that $m\Phi$ is a multiple of 2π. It follows that after m sets of rotations the two vectors derived from PA_1 are equal and opposite, and the vectors derived from A_iA_{i-1} are the sides of a closed regular polygon (possibly interlacing); therefore the resultant displacement vanishes, and the point has moved in a closed curve. The problems proposed belong to this case.

(2) $\Phi/(2\pi)$ *is irrational.* No group of vectors can be made to have a zero sum; and it will be found that the path cannot be closed (except for special positions of P). The path however lies in a finite region, since each group of vectors has a sum not greater than the diameter of its corresponding circle; and it may be shown that the path returns to positions indefinitely near to the initial point.

(3) $\Phi/(2\pi)$ *is an integer.* On taking $m = 1$, A_1P and PA_1 again destroy one another; but the remaining vectors, being now in groups of one, have not generally a zero sum. The resultant transformation is a translation of the entire plane, which by repetition (unless it happens to vanish) carries every point to infinity.

2945. This is completely solved by C. F. Gummer's solution to 2899, this MONTHLY, reprinted here.

3223 [34, 441]. *Solution by R. H. Sciobereti*

Let Oz be the axis of the torus; OX and Ox, the traces of the bitangent plane Σ and of any meridian on the equator, respectively. Let us denote by M one of the points of intersection of the plane Σ with the meridian circle S' of center C so that $OC = a, CM = b$. Let P and Q be the orthogonal projections of M on Ox and OX respectively and R the projection of point O on CM. Then we shall have angle $PQM = \theta$, where θ denotes the angle between the bitangent and the equatorial planes; hence,

(1) $$\sin PQM = \sin \theta = b/a = PM/QM = CM/OC.$$

On the other hand the area of the triangle OCM evaluated in two different ways gives

$$OC \cdot PM = CM \cdot OR, \text{ or } CM/OC = PM/OR,$$

and from a comparison with (1) it follows that $QM = OR$; hence $MR = OQ$, since the two right triangles MRO and OQM are congruent.

Let us now consider a point D on OX at a distance b from O, and such that the three points O, Q, D will be in the same order as the points M, R, C, so that $QD = OD - OQ = CM - MR = RC$; hence, the two right triangles DQM and CRO are congruent and consequently $DM = CO = a$. When the meridian circle S' is rotated about Oz, point M will describe a circle of center D and of radius a. A similar reasoning shows that the locus of the second point of intersection M' of the plane Σ with the meridian circle S' is a circle equal to the first one, but whose center is the symmetrical point of D with respect to O.

REMARK I. These two circles, known as the Villarceau circles, intersect the parallels of the torus at a constant angle. This property may be shown by means of elementary geometry as follows: Let MT be the tangent to the Villarceau circle passing through M, and let MV be the tangent to the parallel through M; MT lies in the plane ODM and is orthogonal to DM, whereas MV is perpendicular to the plane OCM. Since the skew quadrilateral $ODMC$ has its opposite sides equal to each other, it follows that the two triangles ODM and MCO are congruent; hence $DH = CH$, where H denotes the point bisecting OM. In a similar manner it may be shown that if K is the mid-point of DC, $OK = MK$. Hence the line HK is the common perpendicular to OM and DC at their mid-points. This line HK may, therefore, be considered as an axis of symmetry for the skew quadrilateral. Now the symmetrical line of the tangent MT is the perpendicular to CO at the point O in the plane COM; hence it is Oz. The symmetrical line of the tangent MV is a line OU through O orthogonal to the plane DOM; then since angle $VMT =$ angle $UOz = \theta$, it follows that the Villarceau circles intersect the parallels of the torus at a constant angle θ.

REMARK II. The projection of the Villarceau circle on the equator is an ellipse with one focus at point O. In fact its center is at D and its major axis, which is along OX, is $2a$; the minor axis is $2a \cos \theta$; hence, the focal distance is $a \sin \theta = b$ which proves the proposition.

See also Problem 3546, this MONTHLY [40, 183].

3325 [36, 338]. *Solution by E. G. Olds*

Extend AO_2 to meet S_2 at T_2 and AO_1 to meet S_1 at T_1. Then $AK_2 T_2$ and $AK_1 T_1$ are right triangles and are similar, having an acute angle of the one equal to an acute angle of the other. Therefore,

$$AK_2 / AK_1 = AT_2 / AT_1 = 2AO_2 / 2AO_1 = AO_2 / AO_1.$$

Also $\angle K_2 AK_1 = \angle O_2 AO_1$. Therefore the triangles $K_2 AK_1$ and $O_2 AO_1$ are similar.

3482 [39, 364]. *Solution by Wallace Smith*

Sphere S is circumscribed about and sphere s is inscribed in the tetrahedron $ABCD$, each having the center O. Since planes ABD, BDC, ADC, and ABC are tangent to sphere s, they are at equal distances from the center of sphere S. Planes equidistant from the center of a sphere intersect the sphere in equal circles. Therefore, triangles ABD, ABC, ADC, and BCD are inscribed in equal circles. The chords AD, AB, and BD are each common to two equal circles. It follows at once that $\angle ACD = \angle ABD$, $\angle BCA = \angle BDA$, and $\angle BCD = \angle BAD$. Therefore, $\angle BCD + \angle BCA + \angle ACD$ equals $\angle BAD + \angle BDA + \angle ABD$ equals two right angles. A similar proof applies to the other three vertices. Therefore, the sum of the face angles of each trihedral angle of the tetrahedron is equal to two right angles.

See also Problems 3512 and E86, this MONTHLY, [*39*, 552] and [*41*, 520] respectively.

3565 [*40*, 372]. *Solution by F. Underwood*

Let PQR be the given triangle in a plane α. We can project orthogonally upon another plane α' in such a way that the corresponding points $P'Q'R'$ will be the vertices of an equilateral triangle. (See solution of problem 2895 [1933, 274]; or Müller's *Darstellende Geometrie*, vol. I, p. 97). Let C' be the circle circumscribed to $P'Q'R'$, and let E be the corresponding ellipse circumscribed to PQR in the plane α. The ratio of the area of the circle C' to the inscribed equilateral triangle $P'Q'R'$ is $r = 4\pi/3\sqrt{3}$; and, the equilateral triangle being the maximum triangle that can be inscribed in a given circle, the ratio of the area of any circle to that of any triangle inscribed in it can never be less than r. Then since orthogonal projection leaves ratios of areas unchanged, and since any ellipse can be projected orthogonally into a circle, it follows that the ratio of the area of *any* ellipse to that of *any* triangle inscribed in it can never be less than r. But it also follows that the ratio of the area of the ellipse E to that of the triangle PQR is r, and hence E must be the required ellipse of minimum area.

To construct E in the plane α, it is only necessary to observe that an orthogonal projection preserves parallelism, and hence E must be the ellipse circumscribing the triangle PQR and tangent at each vertex to a line parallel to the opposite side. This ellipse can be constructed by well-known methods, such as are given in Minchin and Dale's *Mathematical Drawing*.

Solutions to elementary problems from
THE AMERICAN MATHEMATICAL MONTHLY, since 1933

E16 [*40*, 296]. I. *Solution by Mannis Charosh*

Let the equation of the parabola be $y^2 = 8ax$. The coordinates of the focus F are $(2a, 0)$. The problem is now equivalent to proving that each of the given circles is tangent to the y-axis.

If the point P $(2b, 2c)$ is any point on the parabola, then $c^2 = 4ab$. The coordinates of M, the midpoint of FP, are $(a + b, c)$. The equation of the circle centered at M and with radius FM is therefore

$$[x-(a+b)]^2+[y-c]^2=[a-b]^2+c^2$$
$$=[a-b]^2+4ab$$
$$=[a+b]^2.$$

If $x=0$, the resulting equation has the double root, $y=c$, so that this circle is tangent to the y-axis for each point P on the parabola.

Since each circle goes through the focus F, it too is part of the envelope.

II. *Solution by Simon Vatriquant*

Let F be the focus of the parabola, V the vertex, and P any point on the curve. Let the tangents at P and V meet at T, which is thus the pole of VP. Then the diameter through T bisects VP, and also bisects FP at M (since it is parallel to the base VF of the triangle PVF). Since the projection of the focus of any parabola upon any tangent to that parabola, is a point on the tangent at the vertex, we know that FT is perpendicular to TP. Then the circle on the diameter FP is centered at M, goes through T, and is tangent to VT, since VM is a diameter of the parabola. Since the tangent to the parabola at V is thus tangent to every one of the designated family of circles, it forms an envelope of that family.

E113 [42, 174]. *Solution by M. J. Turner*

Denote the centers of the three circles by A, B and C, and let their radii be P, Q and R respectively. Let L, M and N be the respective intersections of pairs of external tangents to circles centered at B and C, C and A, and A and B. Then from similar triangles, $AN/BN=P/Q$, $CM/AM=R/P$, and $BL/CL=Q/R$. Consequently $(AN/BN)(CM/AM)(BL/CL)=1$, so that the theorem of Menelaus requires that L, M and N must be collinear.

E150 [42, 568]. *Solution by W. B. Clarke*

Let DF cut BC at G. Through E and F draw lines parallel to BC cutting AB at H and J respectively.

Then EH is $CG/2$; FJ is $2CG$, and FJ is $4EH$.

Triangles DEH and DFJ are similar, so that $DF:DE::FJ:EH::1:4$. Consequently, EF is $3DE$.

E337 [46, 237]. *Solution by the Proposer*

Let A and B be the ends of a diameter of the circle of intersection of (S) and (T), and let (A) be the sphere centered at A, with radius AB. Transform the given configuration by inversion with respect to the sphere (A). The spheres (S) and (T) become the planes M and N respectively, each passing through B, and respectively perpendicular to the lines AS and AT. The spheres (X) and (Y) become two new spheres (X') and (Y'), tangent to the planes M and N, and tangent to each other at Q, the transform of P. But Q lies on one of the two perpendicular planes bisecting the dihedral angles formed by planes M and N. Consequently P, the inverse of Q, must lie on one of the two orthogonal spheres obtained by inverting these two perpendicular planes with respect to the sphere (A). These orthogonal spheres pass through A and B, but since AB was any diameter of the circle of intersection of (S) and (T), these orthogonal spheres intersect in that same circle.

In case (S) and (T) are tangent, we call their point of tangency, A, and choose any convenient radius for sphere (A). Now the planes M and N are parallel, and the real locus of Q is the single plane midway between them. The locus of P is then a single sphere, tangent to (S) and (T) at A, and containing the smaller of them. (The second plane on which Q moved in the general case becomes the plane at infinity in this case, and its transform, the second orthogonal sphere, shrinks to the point A.)

In case (S) and (T) have no real points in common, the sphere (A) is imaginary, and so are the inversion transforms of all the spheres in our original configuration. But M and N are still planes, though imaginary, and the locus of Q is two perpendicular, imaginary planes, which invert back into two orthogonal spheres, one of which is real and one imaginary. This real sphere is the desired locus of P, and contains the smaller of the spheres (S) and (T).

E591 [51, 348]. I. *Solution by J. S. Guérin*

The following equivalent theorem will be proved. If tangents be drawn to two circles (A_1) and (A_2) from any point on their radical axis, the line joining either point of contact on (A_1) to either point of contact on (A_2) passes through one of the centers of similitude of (A_1) and (A_2).

Let C_1 on (A_1) and C_2 on (A_2) be two such points of contact. If the line C_1C_2 intersects (A_2) in a second point D_2, the triangle $C_2A_2D_2$ is isosceles, and

$$\angle A_2D_2C_2 = \angle A_2C_2D_2.$$

But, since B lies on the radical axis of the two circles, the triangle BC_1C_2 is isosceles; thus

$$\angle BC_1C_1 = \angle BC_2C_1 = \tfrac{1}{2}\pi - \angle A_2C_2D_2 = \tfrac{1}{2}\pi - \angle A_2D_2C_2,$$

and

$$\angle BC_1C_2 + \angle A_2D_2C_2 = \tfrac{1}{2}\pi.$$

Hence A_2D_2 is perpendicular to BC_1, and the radii A_1C_1, A_2D_2 are parallel. In other words, the line $C_1C_2D_2$ passes through a center of similitude of the circles.

II. *Solution by Augustus Sisk*

(Numbers in parentheses refer to articles in N. Altshiller Court's *College Geometry*.) By (325), B lies on the radical axis of (A_1) and (A_2). By (331), the points of contact of two tangents drawn from B to the respective circles are antihomologous points. But lines joining antihomologous points on two circles pass through one of the centers of similitude of the circles.

E630 [*52*, 160]. *Solution by Howard Eves*

Neglecting differentials of higher order, we easily see (from a figure showing two adjacent normals) that

$$ds' = ds + kd\phi,$$
$$d(A' - A) = \tfrac{1}{2}k(ds + ds'),$$

where $d\phi$ is the angle between the two adjacent normals. Integrating for a complete circuit, these equations become

$$s' = s + 2\pi k,$$
$$A' - A = \tfrac{1}{2}k(s + s') = \tfrac{1}{2}k(2s + 2\pi k) = sk + \pi k^2.$$

E661 [*52*, 517]. *Solution by the Proposer*

We adopt the convention that angular directions on p (or p') are positive if they are counterclockwise when p (or p') is viewed from point L. We now make the

DEFINITIONS. A point on p will be called a *positive isocenter* or a *negative isocenter* on p according as all the angles on p having the point for vertex project respectively into equal and similarly directed or equal and oppositely directed angles on p'.

By an elementary synthetic treatment we shall show that when p and p' are nonparallel there is one and only one positive isocenter and one and only one negative isocenter on p. We shall also geometrically locate these isocenters.

THEOREM 1. *If P is a positive (or negative) isocenter on p, then its image, P', is a positive (or negative) isocenter on p'.*

THEOREM 2. *If p and p' are parallel, then every point on p is a positive or a negative isocenter according as L does not or does lie between p and p'.*

THEOREM 3. *If p and p' are not parallel, then p cannot have two positive isocenters.*

For suppose P and Q are two positive isocenters on p. Draw any circle s on p passing through P and Q, and let R be any other point on s. Then, since $\angle QPR = \angle Q'P'R'$ and $\angle PQR = \angle P'Q'R'$, it follows that $\angle PRQ = \angle P'R'Q'$, whence s projects into a circle s'. Thus all circles on p through P and Q project into circles on p' through P' and Q'. But this is impossible. For consider a circle on p, passing through P and Q and cutting the vanishing line on p. This circle must project into a hyperbola. Thus the original assumption of the existence of two positive isocenters is incorrect.

THEOREM 4. *Let p and p' be nonparallel and let O be the foot of the perpendicular from L onto p, and let V on p be the image of V', the foot of the perpendicular* *from L onto p'. Then any line on p perpendicular to VO is parallel to any line on p' perpendicular to V'O'.*

Draw VQ on p perpendicular to VO. Then

$$LQ^2 = LO^2 + OQ^2 = LO^2 + VO^2 + VQ^2 = LV^2 + VQ^2.$$

Therefore VQ is perpendicular to LV, and therefore to plane VLO. Hence any line on p perpendicular to VO is perpendicular to plane VLO. Similarly, any line on p' perpendicular to $O'V'$ is perpendicular to plane VLO. This proves the theorem.

THEOREM 5. *If p and p' are nonparallel, then p has exactly one positive isocenter, and it is the intersection with p of the bisector of the angle formed by the perpendiculars dropped from L onto p and p'.*

We adopt the notation of theorem 4. Let P be any point on p and let W be the foot of the perpendicular from P on VO. Then W' is the foot of the perpendicular from P' on $V'O'$ and, by Theorem 4, PW is parallel to $P'W'$. Let I be the intersection with VO of the bisector of angle VLO. Draw WK perpendicular to LV to cut LV in K and LI in J. Then

$$IW/I'W' = JW/I'W' = LW/LW' = WP/W'P'.$$

Hence $\angle WIP = \angle W'I'P'$. Similarly, if Q is any other point on p, $\angle QIW = \angle Q'I'W'$. Hence $\angle QIP = \angle Q'I'P'$, and I is a positive isocenter on p. By theorem 3, I is the only positive isocenter on p.

THEOREM 6. *If p and p' are nonparallel, then p has exactly one negative isocenter, and it is the intersection with p of the external bisector of the angle formed by the perpendiculars dropped from L on p and p'.*

We may give a proof of this theorem analogous to that given for theorem 5.

Note. The positive isocenter has long been known and employed in photogrammetry (the science of surveying by means of photographs), but the negative isocenter seems to have escaped notice. For an analytical treatment of the above and for an application to photogrammetry see "Analytical and Graphical Rectification of a Tilted Photograph," by Howard Eves, *Photogrammetric Engineering*, April-May-June issue, 1945.

E665 [*52*, 521]. *Solution by R. A. Rosenbaum*

Problem E 630 can be easily generalized so as to include the present problem. The generalization of E 630 is:

For any point P of a given closed convex curve C, let P' be that point on the normal to C at P for which $PP' = k$, a constant, taken as positive or negative according as P' lies on the exterior or the interior normal. The locus of P' is a curve C'. Let s, s' be the respective lengths of C, C', and A, A' the areas of these curves. Then

$$s' = s + 2\pi k,$$
$$A' = A + sk + \pi k^2.$$

The proof as given for E 630, reprinted here, applies to this generalization, and the present problem is seen to be a special case.

E741 [54, 282]*. *Solution by L. M. Kelly*

The announced relation follows quite nicely from a theorem of L. M. Blumenthal: *If a spherical triangle is reproduced congruently in the plane (that is, with length of sides preserved), each angle of the plane triangle is less than the corresponding angle of the spherical triangle.* In the particular case at hand, reproduction of the right spherical triangle in the plane yields a plane triangle in which angle C is acute, from which it follows, by the cosine law, that $c^2 < a^2 + b^2$.

It might be well to indicate a simple proof of Blumenthal's theorem which, it would seem, should be better known than is the case.

Let ABC be the spherical triangle and $A'B'C'$ the corresponding plane triangle, where $BC = B'C' = a$, $CA = C'A' = b$, $AB = A'B' = c$. Now

$$\sin \frac{A'}{2} = \sqrt{\frac{(s-b)(s-c)}{bc}},$$

$$\sin \frac{A}{2} = \sqrt{\frac{\sin (s-b) \sin (s-c)}{\sin b \sin c}}.$$

Since $A/2$ and $A'/2$ are certainly less than or equal to $\pi/2$, the theorem will be proved if we can show that the right member of the first equation is less than that of the second. This we can do in the following way. First note that $(\sin x)/x$ is a montone decreasing function for $0 < x < \pi/2$. Furthermore, since $a < b + c$, $s - b < c$. Similarly $s - c < b$. Thus

$$\frac{\sin (s-b)}{s-b} \cdot \frac{\sin (s-c)}{s-c} > \frac{\sin b}{b} \cdot \frac{\sin c}{c},$$

or

$$\frac{\sin (s-b) \sin (s-c)}{\sin b \sin c} > \frac{(s-b)(s-c)}{bc}.$$

In a precisely analogous fashion we may prove that reproduction in the plane of a pseudo-spherical triangle (a geodesic triangle on the pseudo sphere) results in correspondingly larger angles. These results may be phrased, of course, in terms of congruent imbedding in spherical and hyperbolic spaces. Furthermore, these two results have been extended by Blumenthal as follows: *If a "shortest distance" triangle on a surface having everywhere positive curvature be reproduced congruently in the plane, the angles of the plane triangle will each be less than the corresponding angle of the surface triangle. A similar result holds for sur-*

faces of everywhere negative curvature. It is necessary to make a careful distinction between geodesic and shortest distance triangles, since they are not always the same.

E774 [*57*, 484]. *Solution by Vern Hoggatt*

The following solution is largely compounded from the partial results sent to the editor by Free Jamison and C. S. Ogilvy. It is a pleasure to acknowledge the assistance furnished by these results.

Using oblique cartesian coordinates let the vertices A, B, C of the triangle be $(0, b)$, $(a, 0)$, $(0, 0)$. Choose the median through A and consider first a line that intersects the side to which the median is drawn and one of the other sides, say the y-axis. The equation of the chosen median is

$$(1) \qquad x = (ab - ay)/2b.$$

The equation of the line passing through $(ar, 0)$ and $(0, b/3r)$, and consequently cutting off one-third of the area, is

$$(2) \qquad x = (arb - 3ar^2y)/b.$$

The equation of another line cutting off one-third of the area is

$$(3) \qquad x = (2arb - 3ar^2y)/2b.$$

Solving (1) and (2) simultaneously we find $y/b = (1 - 2r)/(1 - 6r^2)$, which has a maximum of $(3 - \sqrt{3})/6$ when $r = (3 + \sqrt{3})/6$. As r varies from $1/2$ to $(3 + \sqrt{3})/6$ to 1, y/b varies from 0 to $(3 - \sqrt{3})/6$ to $1/5$. Thus, if z is the fractional part of the median from the vertex to the intersection of (1) and (2), we have $z = 1 - y/b$, and for $z = (3 + \sqrt{3})/6$ there is one line, for $(3 + \sqrt{3})/6 < z \le 4/5$ there are two lines, and for $4/5 < z \le 1$ there is one line. The same results may be obtained by letting the line intersect the third side of the triangle, whence the above number of lines under each condition is to be doubled.

Solving (1) and (3) simultaneously we find $y/b = (1 - 2r)/(1 - 3r^2)$, which is a single-valued function of r with no turning-point maximum or minimum between $r = 2/3$ and $r = 1$. The end-point minimum at $r = 1$ gives $y/b = 1/2$, and $r = 2/3$ gives $y/b = 1$. Thus for $0 \le z \le 1/2$ there are two lines.

Now choose another oblique cartesian frame of reference so that the vertices A, B, C are represented by $(0, 0)$, $(c, 0)$, $(0, b)$. The equation of the median through A is now

$$(4) \qquad x = cy/b,$$

and we have, for lines analogous to lines (2) and (3),

$$(2') \qquad x = (crb - 3cr^2y)/b,$$

$$(3') \qquad x = (2crb - 3cr^2y)/2b.$$

We now have $z = 2y/b$.

Solving (4) and (2′) simultaneously we find $2y/b = 2r/(1+3r^2)$, which has a maximum of $\sqrt{3}/3$ at $r = \sqrt{3}/3$, since r must be non-negative. At $r=1$ and $r=1/3$, $2y/b = 1/2$, and consequently at $z = \sqrt{3}/3$ there is one line, and for $1/2 \leqq z \leqq \sqrt{3}/3$ there are two lines. The lines corresponding to $z = 1/2$ are the same as those found previously.

Solving (4) and (3′) simultaneously we find $2y/b = 4r(2+3r^2)$, which has a maximum of $\sqrt{6}/3$ at $r = \sqrt{6}/3$ and has a value of $4/5$ at $r=1$ and $r=2/3$. So we may conclude that for $z = \sqrt{6}/3$ there is one line, and for $4/5 \leqq z \leqq \sqrt{6}/3$ there are two lines. The lines corresponding to $z = 4/5$ are the same as those found previously.

Combining results we have:

value of z	number of lines
$0 \leqq z < \sqrt{3}/3$	2
$z = \sqrt{3}/3$	1
$\sqrt{3}/3 < z < (3 + \sqrt{3})/6$	0
$z = (3 + \sqrt{3})/6$	2
$(3 + \sqrt{3})/6 < z < \sqrt{6}/3$	4
$z = \sqrt{6}/3$	3
$\sqrt{6}/3 < z \leqq 1$	2

See also Problem E1228, this MONTHLY, reprinted here, Problem 58, MATHE-MATICS MAGAZINE, [*24*, 167], and "Forbidden Area", by V. Hoggatt, this MONTHLY, [*69*, 98].

E788 [*55*, 365]. *Solution by the Proposer*

We first prove by induction that such a map may be colored with two colors, so that no two countries with common boundary have the same color. This is certainly true for $n=1$. Suppose it is true for $n-1$. When we add another great circle, it is only necessary to reverse all the colors to one side of this circle to obtain a coloring for n. Also, if we use this coloring, it is clear that for n even, diametrically opposite countries will have the same color. Hence, in this case, there will be an even number of countries of each color. We next show that these two even numbers cannot be equal. If they were equal, then the total number of countries, F, would be a multiple of four, whereas we shall show that $F = 4k + 2$.

Since each circle intersects each other circle twice, the number of vertices, V, is $n(n-1)$. Now every vertex belongs to four edges, and every edge has two vertices. Therefore the number of edges, E, is $2V$. Using Euler's theorem, $V - E + F = 2$, we get $F = n(n-1) + 2$, and if n is a multiple of four, F will be of the form $4k + 2$.

Finally, we note that in any trip on a map colored by our method, one must change color with every crossing from country to country, and since the number of countries of one color differs from the number of countries of the other color by at least two it will be impossible to reach them all.

For other problems in great circles, see Problems E1667 and E1756, this MONTHLY, [72, 81] and [73, 204] respectively, and Problem 38, MATHEMATICS MAGAZINE, reprinted here.

E888 [57, 339]. I. *Solution by C. S. Ogilvy*

The projection of a unit cube on a plane perpendicular to a main diagonal is a regular hexagon whose inscribed circle has diameter $\sqrt{2}$. Hence any unit square hole (prism) whose axis is this main diagonal has its cross section circumscribed by the circle. Therefore such a hole is completely surrounded by material of the cube, provided only that it is not oriented with a diagonal of its cross section coinciding with a line joining the midpoints of two opposite edges of the cube.

II. *Solution by F. Bagemihl*

Let the vertices of the given cube have rectangular coordinates $(0,0,0),(1,0,0)$, $(1, 1, 0)$, $(0, 1, 0)$, $(0, 0, 1)$, $(1, 0, 1)$, $(1, 1, 1)$, $(0, 1, 1)$. The points $A(0,.9,1), B(.9,0,1), C(1,.1,0), D(.1,1,0)$ are on the cube, and are the vertices of a rectangle with $AB = CD = \sqrt{1.62} > 1, AD = BC = \sqrt{1.02} > 1$, so that a square, s, can be drawn wholly interior to the rectangle $ABCD$, each side of the square having a length greater than unity. Through the given cube cut a hole with a square cross section, perpendicular to $ABCD$, and intersecting the latter in s. A unit cube can pass through this hole if four faces of the cube are parallel, respectively, to the four faces of the hole.

E929 [58, 261]. *Solution by B. D. Roberts*

The double reflection $T_1 T_2$ in the two lines l_1, l_2 (whether intersecting or not) is a rigid sense-preserving transformation involving a rotation through twice the angle from l_1 to l_2. But T^2 is the product of three such double reflections, the sum of the angles involved being a multiple of π. The total rotation due to T^2 is then through an angle which is a multiple of 2π. This means that T^2 produces a translation only. It is easy to show that this translation reduces to the identity transformation if and only if l_1, l_2, l_3 are concurrrent or parallel.

Editorial Note: An interesting consequence of the problem, pointed out by the proposer, is that a plane point set with three non-concurrent and non-parallel axes of symmetry contains an infinite lattice-work of points.

See also Problem E2140, this MONTHLY, [76, 1067].

E1224 [64, 113]. *Solution by N. J. Fine*

Let the group elements T_i be given by $T_i(x) = A_i x + b_i (i = 1, ..., n)$, where the A_i

are linear. Let c be an arbitrary vector, $x = (1/n)\Sigma_{j=1}^{n} T_j(c)$. Then

$$T_i(x) = A_i x + b_i = A_i\left[(1/n) \sum_{j=1}^{n} T_j(c)\right] + b_i$$

$$= (1/n) \sum_{j=1}^{n} \left[A_i T_j(c) + b_i\right] = (1/n) \sum_{j=1}^{n} T_i\left[T_j(c)\right]$$

$$= (1/n) \sum_{k=1}^{n} T_k(c) = x.$$

Hence x is invariant under the entire group.

Editorial Note: One cannot help but notice the following property. Let c and d be any two points and let c_i and d_i be the maps of c and d, respectively, under the transformation T_i. Then, if there is only one fixed point of G, the centroids of the two sets of points $c_1, \ldots, c_n$ and $d_1, \ldots, d_n$ coincide. Examples, like the octic group of the symmetries of a square, are abundant.

E1228 [*64*, 198]. *Solution by the Proposer*

The theorem is a consequence of the facts that $n(P)$ and the ratio of areas are affine invariants and that any triangle is affinely equivalent to an equilateral triangle.

For an equilateral triangle the required region is bounded by three arcs of hyperbolas, each of which has two sides of the triangle for asymptotes, and pairs of which are tangent to one another at the midpoints of the medians. Taking the side of the equilateral triangle as the unit length, the rectangular equation of one of the hyperbolas referred to one side and one vertex of the triangle as x-axis and origin is $y(\sqrt{3}\, x - y) = 3/16$. The area of the region can be obtained by simple integration as $\sqrt{3}\,(\ln 8 - 2)/16$, and hence the required constant is $c = (\ln 8 - 2)/4 \approx 0.0198$.

See also E774, this MONTHLY, reprinted here.

E1423 [*68*, 179]. *Solution by W. C. Waterhouse*

A proof is given in Selection 10 of *The Enjoyment of Mathematics*, by H. Rademacher and O. Toeplitz (translated into English by H. Zuckerman), and is roughly as follows.

We want to show that between successive passages through a given double point an even number of double points are passed through. Call the part of the curve traced (itself a closed curve) B, and the rest of the curve (also a closed curve) C. All double points of B are certainly passed through twice, and we need consider only the intersections of B and C. But C can be replaced by a regular curve without changing its intersections with B, and then the Jordan Curve Theorem shows that there are an even number of intersections of B with C.

E1515 [70, 95]. *Solution by L. F. Meyers*

First we define the following point sets I, B, C of the Euclidean plane:

$$I = \{(x,y) : 0 \leqq x < 1, 0 \leqq y < 1\},$$
$$B = \bigcup_{m=2}^{+\infty} \{(x,y) : 0 < x < 1, y = (1-x)/m\} \cup \{(0,0)\},$$
$$C = I \setminus B.$$

Thus I is the half-closed unit square; B (a subset of I) is a union of certain open segments together with the limit point $(0,0)$; C is the complement of B with respect to I. Next we introduce a convenient notation: if T is a point set of the Euclidean plane, and a and b are integers, then by $T_{a,b}$ we mean the point set obtained by translating the set T through the vector whose components are a and b. Now we define

$$R = \left(\bigcup_{k=-\infty}^{0} I_{0,k} \right) \cup \left(\bigcup_{k=1}^{n-1} C_{0,k} \right) \cup \left(\bigcup_{k=n}^{+\infty} I_{0,k} \right) \cup \left(\bigcup_{k=1}^{n-1} B_{k,0} \right),$$

and then

$$S^{(0)} = \bigcup_{k=-\infty}^{+\infty} R_{kn,-kn},$$

and finally

$$S^{(i)} = S^{(0)}_{i,-i} \text{ for } i = 0, 1, \ldots, n-1.$$

It is not a difficult matter to show that the n congruent sets $S^{(i)}$ are individually connected and mutually disjoint, and that together they exhaust the whole Euclidean plane.

If $n = 2$, then a simple solution (polygonally connected) can be obtained if congruence by reflection is permitted. The two sets are the upper and lower open half-planes, respectively, together with portions of their boundary. The portions of the boundary will be intervals open on the left and closed on the right, all of equal length, and belonging alternately to the two sets.

Solutions to advanced problems from
THE AMERICAN MATHEMATICAL MONTHLY, since 1933

3829 [46, 657]. *Solution by H. E. Vaughan*

The proof given here holds for any simple closed curve C, rectifiable or not, lying in a plane S. Denote by D_1 and D_2 the interior and exterior of C and let P be any point of D_1. Let d be a fixed line through P and denote by C_0 and C_π the intersections of C with the two rays into which d is separated by P. Denote by C', C_0' and C_π' the sets obtained from C, C_0, and C_π by rotating S about P through an angle π.

We wish to show that C and C' have at least one point in common, since any such point will be admissible as the point A of the theorem, while the point mapped into it by the above rotation may be used for B. To prove the existence of such a point we shall show that either (1) C_0' and C_π have a point in common or (2) C' contains points of both D_1 and D_2. If (1) holds, the theorem will be proved, since C_0' and C_π are subsets of C' and C respectively, so that a point common to them will be common to the latter pair also. If (2) holds, the connected set C' contains points of both domains complementary to C and hence contains at least one point of C, so that the theorem also follows in this case. (We make use here of the theorem that a connected set which contains points interior and exterior to a set contains at least one boundary point of the set. We might also use the more sophisticated Jordan Curve Theorem.)

Let us suppose, then, that C_0' and C_π have no points in common. Then any point of C_0' is in D_1 or D_2. Without loss of generality we may assume that some point of C_0' is in D_1. If, in addition, there is a point of C_0' in D_2, we have case (2) and the theorem is proved. Since C_0' is a closed subset of the ray issuing from P and containing C_0' and C_π, it has a last point, R'. If no point of C_π lies beyond this, R' lies in D_2 and the theorem is proved. If there is a point Q of C_π beyond all points of C_0' on the ray in question, the corresponding point Q' of C_π' lies, on the opposite ray, beyond all points of C_0 and is a point of C' in D_2.

As a matter of fact, the following more general theorem has been proved: Let C be any plane continuum, D a bounded domain complementary to C, and P an arbitrary point of D. Then there exist two points A and B of the boundary of D such that P bisects the segment AB. It is easily seen that the property need not hold in case D is not bounded.

Second Solution. This solution is less elementary than the preceding but seems to be of interest in view of its purely topological character. Suppose that C is given in rectangular coördinates in S by the equations $x = f(t)$, $y = g(t)$, $0 \leq t \leq 1$. The locus M of midpoints of chords of C is given by the equations

$$(1) \qquad x = \tfrac{1}{2}[f(t_1) + f(t_2)], \qquad y = \tfrac{1}{2}[g(t_1) + g(t_2)]$$

where (t_1, t_2) runs over the unit square in an auxiliary (t_1, t_2)-plane, and it is sufficient for the proof of the theorem to show that M covers the interior of C. Due to symmetry the range of (t_1, t_2) may be restricted to the triangle $t_1 \leq t_2$, $0 \leq t_2 \leq 1$, and also, for each number s, $0 \leq s \leq 1$, the points $(0, s)$ and $(s, 1)$ may be identified, since the functional values x and y are the same for both points. In this process the points $(0, 0)$ and $(1, 1)$ are also identified since each is identified with $(0, 1)$. This identification introduces a "twist" so that the resulting surface is a cross-cap (topologically equivalent to a Moebius band). Its boundary mod 2 is the line $t_2 = t_1$ which, through the identification of its endpoints, has become a simple closed curve. Now, by the mapping (1), this curve is mapped into C and the cross-cap as a whole goes into the set M. In the language of combinatorial topology we may say that the singular cycle determined by the mapping on C bounds, mod 2, the singular 2-chain determined by the mapping on M. But this necessitates that M cover the interior of C because of a theorem whose intuitive content is that if P is a point interior to C, C does not bound in $S-P$.

See also Problem 4325, this MONTHLY [57, 423].

4036 [50, 397]. I. *Solution by Fritz John*

Let $p(\alpha)$ denote the "function of support" of C_1, *i.e.*, $p(\alpha)$ shall be the distance of that tangent of C_1 from the origin, whose normal forms the angle α with the x-axis (See Courant: *Calculus*, II, p. 213). Then

$$x = p\cos\alpha - p'\sin\alpha, \qquad y = p\sin\alpha + p'\cos\alpha$$

is a parametric representation for C_1. The radius of curvature of C_1 is given by $R = p + p''$, the enclosed area by

$$A_1 = \frac{1}{2}\int_0^{2\pi}(xy' - yx')d\alpha = \frac{1}{2}\int_0^{2\pi}(p^2 + pp'')d\alpha = \frac{1}{2}\int_0^{2\pi}(p^2 - p'^2)d\alpha.$$

Similarly the parametric representations of C_2 and C_3 are respectively

$$x = (2p + p'')\cos\alpha - p'\sin\alpha, \qquad y = (2p + p'')\sin\alpha + p'\cos\alpha,$$

and

$$x = (p + p'')\cos\alpha, \qquad y = (p + p'')\sin\alpha;$$

hence the areas enclosed by C_2 and C_3 are easily found to be

$$A_2 = \frac{1}{2}\int_0^{2\pi}(4p^2 - 7p'^2 + 3p''^2)d\alpha$$

$$A_3 = \frac{1}{2}\int_0^{2\pi}(p^2 - 2p'^2 + p''^2)d\alpha.$$

Consequently $A = A_2 - A_1 = 3A_3$, which is the first statement.

Now "Wirtinger's inequality" states, that for a function $f(\alpha)$ of class C' with $\int_0^{2\pi}f(\alpha)d\alpha = 0$

$$\int_0^{2\pi}f'^2(\alpha)d\alpha > \int_0^{2\pi}f^2(\alpha)d\alpha,$$

unless f is of the form $f(\alpha) = a\cos\alpha + b\sin\alpha$; (see Hardy-Littlewood-Polya: *Inequalities*, pp. 185–187). For $f = p'$ it follows that $\int_0^{2\pi}p''^2(\alpha)d\alpha > \int_0^{2\pi}p'^2(\alpha)d\alpha$, and hence $A_3 > A_1$, unless $p' = a\cos\alpha + b\sin\alpha$; in the latter case $p = c + a\sin\alpha - b\cos\alpha$, and C_1 is a circle of radius c. The isoperimetric inequality (which may be based on Wirtinger's inequality), yields

$$L_2^2 \geqq 4\pi A_2, \qquad L_3^2 \geqq 4\pi A_3;$$

hence

$$L_2L_3 \geqq 4\pi\sqrt{A_2A_3} = 4\pi\sqrt{(3A_3 + A_1)A_3} > 4\pi\sqrt{4A_1^2} = 8\pi A_1$$

unless C_1 is a circle.

In the case where C_1 is a circle of radius c, C_2 is a circle of radius $2c$, and C_3 a circle of radius c, so that $L_2L_3 = 8\pi A_1$.

II. *Solution by the Proposer*

We consider two normals to C_1 corresponding to the directions ϕ and $\phi + d\phi$; a point on a normal whose distance to C_1 is a constant equal to λ will describe a curve whose arc s^* satisfies

$$ds^* = (R + \lambda)d\phi.$$

The area A will be then

$$A = \int\int ds^* d\lambda = \int_0^{2\pi} d\phi \int_0^R (R + \lambda)d\lambda = \frac{3}{2} \int_0^{2\pi} R^2 d\phi = 3A_3$$

which proves (*a*).

We have also, if s_2 is the arc of C_2,

(1) $$ds_2 = \sqrt{4R^2d\phi^2 + dR^2} = \sqrt{4R^2 + R'^2}\, d\phi$$

where R' represents the derivative with respect to ϕ. We have also

(2) $$ds_3 = \sqrt{R^2 + R'^2}\, d\phi.$$

From (1) and (2) we deduce, representing by s_1 the arc of C_1

$$ds_2 \geqq 2R d\phi = 2ds_1 \quad \text{and} \quad L_2 \geqq 2L_1$$
$$ds_3 \geqq R d\phi = ds_1 \quad \text{and} \quad L_3 \geqq L_1.$$

This gives us

(3) $$L_2L_3 \geqq 2L_1^2.$$

But it is known that for every plane closed curve we have $L_1^2 - 4\pi A_1 \geqq 0$; so this inequality and (3) proves the last part (b).

The equality in (b) is valid only if $R' = 0$, and then the radius of curvature is constant and the closed curve must be a circle.

4087 [*51*, 593]. *Solution by Howard Eves*

(1) This part follows from the facts that the ratio of two segments on a line is an absolute invariant, and the area of a polygon is a relative invariant, with respect to the group of affine transformations.

(2) Since any three non-collinear finite points may be carried affinely into any other three non-collinear finite points it follows that any triangle is affinely equivalent to an equilateral triangle. But it is easy to show that, for an equi-

lateral triangle,

$$R(k) = (k - 2)^2/(k^2 - k + 1).$$

By (1) this relation is then true for all triangles.

(3) As in (2), and since parallel lines carry affinely into parallel lines, it follows that any parallelogram is affinely equivalent to a square. It is easy to show that, for a square,

$$R(k) = (k - 1)^2/(k^2 + 1).$$

By (1) this relation is then true for all parallelograms.

(4) To show that $R(k)$ does not have the same value for all quadrilaterals, consider a quadrilateral, S, three of whose vertices are the vertices of an equilateral triangle, and whose fourth vertex, A_4, is near one of the other vertices, say A_3. Then it is readily seen from a figure that, as $A_4 \to A_3$, $R(2) \to 1/6$. But for a parallelogram $R(2) = 1/5$.

Editorial Notes. A solution by the proposers is similar to the above. For (2) the triangle $(0,0), (k,0), (0,k)$ is used; for (3) the square with three vertices as above and the fourth (k,k) is used; and for (4) a trapezoid with one base twice the length of the other gives $R(2) = 156/775$, whereas for a parallelogram $R(2) = 1/5$. For (2) Butchart used vectors for a triangle $A_1A_2A_3$ and eliminated the vector for A_3 from the vector equations for B_2 and B_3 and thus obtained $A_2C_1/C_1B_3 = k/(k-1)^2, A_2C_2/C_2B_3 = k(k-1)$. Additional combinations give the desired result. For (3) the square with side k is used, then Q is a square with side $C_1C_2 = k(k-1)/(k^2+1)^{1/2}$ and this gives the desired result for $R(k)$.

See also Problem 282, MATHEMATICS MAGAZINE [14, 109] for the case $n = 3$.

4262 [56, 270]. *Solution by the Proposer*

Let (x,y) be the coördinates of the variable center of a circle of constant radius ρ. Let $N \equiv N(x,y)$ be the number of common points of this circle and the curve C for each position of (x,y). Then the Proposer has shown, in the paper already cited, that the following integral formula holds;

$$\iint N \, dx \, dy = 4L\rho.$$

On the other hand, if $\rho \geq R$, the area covered by the points (x, y) which are centers of circles of radius ρ and which cut the given circle of radius R, has the value

$$\pi(\rho + R)^2 - \pi(\rho - R)^2 = 4\pi R\rho.$$

Consequently the mean value of N is

$$\overline{N} = \frac{\iint N \, dx \, dy}{\iint dx \, dy} = \frac{L}{\pi R}.$$

As the mean value of a function is not greater than its maximum value, the inequality (1) is established.

If $\rho < R$, the area covered by the centers (x, y) of circles of radius ρ which cut the given circle of radius R or are contained in its interior is $\pi(\rho+R)^2$. Consequently $\overline{N} = 4L\rho/\pi(\rho+R)^2$ and inequality (2) holds.

4739 [65, 214]. *Solution by Harley Flanders*

The answer is yes. For suppose $C_1 \cap C_2$ and C_3 are disjoint. We note that C_3 is the intersection of the eight closed half-spaces determined by its faces, consequently there must be a plane P which contains a face of C_3 such that $C_1 \cap C_2$ lies entirely in the open half-space determined by P which does not contain C_3. We now choose any plane M perpendicular to P and form the orthogonal projections R_1, R_2, R_3 of C_1, C_2, C_3 on M. The R_i are rhombuses obtained from each other by translation and evidently lie in the impossible configuration in which each two intersect, but $R_1 \cap R_2$ and R_3 are disjoint.

The corresponding result for rhombuses in the plane reduces to a statement about disjoint segments on a line via oblique projection.

The same proof applies when the regular octahedra are merely homothetic.

Editorial Note: Although the proposer had originally an independent argument, he finds the result in a recent paper by Hanner, *Math. Scand.*, vol. 4, 1956, pp. 65–87. If the three-dimensional cartesian space E is metrized by means of the norm $N(x) = |x_1| + |x_2| + |x_3|$ then its unit cell is an octahedron. From the result stated above and another of Hanner's results, it follows that E is 4-hyperconvex but not 5-hyperconvex, in the sense of Aronszajn and Panitchpakdi (*Pacific J. Math.*, vol. 6, 1956, pp. 405–440). This solves part of their Problem 1, p. 437. From another result in Hanner's paper it follows that if a finite-dimensional Banach space is 5-hypercomplex, it is hypercomplex.

4927 [68, 809]. *Solution by Robert Brown*

Let A be the image of n-space under f. From the inequality in the hypothesis it follows immediately that f is one-to-one and that f^{-1} is a continuous transformation of A onto n-space. Hence, A is homeomorphic to n-space, and by Brouwer's theorem on the invariance of domain, A is open in n-space. Let a be an accumulation point of A, and let $f(x_1), f(x_2), \ldots, f(x_n), \ldots$ be a sequence in A converging to a. This sequence is a Cauchy sequence, and by the inequality, the sequence $x_1, x_2, \ldots, x_n, \ldots$ is also Cauchy, say with limit x. Then $a = f(x) \in A$. Hence, A is closed, as well as open, and must be the entire n-space.

5527 [76, 311]. I. *Solution (to first part) by Harley Flanders*

Let S be the unit circle and $f: S \to E^3$ a one-one C^1 mapping defining the curve. We define a continuous mapping g on the torus $T = S \times S$ into the projective plane P as follows:

Identify P with the plane at infinity. If $x \neq y$, the secant $f(x) \cup f(y)$ meets P in $g(x,y)$. If $x = y$, the tangent line t_x to the curve at x meets P in $g(x,x)$.

The continuity of g follows from the continuous differentiability of f.

The mapping g cannot be one-one. Suppose it were: then $g: T \to P$ is one-one on a compact space into a Hausdorff space, hence is a homeomorphism on T into P, actually *onto* P since T and P are both two-manifolds and T is compact (invariance of domain). This is impossible since T and P have different homology groups.

Each of the various cases in which g sends two distinct points of T to the same point of P yields a direction of projection onto a plane curve with the desired singularities. Notation: distinct letters = distinct points.

(In each case the minimal possible singularity is listed.)

(i) $g(x,y) = g(z,w)$ Two double points.
(ii) $g(x,y) = g(x,z)$ Triple point.
(iii) $g(x,x) = g(y,z)$ Cusp and double point.
(iv) $g(x,x) = g(y,y)$ Two cusps.

II. *Solution (to second part) by the Proposer*

An example of a curve all of whose plane projections have at most two double points is given parametrically by

$$x = \cos t, \qquad y = \sin t, \qquad z = \sin 2t + \sin 3t.$$

There are no three parallel lines each meeting the curve twice. Since all lines parallel to the z-axis meet the curve only once, we can use azimuth θ and slope relative to the xy-plane to distinguish the direction of lines. The strategy is to take any fixed θ, and for each t compute the slope u of the unique line of azimuth θ meeting the curve in two points, one of which corresponds to the parameter value t, and show that u takes any value for at most two values of t. To do this, first turn the curve around the z-axis by an angle θ, i.e. replace t by $t-\theta$ in z, then compute the slopes of chords parallel to $y = 0$. We get

$$u = \frac{\sin 2(t - \theta) + \sin 3(t - \theta) + \sin 2(t + \theta) + \sin 3(t + \theta)}{2 \sin t}$$

$$= 2 \cos 2\theta (\cos t) + \cos 3\theta (1 + 2 \cos 2t).$$

For each θ we must consider this function on the interval $0 \leq t \leq \pi$. Note that the coefficients $2 \cos 2\theta$ and $\cos 3\theta$ are never simultaneously zero. The derivative

$$\frac{du}{dt} = -2 \sin t (\cos 2\theta + \cos 3\theta \cos t)$$

vanishes at the ends of the interval, and otherwise only at most once. Therefore u is monotone on each of two complementary subintervals, giving the desired result.

Solutions to problems from MATHEMATICS MAGAZINE

159 [12, 246]. *Solution by George Dantzig*

Let A, B, C and A', B', C' be the vertices of the two triangles which lie respectively in the planes P and P'; let the intersections of the lines BC and $B'C', CA$ and $C'A', AB$ and $A'B'$ be the points U, V, W, respectively; and let the intersections of the planes P and P' be the line p. The points U, V, W must lie on the line of intersection of the planes, hence p is the axis of perspective of the two triangles. Let Q be the center of perspective of the two triangles. If we keep the plane P fixed and turn the plane P' about the line p, the axis of perspective of the two triangles remains fixed but the center of perspective Q, however, will change. We will show that the locus of Q is a circle whose plane is perpendicular to p.

In this rotation the points A', B', C' will describe circles about p. Let the centers of these circles be A'', B'', C'' and their radii be a, b, c, respectively. Call the intersections of the lines AA'' and BB'' the point K; we shall show that the line QK is of constant length and is perpendicular to the line p. Consider the configuration after revolution through any fixed angle. In the planes $AA'A''$ and $BB'B''$ the lines $A'A''$ and $B'B''$ are parallel and both are perpendicular to the line p. Hence the intersection of these two planes, i.e. QK, is parallel to $A'A''$ and $B'B''$ and perpendicular to p. Therefore, triangles $AA'A''$ and QKA are similar and we have the proportion $A'A'' / QK = AA'' / KA$. But AA'', KA, and $A'A'' = a$ are fixed lengths independent of the angle of rotation. We conclude therefore that QK is also constant. Moreover QK is perpendicular to the line p; thus the point Q describes a circle with K as center.

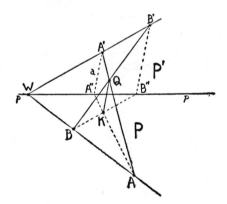

See also Problem 206, this MAGAZINE [15, 256].

214 [12, 416]. *Solution by Henry Schroeder*

The line that joins the intersections of the opposite sides of a quadrilateral is divided harmonically by the diagonals produced.* Thus $(BCDH)$ is a harmonic range and *FB, FD, FC, FH* is a harmonic pencil. Since any line cutting a harmonic pencil is divided harmonically by it, then (*BPIE*) is also a harmonic range. Moreover, if two conjugate rays of a harmonic pencil are perpendicular they bisect the angles formed by the other two rays. Accordingly, *DP* bisects angle *IDE*.

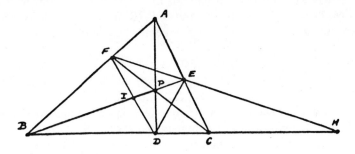

As *P* moves up and down *AD*, the points *B, C*, and *D* remain fixed. Thus the fourth point *H* of the harmonic range is fixed.

See also this MAGAZINE, vol. 4, issue 6, pages 18–20.

391 [*15*, 430]. *Solution by R. V. Sweeney*

Extend *AB* to *R*, so that *NR = AN*. Complete the triangle *AMR* which is isosceles. Next, complete the triangle *MCR* in which *MC = MR*. Angle *BCM* equals angle *BAM* and is therefore equal to the other base angle of the triangle *AMR*. Accordingly, triangle *CRB* is isosceles. It follows that *RB = CB* and the theorem is evident.

38 [*23*, 207]. *Solution by Dewey Duncan*

Three great circles intersecting in distinct points divide a spherical surface into 8 spherical triangles occurring in symmetric pairs which lie in opposite hemispheres. Let 2 great circles intersect at *O* and *O'*, and let them meet a 3rd great circle in points *A, A'* and *B, B'*, respectively. On the hemisphere bounded by the great circle through *A, B, A', B'* and containing point *O*, examine the configuration obtained by drawing a 4th great circle. This 4th great circle meets the 3rd great circle at the diametrically opposite points *C, C'*, the semicircle *AOA'* at *P*, and the semicircle *BOB'* at *Q*. Thus a triangle *OPQ* is formed and is surrounded cyclically by triangle *AOB*, quadrilateral *BOPC*, triangle *CPA'*, quadrilateral *A'PQB'*, triangle *B'QC'*, and quadrilateral *C'QOA*. Hence 4 great circles having all intersection points distinct must always yield 8 spherical triangles and 6 spherical quadrilaterals on a sphere.

Consider again the hemisphere bounded by the original 3rd great circle and containing the aforesaid configuration. Let the 5th great circle cut the 3rd great circle in points *D* and *D'*, one of which must lie on a side of a triangle, the other on a side of a quadrilateral. Let *D* lie on side *AB* of *AOB* and *D'* lie on side *A'B'* of

$A'PQB'$. Observe that the sides of the triangle OPQ are also sides of quadrilaterals. Now DD' will or will not enter triangle OPQ.

If DD' does enter triangle OPQ it must cut 2 of its sides, thereby dividing triangle OPQ into a triangle and a quadrilateral. Semicircle DD' must also divide AOB into a triangle and a quadrilateral, the quadrilateral from which it enters OPQ into a triangle and a pentagon, and $A'PQB'$ into 2 quadrilaterals. Hence the 5th circle yields a configuration on the hemisphere consisting of 5 triangles, 5 quadrilaterals and one pentagon.

If DD' does not enter OPQ it must pass through four consecutive peripheral polygons, thereby dividing AOB into a triangle and a quadrilateral, the adjacent quadrilateral into 2 quadrilaterals, the next adjacent triangle into a triangle and a quadrilateral, and $A'PQB'$ into a triangle and a pentagon. Again, we have 5 triangles, 5 quadrilaterals and one pentagon.

Hence, the configuration formed on a sphere by 5 great circles, no 3 of which are concurrent, consists of *exactly* 10 triangles, 10 quadrilaterals and 2 pentagons. Furthermore, a 6th great circle will either leave the pentagon intact, will divide it into a pentagon and a quadrilateral, or will divide it into a triangle and a hexagon. Hence, if on the sphere there are n great circles ($n \geq 5$), no 3 of which are concurrent, there will always be at least one spherical polygon with 5 or more sides.

See also Problem 130, this MAGAZINE [*43*, 233], and Problem E788, AMERICAN MATHEMATICAL MONTHLY, reprinted here.

146 [*26*, 222]. *Solution by Leon Bankoff*

Consider the traces of the plane of the circle on the three mutually perpendicular planes, which are taken as a coordinate system. The radii at the three points of tangency are each perpendicular to one of the traces, as are the projections of these radii on the respective tangent coordinate planes. Consequently, the three radii and their corresponding projections form angles which determine the orientation of the plane in the same manner as the direction angles of the normal do. Indeed, the three dihedral angles between the cutting plane and the coordinate planes are equal to corresponding direction angles of the normal. We therefore may call the dihedral angles made with the $YZ, ZX,$ and XY planes, α, β, and γ, respectively, and the projections upon these planes of the radii to the contact points, j, k, and l, respectively.

Since $\cos^2\alpha + \cos^2\beta + \cos^2\gamma = 1$, we have

$$j^2/r^2 + k^2/r^2 + l^2/r^2 = 1 \text{ or } j^2 + k^2 + l^2 = r^2.$$

Now the center of the circle is $(\sqrt{r^2-j^2}, \sqrt{r^2-k^2}, \sqrt{r^2-l^2})$ so

$$x^2 + y^2 + z^2 = 3r^2 - (j^2 + k^2 + l^2) = 2r^2.$$

Accordingly, the center of the circle lies on the surface of a sphere with radius $R = r\sqrt{2}$. Since the distance of the center of the circle to any of the coordinate planes may equal but not exceed r, the desired locus is the surface of this sphere

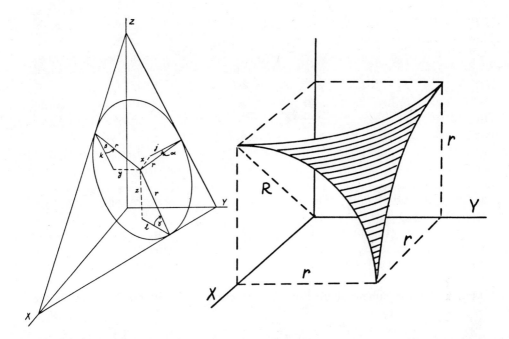

included between the planes $x = r$, $y = r$, $z = r$, together with congruent surfaces in the other seven octants.

The analogous problem in two dimensions requires the locus of the midpoint of a line of length c which slides so that its extremities move along two perpendicular lines. This locus is a circle with radius $c/2$.

AUTHOR INDEX

This index includes the authors of articles from the AMERICAN MATHEMATICAL MONTHLY and MATHE-MATICS MAGAZINE which are reprinted or cited in this volume.

(EN) indicates that an article is mentioned in an Editorial Note.
(#xx) designates the number of an article listed in the Supplementary Bibliography.

Abhyankar, S. S., 177 (EN), 288 (EN)
Alexander, J. C., 295 (#1)
Allendoerfer, C. B., 136
Amur, K., 216
Appel, K., 28 (EN)
Atneosen, G. H., 295 (#2)

Bagemihl, F., 295 (#3)
Ball, N. H., 295 (#4)
Ballantine, J. P., 295 (#5)
Banchoff, T. F., 192
Barnett, J., 50
Beckenbach, E. F., 233 (EN)
Bennett, A. A., 295 (#6,7)
Berge, C., 295 (#8)
Berlekamp, E. R., 128
Bliss, G. A., 295 (#9)
Botts, T., 233 (EN)
Brahana, H. R., 20
Brown, B. H., 29
Brown, R. F., 295 (#10)
Brown, R. L. W., 295 (#11)
Brown, W. S., 177
Bruck, R. H., 295 (#12)
Buck, E. F., 295 (#13)
Buck, R. C., 295 (#13)
Busemann, H., 51

Callahan, J., 295 (#14)
Cell, J. W., 295 (#15)
Cerutti, E., 295 (#16)
Chakerian, G. D., 292, 295 (#17–19)
Chernoff, P. R., 167
Cohen, M., 295 (#20)
Coleman, A. J., 295 (#21)
Conrad, S. R., 296 (#22)
Coolidge, J. L., 296 (#23)
Courant, R., 40
Coxeter, H. S. M., 296 (#24, 25)

Daus, P. H., 296 (#26)
Davis, P. J., 295 (#16)
deBruijn, N. G., 296 (#27)
Deo, N., 202
Dickson, L. E., 296 (#28,29)
Dieudonné, J., 248
Dines, L. L., 233 (EN)
Dowling, L. W., 296 (#30)
Duke, R. A., 296 (#31)

Echols, W. H., 13

Finney, R. L., 206
Flanders, H., 154
Fulton, C. M., 296 (#32)
Fusaro, B. A., 296 (#33)

Gans, D., 296 (#34–37)
Gerriets, J., 296 (#38)
Gilbert, E. N., 128
Gilbert, G., 127
Givens, W. B., 296 (#39)
Gluck, H., 143, 148 (EN)
Goldberg, M., 296 (#40)
Graves, L. M., 296 (#41)
Guggenheimer, H., 204 (EN)

Haken, W., 28 (EN)
Halsted, G. B., 1, 3 (EN), 296 (#42)
Herda, H., 294 (EN)
Horn, R. A., 214
Huff, G. B., 77

Jackson, S. B., 296 (#43)
James, R. C., 79
Johnson, H. H., 288
Johnson, R. A., 296 (#46)
Johnston, L. S., 296 (#44, 45)

Kalish, A., 296 (#48)
Kasner, E., 296 (#47, 48)
Kelly, L. M., 296 (#49)
Kelly, P. J., 296 (#50)
Kenelly, J. W., 296 (#51)
Kershner, R. B., 296 (#52)
Keyser, C., 297 (#53)
Klamkin, M. S., 202, 295 (#18), 297 (#54, 55)
Klee, V., 218

Lefschetz, S., 168, 297 (#56)
Leisenring, K., 67
Lenzen, V. F., 297 (#57)
Levin, M., 128 (EN)
Levow, R. B., 28 (EN)
Lewis, F. P., 6
Lovett, E. O., 297 (#58)

MacDonnell, D., 127
MacDuffee, C. C., 297 (#59)
Mac Lane, S., 114
Miller, G. A., 297 (#60)
Millman, R. S., 297 (#61)
Milnoi, J., 180
Moran, D. A., 204
Morley, F. V., 297 (#62)

Nitsche, J, C. C., 47 (EN), 297 (#63)
Niven, I., 149

Osaka, J., 288
Osborn, R., 297 (#64)
Ostrom, T. G., 297 (#65)
Ott, E. R., 297 (#66)

Patterson, B. C., 297 (#67)
Pavlick, F., 297 (#68)
Pedoe, D., 297 (#69)
Pfeffer, W. F., 246
Pólya, G., 74
Poole, G., 296 (#38)

Ramaley, J. F., 297 (#70)
Roever, W. H., 3, 4 (EN)

Rutt, N. E., 297 (#71)

Saaty, T. L., 28 (EN)
Schaaf, W. L., 297 (#72)
Scherk, P., 297 (#73)
Schultz, R., 234
Sinden, F. W., 128
Sobczyk, A., 296 (#51)
Stehney, A. K., 297 (#61)

Tan, K., 297 (#74)
Tóth, L. F., 48

Valentine, F. A., 297 (#75, 76)

Weaver, J. H., 297 (#77)
Wegner, B., 204 (EN)
Winger, R. M., 297 (#78)
Wong, R. E. M., 297 (#79)

Yates, R. C., 297 (#80, 81)

Zirakzadeh, A., 297 (#82)
Zuckerman, H. S., 149

INDEX OF KEY WORDS

(EN) indicates that the topic is mentioned in an Editorial Note.
(#xx) designates the number of an article in the Supplementary Bibliography.

Abelian integral, 171, 255 f, 264
Abel's theorem, 171, 255 f
Adjoint of a curve, 175
Algebraic geometry, 168 ff, 248 ff
Angle, 296 (#39, 46)
  solid, 295 (#15)
Angle-sum formulas, 137 ff
Angular deficiency, 67
Annulus conjecture, 237
Antipodal map, 247
Apollonius, problem of, 296 (#25)
Archimedes, axiom of, 68
Arcs, 296 (#38)
Area, 29 ff, 67 ff, 74(EN), 74 ff, 149 ff, 154,
  157 ff, 164, 167, 296 (#33)
Art, 297 (#72)
Axiom, Archimedes, 68
  continuity, 121 ff
  See also Postulate
Azimuthal equidistant projection, 182 ff

Betti number, 101, 296 (#31)
Birational geometry, 168, 255 ff
Birkhoff, G. D., 24, 28, 114
Bolyai, J., 1, 9
Bolyai, W., 8
Bolyai-Lobachevski geometry, see Hyperbolic
  geometry
Boundary, 149 ff, 234
Branch point, 103
Breadth of a convex body, 223 f
Brill-Noether theory, 175, 262 ff
Brilliant point, 4
Brouwer-Schauder-Tychonov fixed point
  theorem, 230
Brunn-Minkowski inequality, 160, 223
Buffon problem, 297 (#70)
Butterfly problem, 296 (#22)

Caratheodory's theorem, 224 ff
Cartography, 20 ff, 29 ff, 180 ff
Catenary, 297 (#81)
Cayley, A., 1, 20
Centroid, 202 ff, 296 (#32)
Chain, 22 ff
Chasles' correspondence principle, 253
Chebyshef theorem, 190
Chord, diametral, 292
Circle, 167, 246, 295 (#10), 297 (#75)
Circular horn triangle, 296 (#48)
Clifford, W. K., 1
Collar neighborhood theorem, 235

Collinear, 295 (#17), 296 (#24, 26)
Coloring problems, 20 ff, 28 (EN)
Combinatorial topology, 79 ff
Cone, 221
Conformal geometry, 297 (#73)
  map, 29 ff, 188
Conic section, 29 ff, 77 ff, 296 (#28)
Connected sum, 295 (#1)
Connections on manifolds, 297 (#61)
Connectivity number, 101
Continuity axiom, 121 ff
Convex body, 165, 222
  combination, 221
  curve, 154 ff, 167
  hull, 221, 225
  hypersurface, 216
  neighborhood, 197
  polygon, 128
  region, 296 (#38)
  set, 218 ff, 233 (EN), 295 (#13)
Covering surface, 102 ff
Cremona transformation, 175
Critical point theorem, 192 ff, 204
Crofton, formula of, 293
Crosscap, 79 ff
Cuff, 79 ff
Curvature, 192 ff
  centroid, 202
  of a curve, 143 ff
  Gaussian, 77, 197, 216
  mean, 216
  principal, 216
  total, 165, 197 f, 214
Curve, 39 (EN) 297 (#59, 62, 63)
  adjoint of a, 175
  boundary, 40 ff
  closed, 214, 292, 294 (EN), 295 (#19)
  convex, 154 ff, 167
  covers for a, 295 (#19)
  curvature of a, 143 ff
  function field of a, 169
  involution of a, 246 ff, 292 ff
  Jordan theorem, 182, 296 (#43)
  Jordan-Schoenflies theorem, 235 f, 296 (#43)
  normal to a, 202
  points on a, 262, 297 (#56)
  on a surface, 295 (#21)

Deficiency, angular, 67
DeMorgan, A., 2, 8, 20
DeRham theorem, 268
Diametral chord, 292

Dirichlet principle, 258
Distance, 295 (#18)
Distortion, 181
Divisor, 261, 267, 276
Dodecahedron, great, 107
Duplication problem, 297 (#77)

Ellipsoid, 4
Elliptic function, 170
   geometry, 8, 48, 57, 67, 296 (#35)
Equiareal map, 29 ff
Equichordal point, 224
Equidistant lines, 3, 7, 67
Equipartition of convex sets, 295 (#13)
Euclid, 1 ff, 6 ff, 114 ff, 297 (#71)
Euclidean geometry, 8, 50, 114 ff, 206 ff, 295
   (#12), 296 (#34, 36), 297 (#68)
   See also Plane geometry
Euclidean n-space, 50, 143 ff
Euler, L., 40
   formula, 100
   theorem, 205, 225
Euler-Poincaré characteristic, 99 ff, 153, 173,
   194 ff
Extreme point, 221

Face, 221
Fagnano-Euler theorem, 255 f
Farkas lemma, 225
Fenchel's theorem, 214 ff
Fifth postulate, see Parallel postulate
Finite geometry, hyperbolic, 296 (#41), 297
   (#65)
   projective, 297 (#66)
Fixed point theorem, 230, 246 ff
Foliations of manifolds, 295 (#20)
Formac, 295 (#16)
Foundations of geometry, 295 (#12), 297 (#71)
Four-color problem, 20 ff, 28 (EN)
Four normals theorem, 202 ff
Fourth dimension, 295 (#7), 297 (#53)
Frenet frame, 143
Function, elliptic, 170
   field, 169
   gauge, 227
   zeta, 273 ff

Galileo, 50
Galois imaginary, 27
Gauge function, 227
Gauss, C. F., 2, 8 ff, 51
   Theorema egregium, 77, 193, 200 f
Gauss-Bonnet theorem, 74, 192, 197 ff
Gaussian curvature, 77, 197, 216
Genus, 173, 258, 265, 296 (#31)
Geometric probability, 223
Geometric thought, 297 (#69)
Geometry, see Algebraic, Birational, Conformal,
   Elliptic, Euclidean, Finite, Hyperbolic,
   Non-Euclidean, Plane, Projective, Solid,
   Spherical, Synthetic Geometry

Graph, 295 (#8), 296 (#31)
   intersection, 224 ff
   interval, 224 ff
   linear, 21 ff
Group, Lie, 297 (#58)
   symmetry, 296 (#47), 297 (#60)
   transformation, 127 (EN), 297 (#58)

Hahn-Banach theorem, 228
Handle, 79 ff, 241 f
Hauptvermutung, 240
Helly's theorem, 224
Helmholtz, 1
   mobility principle, 51 ff
Hilbert, D., 10, 114
History, 1 ff, 6 ff, 168 ff, 248 ff, 296 (#23), 297
   (#64, 71)
Hodge theory, 269 f
Homology, 295 (#2)
Hyperbolic geometry, 8, 57, 64 ff, 67 ff, 296
   (#37)
   finite, 296 (#41), 297 (#65)
Hyperplane, 219 f
Hypersphere, 296 (#33)

Icosahedron, great, 107 f
Imaginary element, 297 (#78)
   point, 295 (#6)
Index, rotation, 215
Indicatrix, tangent, 214 ff
Inequality, 295 (#18)
   Brunn-Minkowski, 160, 223
   Isoperimetric, 154, 165, 288
   Jung, 222
   Lipschitz, 182
   Minkowski, 154 ff
   Riemann, 174
Infinite elements, 297 (#78)
Integral, 170
   abelian, 255 f, 264
Intersection graph, 224
   multiplicity, 254, 271
Interval graph, 224 ff
Inversive plane, 297 (#67)
Involution of a curve, 246 ff, 292 ff
Isoperimetric inequality, 154, 165, 288
   problem, 40 ff

Jacobi theorem, 171
Jordan curve theorem, 182, 296 (#43)
Jordan-Schoenflies curve theorem, 235 f, 296
   (#43)
Jung's inequality, 222

Kähler variety, 267 ff
Klein, F., 253
   bottle, 101, 108, 112, 295 (#1, 11)
Krein-Milman theorem, 229

Lagrange, J. L., 8, 29 ff
Lambert, J. H., 1, 29 ff

Laplace, P. S., 8, 40
Lattice point, 149 ff
Laws of sines and cosines, generalized, 137 ff
Legendre, 1, 7 ff, 170
Lie, S., 1
    group, 297 (#58)
Liebmann-Süss theorem, 216
Line bundle, 277
Linear programming, 226
Lines, equidistant, 3, 7, 67
Lipschitz inequality, 182
Lobachevski, N. I. (Lobatschewsky), 1 ff, 9, 51
Loewner's theorem, 222 f

Malfatti problem, converse, 296 (#40)
Manifold, 234 ff
    combinatorial, 237, 240
    connections on, 297 (#61)
    differentiable, 237 f
    foliations of, 295 (#20)
    piecewise linear, 238 ff
    topological, 234 ff
    triangulations of, 235
Maps, 180
    antipodal, 247
    plane, 295 (#14)
    See also Cartography
Markov-Kakutani theorem, 230
Mass centroid, 202 ff
Mean curvature, 216
Mercator projection, 30 ff
Metric postulate, 114 ff
Minimal surface, 40 ff
Minkowski inequality, 154 ff
    space, 228, 297 (#75)
Mixed volumes, 223
Mobility principle, 51 ff
Möbius, 253
    band, 41, 47, 81, 101
Morse theory, 193 ff, 204

n-space, Euclidean, 50, 143 ff
    projective, 297 (#82)
Napoleon's theorem, 210
Newton's theorem, 297 (#74)
Noether, M., 175, 263
Non-Euclidean geometry, 1 ff, 3 (EN), 6 ff,
    61 ff, 67 ff
    See also Hyperbolic and Elliptic geometry
Non-orientable surface, 82
Normal to a curve, 202

Optical phenomenon, 3, 4 (EN)
Oval, 203, 295 (#4), 297 (#65)

Packing, 48 ff, 222, 296 (#27)
Pappus, 295 (#16)
Parallel postulate, 1 ff, 6 ff, 51 ff
Perimeter centroid, 202 ff
Perspective, 295 (#5)
Physical phenomenon, 3 f, 295 (#9), 297 (#57)
Picard variety, 274

Pick's theorem, 149
Piecewise linear function, 239
    manifold, 238 ff, 246 (EN)
Plane geometry, 1 ff, 6 ff, 13 ff, 114 ff, 127 f,
    128 ff, 154 ff, 295 (#12), 296 (#34, 36, 46,
    52)
    See also Euclidean geometry
Plane, inversive, 297 (#67)
    map, 295 (#14)
    projective, 108, 295 (#1)
Plateau experiment, 40 ff
    problem, 40 ff, 47 (EN)
Plücker, J., 173, 253
    formulas, 254
Poincaré conjecture, 236 ff
    vector field theorem, 101
Point, branch, 103
    critical, 192 ff, 204
    equichordal, 224
    extreme, 221
    fixed, 230, 246 ff
    lattice, 149 ff
    ramification, 258
    "remarkable," on a curve, 297 (#56)
Polar, 226
Polygon, 13 ff, 74 ff, 128 ff, 149 ff, 296 (#39, 43,
    47)
Polyhedron, 192 ff, 225 ff, 295 (#3)
Polytope, 225
Poncelet closure theorem, 253
    continuity principle, 272
Postulate, 295 (#12)
    metric, 114 ff
    parallel, 1 ff, 6 ff, 51 ff
    See also Axiom
Probability, geometric, 223
Problems
    Apollonius, 296 (#25)
    Buffon, 297 (#70)
    Butterfly, 296 (#22)
    Duplication, 297 (#77)
    Four-Color, 20 ff, 28 (EN)
    Malfatti, 296 (#40)
    Plateau, 40 ff, 47 (EN)
    Trisection, 297 (#80)
Proclus, 6 ff
Projection, 296 (#26)
    azimuthal equidistant, 182
Projective geometry, 252 ff, 296 (#23, 30, 36),
    297 (#82)
    finite, 297 (#66)
Projective plane, 108, 295 (#1)
Proof, dynamic, 206
    vector, 297 (#55)
Ptolemy, 7
Puiseux expansion, 171, 257
Pythagorean theorem, 115

Quadrilateral, 13 ff, 296 (#29, 50)

Ramification point, 258
Ray, 219

Remainder theorem, 175
Riemann, G. F. B., 1, 40, 255 ff
  equality, 174
  hypothesis, 274
  inequality, 174
  space, 296 (#33)
  surface, 87, 102 ff, 172, 275 ff
Riemann-Roch Theorem, 174, 176, 261 f, 276 ff
Rotation index, 215

Saccheri, G., 1 ff, 7 ff, 12 (EN)
Schäfli-Poincaré theorem, 225
Schemes, 284 ff
Schoenflies theorem, 235 f, 296 (#43)
Schwarz, H. A., 40
Sheaves, 281 ff
Similarity transformation, 127 (EN)
Simplex, 222, 238, 296 (#51)
Singularity, 295 (#14)
Skewsquare, 13 ff
Soap film, 40 ff
Soddy, F., 177
Solid geometry, 295 (#15), 297 (#55)
Sphere, 20 ff, 29 ff, 48 f, 50, 67, 180 ff, 216, 297 (#63)
Spheres, tangent, 177
Spherical geometry, 52 ff
  See also Elliptical geometry
Spheroid, 4
Steiner-Lehmus theorem, 127 f
Steinitz theorem, 225
Supporting hyperplane, 220
  line, 292
Surface, area of, 165
  curvature of, 216
  curves on, 295 (#21)
  minimal, 40 ff
  non-orientable, 82
  polyhedral, 192 ff
  of revolution, 297 (#54)
  Riemann, 87, 102 ff, 172, 257 ff

  tangent, 4
  topology of, 41 ff, 79 ff
  See also Gaussian Curvature
Symmetry, 296 (#47), 297 (#60)
Synthetic geometry, 296 (#49)

Tangent indicatrix, 214 ff
Tchebyschef, see Chebyshef
Tetrahedron, 136 ff, 149
Theorema egregium of Gauss, 77, 193, 200 f
Topology of surfaces, 79 ff, 192 ff, 204 f, 234 ff
Torus, 79, 86 ff, 295 (#1)
Total curvature, 165, 197 f, 214
Tractrix, 297 (#81)
Transformation, Cremona, 175
  group, 127 (EN), 297 (#58)
Triangle, 120 ff, 127, 136 ff, 296 (#29, 44, 45, 48)
Triangulation conjecture, 240
  of a manifold, 235, 246
Trisection problem, 297 (#80)

Variety, algebraic, 259 ff
  irreducible, 259
  Kähler, 267 ff
  Picard, 274
Veblen, 0., 114
Vector field theorem, Poincaré, 101
Vector proof, 297 (#55)
  space, normed, 226
Volume, 50, 165, 223
Von Aubel's theorem, 296 (#50)

Wallis, J., 7
Weierstrass, K. W. T., 40, 171
Width of a curve, 167, 292 ff

Zeno, 7
Zeta functions, 273 ff

**OVERDUE CHARGE IS 10 CENTS
A DAY, INCLUDING SATURDAYS
SUNDAYS AND HOLIDAYS.**